D0142724

PRIMER OF GENETIC ANALYSIS

PRIMER OF GENETIC ANALYSIS
A PROBLEMS APPROACH

SECOND EDITION

JAMES N. THOMPSON, Jr.
University of Oklahoma

JENNA J. HELLACK
University of Central Oklahoma

GERALD BRAVER
University of Oklahoma

DAVID S. DURICA
University of Oklahoma

Xavier University Library
New Orleans, LA 70125

575.1076
P953t
1997

PUBLISHED BY THE PRESS SYNDICATE OF THE UNIVERSITY OF CAMBRIDGE
The Pitt Building, Trumpington Street, Cambridge CB2 1RP, United Kingdom

CAMBRIDGE UNIVERSITY PRESS
The Edinburgh Building, Cambridge CB2 2RU, United Kingdom
40 West 20th Street, New York, NY 10011-4211, USA
10 Stamford Road, Oakleigh, Melbourne 3166, Australia

© Cambridge University Press 1997

This book is in copyright. Subject to statutory exception
and to the provisions of relevant collective licensing agreements,
no reproduction of any part may take place without
the written permission of Cambridge University Press.

First edition published in 1987

Printed in the United States of America

Typeset in Stone Sans and Stone Serif

Library of Congress Cataloguing-in-Publication Data

Primer of genetic analysis : a problems approach / James N. Thompson
. . . [et al.]. – 2nd ed.
p. cm.
Includes biographical references.
Previous ed. cataloged under: Thompson, James N.
ISBN 0 521 47312 8 (hbk.). – ISBN 0 521 47891 X (pbk.)
1. Genetics–problems, exercises, etc. I. Thompson, James N.
QH440.3.T48 1996
575.1′076–DC20 96-14267
CIP

A catalogue record for this book is available from
the British Library

ISBN 0 521 47312 8 hardback
ISBN 0 521 47891 X paperback

CONTENTS

PREFACE

Many beginning students find some aspects of genetics much more difficult to grasp than others. Not very surprisingly, this is especially true of the problem solving and calculations involved in Mendelian genetics, pedigree analysis, linkage and mapping, quantitative analysis, and population genetics. We have found that many texts provide little if any feedback to students in these areas of possible weakness. Even if the correct answer is readily available, a student who arrives at the wrong answer may not necessarily know whether the incorrect answer resulted from a minor mathematical error or a more fundamental misunderstanding of the material. This primer of genetic analysis is designed to address some of these difficulties.

In order to help you build confidence in your grasp of the subject and in your ability to apply that knowledge to novel problems, we start with an array of problems that test simple, basic ideas. More challenging problems then follow naturally. In addition, we have supplied questions with detailed explanations of how to work them. You can then work at your own pace through the logical analysis of different types of problems. With this practice and some guidance on how to avoid common pitfalls, you will gain experience and confidence in your ability to solve problems in the major areas of genetics.

Each chapter begins with Study Hints in which we suggest ways of organizing your studies of a topic or take you step-by-step through a problem-solving approach. Practice Tests are provided at the end of the book. Some areas, such as linkage and mapping, are given especially extensive coverage, since they are particularly well suited to practicing data analysis and interpretation. It is important to remember, however, that these hints are not comprehensive. They are intended to complement the material in a general textbook. The hints and problems should also be useful in reviewing for examinations like the Medical Candidacy Aptitude Test (MCAT) or Graduate Record Examination (GRE). Important Terms lists appear at the end of each Study Hints section as a convenient guide for review. Definitions of these and other basic terms are given in the Glossary at the end of the book.

In preparing this and the first edition, we have benefited from the helpful suggestions made by students in our courses. In addition, Fred B. Schnee, William E. Spivey, and Robert R. Tucker made many constructive recommendations during the writing of the first edition. Eric M. Weaver helped us see the material again from the eyes of a dedicated student while we prepared this second edition, and the suggestions by Stanton B. Gray and K. April Sholl are also greatly appreciated. Their suggestions have led to the addition of several "overview" chapters, additional problems, and a significant expansion in the area of molecular biology. We also thank Coral McCallister for her talented help with many of the figures.

We hope that you find this excursion into genetic analysis an enjoyable and profitable one. Good luck!

NOTE ON GENETIC SYMBOLS

It is very important to learn to recognize the meaning of genetic symbols and to use them consistently. Some symbols can be used interchangeably, and we have chosen to use several common versions in this text. Care in defining symbols will help prevent errors in interpreting and solving genetic problems.

The various forms in which a gene can exist (its alleles) are all given the same letter symbol. Uppercase letters (for example, *A* or *Cy*) denote the dominant alleles, and the lowercase letters (*a* or *cy*) denote the recessive alleles. When more than two alleles are known for a particular genetic locus, numerical superscripts are usually used to identify them. For example, a^1, a^2, and *A* could represent three different alleles at the *A* locus. Allele symbols, you will note, are italicized.

Another common convention is to use a plus (+) sign to denote the normal (or "wild-type") allele. If several wild-type alleles are being considered in the same problem or breeding plan, it is often useful to distinguish among them by adding the plus sign, as a superscript, to the mutant symbol. For example, *ri* would be the recessive mutant allele for the wild-type locus ri^+.

In this text, the complete genetic makeup of an individual may not necessarily be known. In those instances, a short dash will be used to indicate that an allele is present but unknown. For example, in the genotype *AaB –,* we know that the individual is heterozygous for the *A* locus and carries at least one *B* allele. We do not, however, know whether the individual is *Bb* or *BB*.

When the linkage relationships of a set of genes are important, the linked combinations will be separated by a slash. In the genotype *R Y U / r y u,* one chromosome carries the dominant alleles of all three genes, and the other chromosome carries the recessive alleles. Other special types of symbols should be clear from the context of the problem or will be defined for a specific situation.

1 OVERVIEW OF GENETIC ORGANIZATION AND SCALE

The genetic material is a molecule called **deoxyribonucleic acid (DNA).** Each **chromosome** contains a single long strand of DNA that encodes the information needed to produce hundreds or even thousands of different proteins. Each species has a characteristic array of chromosomes that carries all the **genes** needed to produce that organism from a single cell. The relationship between the genetic makeup of an organism (the **genotype**) and the developmental effects of these genes (the **phenotype**) can be complex. It is, therefore, useful to begin with a simple overview of these processes. Here we introduce some of the key concepts of genetics using an illustrated guide that begins at the smallest unit of genetic organization within a **nucleus** and ends at the level of the **population.** Some important terms are shown in boldface type, and definitions are given in the Glossary.

- DNA is made up of subunits called **nucleotides** composed of a sugar (S), a phosphate group (P), and a nitrogenous base (B). There are four nucleotides that differ by the nucleotide base they contain: **adenine (A), guanine (G), thymine (T),** and **cytosine (C).** Genetic information is encoded in DNA by the sequence of these four bases.

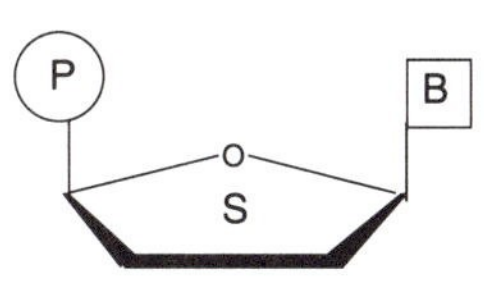

nucleotide

- **Nucleotides** are linked by a bond between the sugar of one nucleotide and the phosphate group of the next.

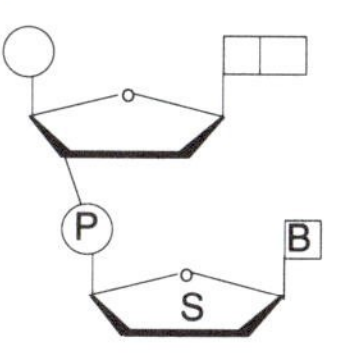

two nucleotides (schematic representation)

- This produces a long chain that can be literally millions of nucleotides long.

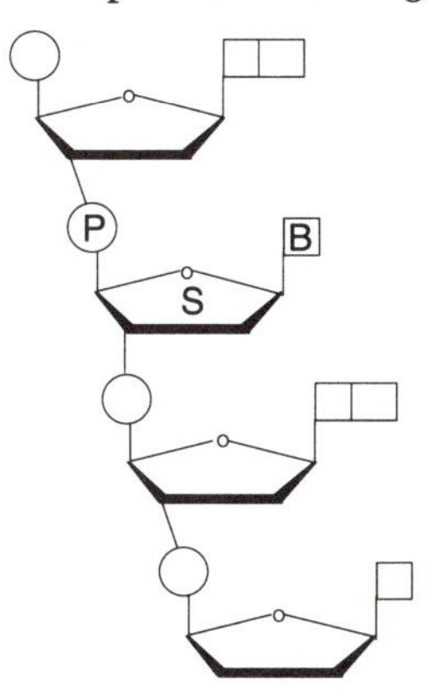

a portion of one strand

- A **single DNA molecule** is composed of two such strands that join together by bonds between the nucleotide bases (A paired with T, and C paired with G). This forms a DNA **double helix.**

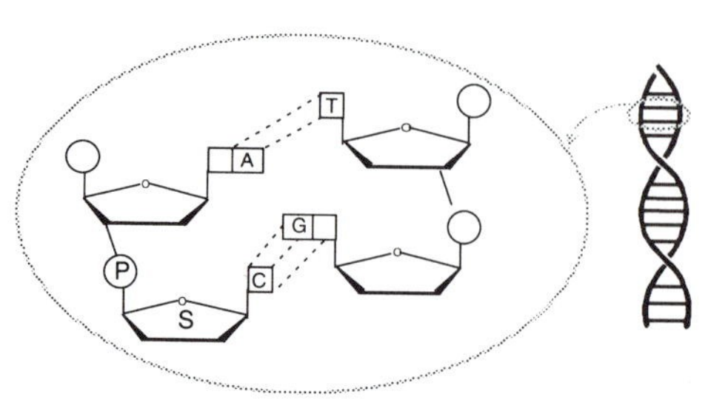

schematic of DNA and double helix

- The DNA is bound to structural proteins (**histones** that make up the **nucleosome**) that help pack the DNA in the nucleus and to regulatory proteins that turn genes on and off during development.

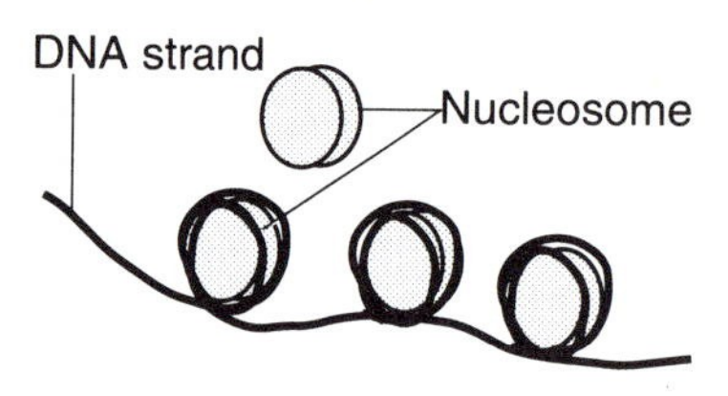

nucleosomes with DNA

- Each **chromosome** is made up of one long DNA molecule (one DNA double helix) and its associated proteins. The **centromere** is the attachment site for the spindle fiber that moves the chromosome during cell division.

Chromosome
Centromere

- Along the length of this DNA molecule are the regions that code for the production of proteins. These regions (or genetic **loci**) are the **genes.** Each gene can be up to a thousand or more nucleotides long, and every chromosome carries as many as several thousand different genes. All of the genes on a given type of chromosome are thus linked on a single DNA molecule. This linked group of genes is called a **linkage group.** The linear order of genes on a chromosome can be mapped to produce a linkage map.

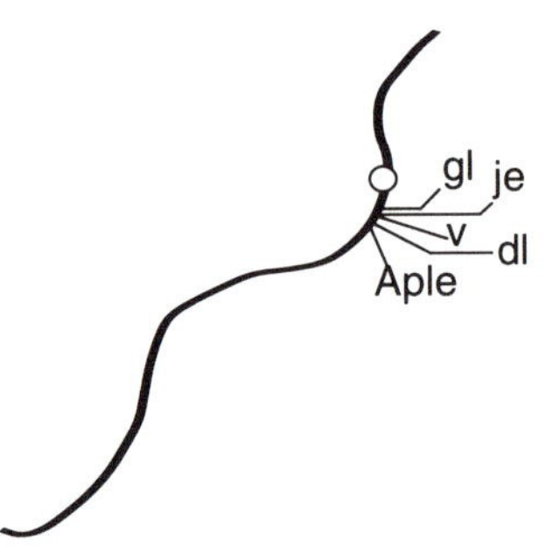

chromosome showing a map of five genes in linear order on its linkage group

- The body cells (**somatic** cells) of most organisms contain two copies of each type of chromosome **(diploid).** These are the homologous chromosomes. Since **homologous** chromosomes carry the same series of genes, they are members of the same linkage group. They can, however, differ in the form that a given gene takes (that is, normal or mutant). The different forms of a gene are called **alleles.**

A *a*

two homologous chromosomes carrying alleles *A* and *a*

- When the nucleus prepares to divide, each DNA molecule **replicates** except for the centromere. This yields two identical copies of the DNA molecule bound at the centromere. At this stage, the two copies are called sister **chromatids.** Since they have not yet divided at the centromere, however, each unit is still considered a single chromosome.

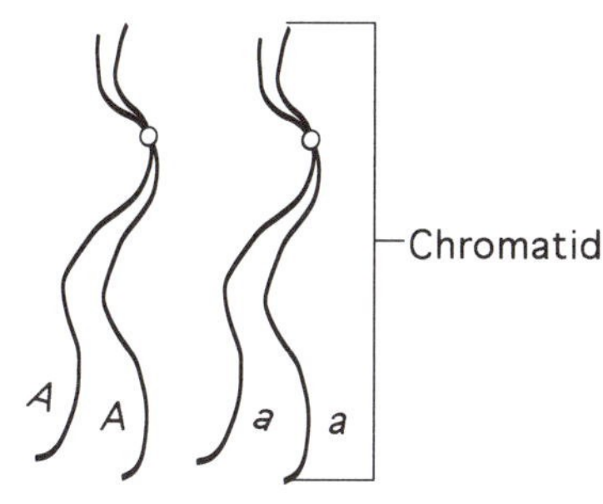

two homologous chromosomes, each with two sister chromatids

- Chromosomes that carry different sets of genes are called **nonhomologous** chromosomes. Every species has its own characteristic number of different chromosomes **(*n*).** **The total number of chromosomes in a somatic cell is, therefore, 2*n*.** All the genes needed to produce that organism will be found somewhere on one of these ***n*** linkage groups. The total **2*n*** genetic makeup is the **genome.** In the figure of the hypothetical cell, there are three pairs of nonhomologous chromosomes in the genome.

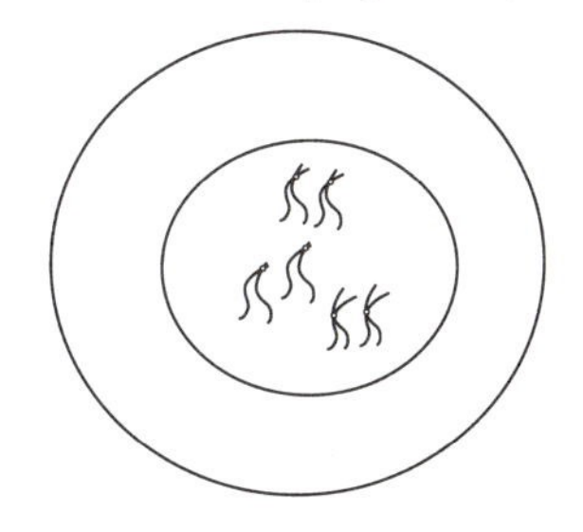

hypothetical cell with $2n = 6$, as seen during nuclear division

- **Mitosis** is a type of nuclear division that yields two identical diploid cells ($2n$). **Meiosis** is a special type of nuclear division found in reproductive **(germinal)** tissue that yields **gametes.** Each gamete carries only one copy of each linkage group and has a haploid (n) number of chromosomes. The diploid chromosome number is re-created at fertilization when the haploid maternal set and the haploid paternal set fuse.

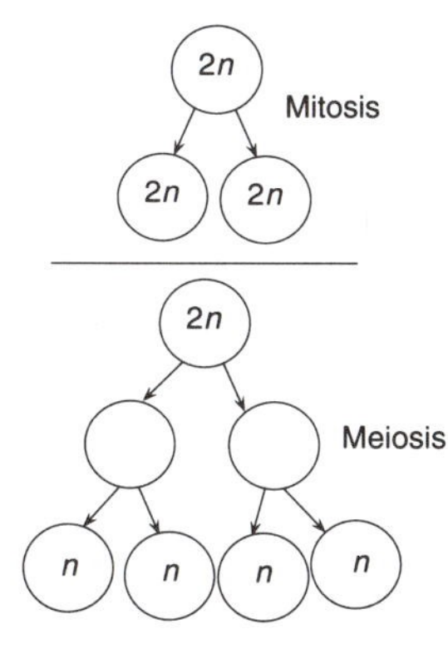

mitosis and meiosis

- If we focus our attention on one gene, the alleles on the two homologous chromosomes can either be the same (*AA* or *aa* = **homozygous** genotypes) or be different (*Aa* = **heterozygous** genotype). These separate **(segregate)** during meiosis to produce haploid gametes.

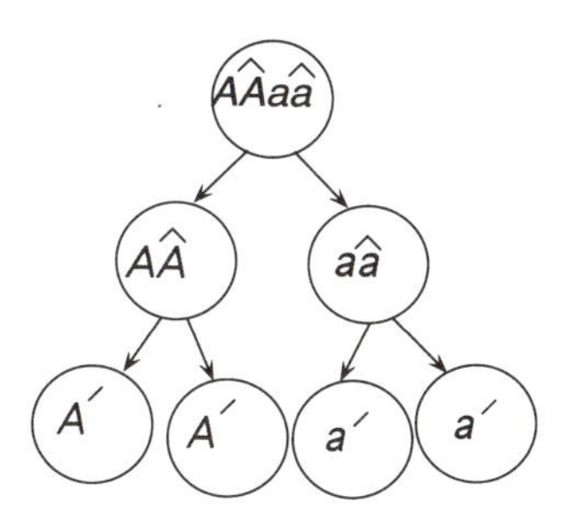

branching diagram to show haploid products

- The products of segregation and fertilization are highly predictable, giving rise to the basic rules of **genetic transmission.** Gregor **Mendel** set the foundation for this area of genetics.

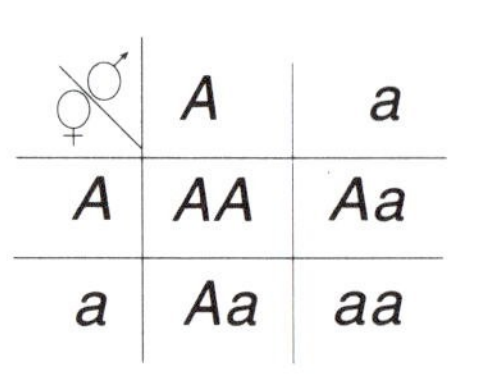

Mendelian cross using Punnett square for two heterozygous parents

- A **pedigree** diagram shows genetic relationships from a series of different Mendelian crosses. Circles indicate females and squares indicate males.

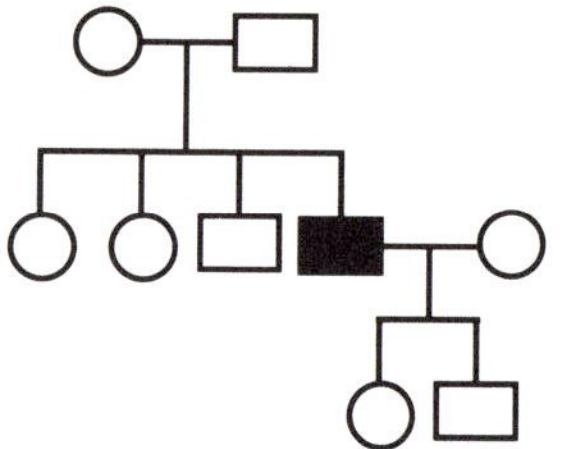

a simple pedigree

- Most genes code for the production of proteins. One of the two strands (the template strand) of a DNA molecule is "read" (through **transcription**) to yield a molecule of messenger **RNA** (mRNA). This then binds with ribosomes, where it defines the sequence of **amino acids** needed to produce the correct **polypeptide** (protein). This is **translation.**

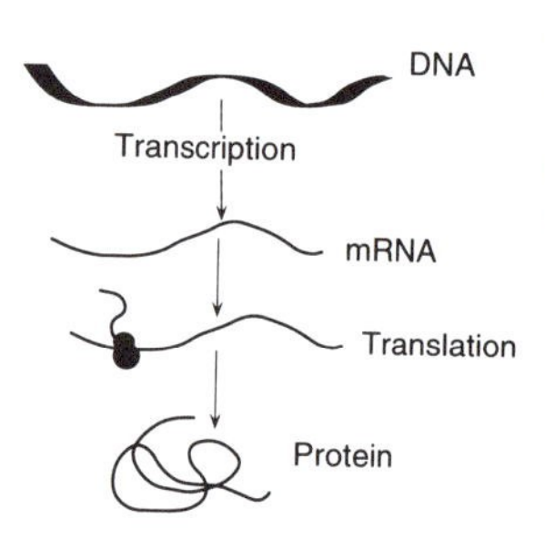

DNA → mRNA → protein

- Many proteins are **enzymes,** which catalyze specific biochemical steps. Thus, genes work by controlling the biochemical activities for growth and function of cells. In this way, the **genome** codes for all of the morphological, physiological, and behavioral characteristics **(phenotypes)** of an animal or plant.

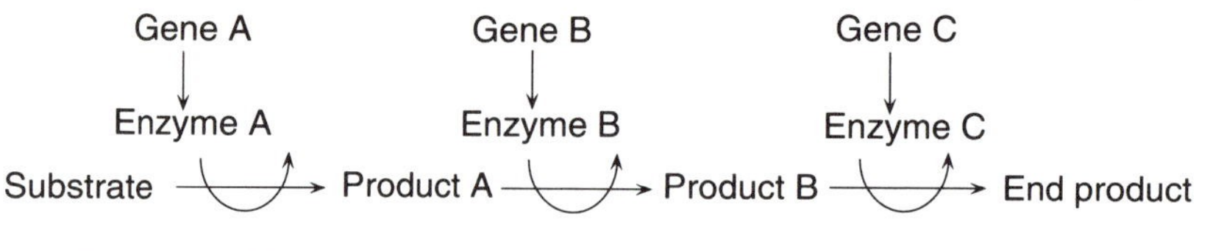

biochemical pathway

- Some characteristics are the result of several genes and environmental factors working together. Their expression is measured on an appropriate scale (such as height in meters). These are **quantitative traits** (**multifactorial** or **polygenic traits**).

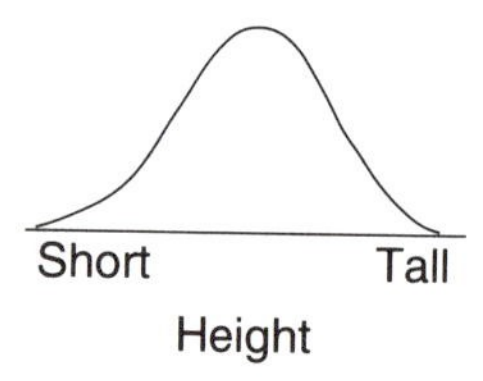

- The genetic makeup of an individual is the **genome,** whereas the total genetic makeup of all individuals in the **population** is the **gene pool.** The gene pool is described in terms of allele frequencies, where ***p*** is the frequency of the *A* allele and ***q*** is the frequency of the *a* allele. By using appropriate assumptions, the genetic makeup of individuals in the population can be predicted.

p = .6
q = .4

a A A a a A A A A a

alleles in a hypothetical gene pool

- **Hardy, Weinberg,** and **Castle** established the foundation for **population genetics** by showing that allele frequencies remain in **equilibrium** unless acted upon by **selection, migration, mutation,** sampling error in **small populations,** or deviations from **random mating.** Population genetics is the study of changes in allele and genotype frequencies that occur when these factors act on animals or plants.

	A (.6)	*a* (.4)
A (.6)	*AA* (.36)	*Aa* (.24)
a (.4)	*Aa* (.24)	*aa* (.16)

Punnett square with allele frequencies

Genetics is a dynamic and exciting field (but, of course, you would expect us to say something like that). But it can also be confusing, since there are so many levels at which you can look at inheritance and the use of genetic information. This introduction to genetic organization and scale is intended as a kind of outline to some key levels and processes. Although they overlap, we can readily see three main perspectives. First is the molecular level of DNA structure and coding (nucleotide—DNA—transcription—translation—protein product). Second are the rules of genetic transmission, which are based on probabilities of inheritance (segregation and independent assortment in meiosis—probabilities and genotypes—genes in families—genes in populations). Third is the way the genotype controls biochemical activities during development to produce the organism's phenotype (proteins—enzyme control of biochemical pathways—gene interactions—development). This primer will investigate these areas of genetics individually. But it is always important to keep in mind that they are really just different ways of looking at the same thing: the coding, transmission, and use of information by cells.

2 MITOSIS AND MEIOSIS

STUDY HINTS

Genes are located on chromosomes, and the stable manner in which chromosomes are first replicated and then distributed to daughter cells during cell division is the basis for genetic inheritance. Since much of genetic theory is based on the behavior of chromosomes and the genes they carry, it is very important to understand clearly how nuclear division occurs. In this way you can predict its consequences and understand the effect of errors that might occur in it. Yet the subject of cell division is complex, with many new terms to memorize and numerous things happening simultaneously. It is a continuous process that has been divided into stages somewhat artificially, so that we can describe it conveniently. All of this makes it rather hard to grasp at the beginning. Do not despair! It is really much simpler than it looks at first. The secret is to learn in stages. First one must understand the "strategies" of mitosis and meiosis, and the differences between them.

Mitosis has evolved as a mechanism to distribute accurately a copy of each chromosome present in the original cell to two new cells. The "goal" of *meiosis* is quite different. Meiosis passes alternate (homologous) copies of each type of chromosome to daughter cells and reduces the total chromosome number by half. These different objectives require slightly different chromosome behaviors. We shall briefly summarize these two processes, keeping in mind the different strategies they represent.

Both processes begin in essentially the same way. The chromosome (and the deoxyribonucleic acid [DNA] molecule it contains) duplicates, forming two identical chromosome strands attached to each other at the centromere. This is accompanied by a physical reorganization (coiling) that greatly reduces the chromosome's apparent length. The transition between these levels is illustrated in the following figure. One of the earliest signs of coiling is the formation of chromomeres, shown in this "magnified" insert to part A in the figure.

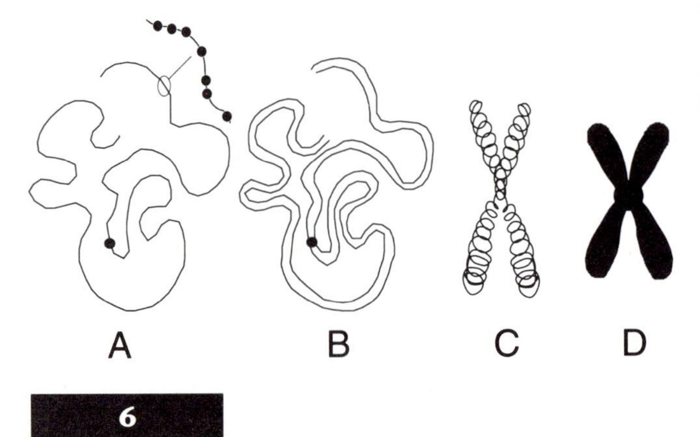

During late interphase, nuclear DNA and the histone proteins in chromosomes (A) duplicate (B), though remaining attached at the centromere. The centromere is really a constriction, but we draw it here as a dark spot so it can be seen easily. The two identical DNA copies are called sister chromatids. During prophase (C and D), these DNA strands coil into discretely identifiable chromosomes.

Each of the identical chromosome strands coils to form one of the two strands (sister chromatids) in a duplicated chromosome. The shape of the chromosome is determined by the position of the centromere. Without such coiling (or condensation), separating chromosomes would be a little like trying to separate a plate of spaghetti into two piles without breaking anything. People sometimes get confused about chromosome number at this point. Just remember that chromosome means "colored body." It is a single structural unit, no matter how much DNA it holds. So when counting chromosomes, count the centromeres, since whatever is attached to a centromere is a chromosomal unit.

In meiosis, the objective is to reduce the chromosome number so that there is only one copy of each kind in a gamete. The most direct way to do this is to pair the chromosomes carrying the same type of information or the same linear array of genes (homologous chromosomes). This is one purpose of synapsis. The first meiotic division, therefore, simply separates (segregates) homologous copies of each chromosome. The second meiotic division, like mitosis, distributes the identical copies (sister chromatids) that originated from chromosome replication.

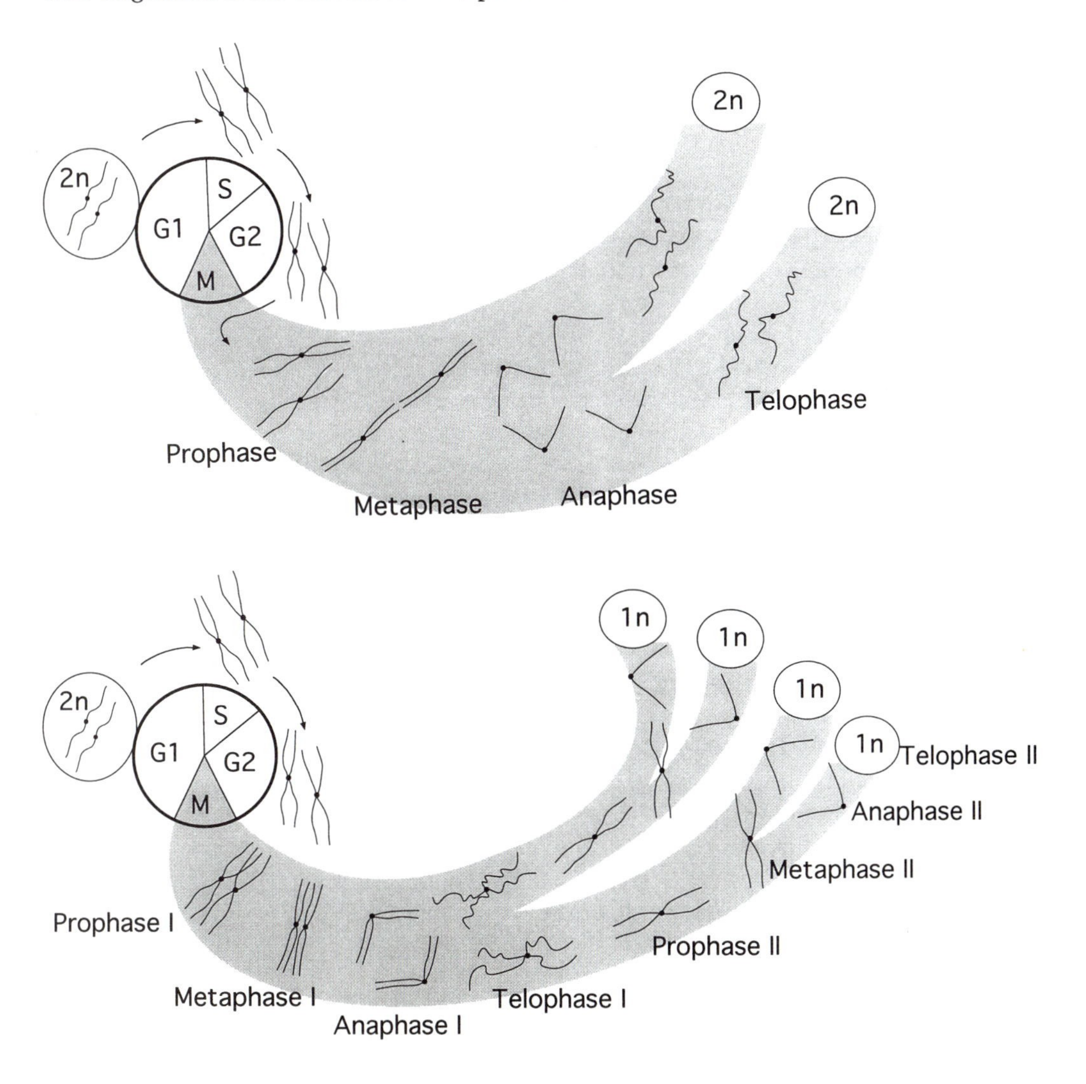

These two strategies are contrasted in the preceding figure, in which the original cell has a chromosome number of $2n = 2$. In the Important Terms section of the chapter we have listed some of the most important terms you need to know to describe these processes. Definitions are given in the Glossary.

The only additional hint we might add concerns prophase I of meiosis. The divisions of this phase of meiosis are described in your text and summarized in Table 2.1. It is sometimes useful to have a mnemonic (memory-assisting) device to help you remember the sequence of events. For the divisions of prophase I, try this phrase, whose first letters are the same as the letters of the terms you need to remember: "Leaping zebras pound down dunes": that is, *leptotene, zygotene, pachytene, diplotene, diakinesis* (see Table 2.1). Also remember these divisions are somewhat artificial, since the process is continuous.

TABLE 2.1 SUBSTAGES OF PROPHASE I OF MEIOSIS

I. **Leptotene**
   A. Chromatin begins condensing and becomes visible.
      1. Chromomeres form as irregular tightly coiled regions of the chromosome.
         a. Location, size, and number of chromomeres are chromosome specific.
         b. G bands on human chromosomes are represented by several chromomeres coiling together.
         c. Homologous chromosomes have the same banding pattern.
   B. Homologous search begins.
   C. Telomere proteins attach the telomeres to nuclear membrane.

II. **Zygotene**
   A. Pairing of homologous chromosomes occurs.
      1. Homologous search and rough pairing take place.
      2. Synaptonemal complex is present but homologues are not tightly synapsed.
         a. The pairs of chromosomes are called *bivalents* at this stage.
         b. The number of bivalents is equal to the haploid number.
   B. Chromosomes continue to coil and shorten.

III. **Pachytene**
   A. Synapsis tightens so homologues are very close together (about 100 Å between the pairs). The term *tetrad* is often used here to indicate the presence of 4 chromatids (2 for each chromosome).
   B. Recombination occurs.
   C. The telomeres lose contact with the nuclear membrane.

IV. **Diplotene**
   A. Synaptonemal complex breaks down.
   B. Homologues begin to move apart.
   C. Chiasmata are present.
   D. Terminalization begins.[a]

V. **Diakinesis**
   A. The homologous chromosomes continue to terminalize.[a]
   B. Nuclear membrane breaks down.
   C. Chromosomes move toward the equatorial plate.

[a] As terminalization is completed, the chromosomes lineup in the center of the cell as the cell enters metaphase I.

IMPORTANT TERMS

Acrocentric
Bivalent
Centriole
Centromere
Chiasma (pl., chiasmata)
Chromatid
Chromatin
Chromomere
Cytokinesis
Diploid
Dyad
Euchromatin
Gamete
Haploid
Heterochromatin
Homologous chromosomes
Linkage group
Meiosis
Metacentric
Mitosis
Nucleolus
Polar body
Primary oocyte
Primary spermatocyte
Secondary oocyte
Secondary spermatocyte
Spindle
Synapsis
Synaptonemal complex
Telocentric
Terminalization
Tetrad

PROBLEM SET 2

General Study Hint: To help reinforce these ideas, you should sketch the cell and its chromosomes before trying to answer these questions.

1. Suppose that you are studying a new species in which the chromosome number is unknown. You are looking at a spread of cells in some unknown stage of cell division, and you discover that there are six chromosomes (three metacentric ones and three telocentric ones) visible in each dividing cell. What type of nuclear division have you found, and what conclusions can be drawn about the chromosome makeup of the species?
2. How many chromosomes are there in a human intestinal cell at anaphase?
3. How many different kinds of gametes can be produced by an individual who has three linkage groups with each chromosome identifiable by the forms of the genes (alleles) it carries?
4. In an organism with a haploid chromosome number of 7, how many sister chromatids are present

   (a) in its mitotic metaphase nucleus?
   (b) in its meiotic metaphase I nucleus?
   (c) in its meiotic metaphase II nucleus?
5. A particular beetle species is found to have a chromosome complement of $2n = 16$ chromosomes. How many linkage groups would the beetle have?
6. Recombination can be seen as chiasmata between homologous chromosomes at what stage of cell division? Sketch a bivalent involving metacentric chromosomes having one chiasma.
7. Is it possible for meiosis to occur in a haploid cell? Is mitosis possible? Please explain your answer to each question.
8. Draw and label a cell at metaphase I of meiosis, where $2n = 8$ metacentric chromosomes.

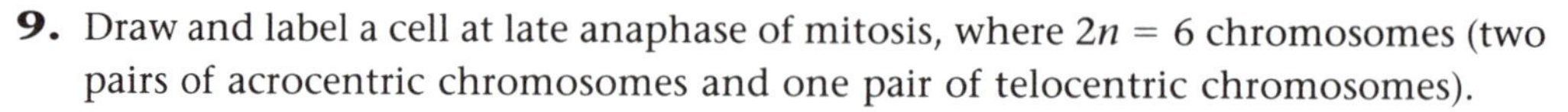

9. Draw and label a cell at late anaphase of mitosis, where $2n = 6$ chromosomes (two pairs of acrocentric chromosomes and one pair of telocentric chromosomes).

10. Draw and label all cells at telophase II that result from meiosis of a primary spermatocyte ($2n = 8$ acrocentric chromosomes).

11. Draw, label, and compare cells at prophase I and prophase II of meiosis in a species characterized by $2n = 6$ acrocentric chromosomes.

12. Draw, label, and compare cells at anaphase of mitosis and anaphase I of meiosis in a species having $2n = 4$ chromosomes (one homologous pair of metacentric chromosomes and one pair of telocentric chromosomes).

ANSWERS TO PROBLEM SET 2

1. This cell must be in a postreductional stage of meiosis (that is, in the second meiotic division), since a mitotic cell would have two of each type of chromosome and could not, therefore, have an odd number of metacentric or telocentric chromosomes. The cell you have found must consequently have a haploid chromosome number of 6 and a diploid complement of $2n = 12$ chromosomes (three pairs of metacentrics and three pairs of telocentrics).

2. Answer: 92. At anaphase of mitosis, the centromeres divide, but cytokinesis has not yet been completed. Thus for a short period the cell has twice as many chromosomes as normal. Remember that we have defined whatever is attached to a single centromere as a single chromosome.

3.

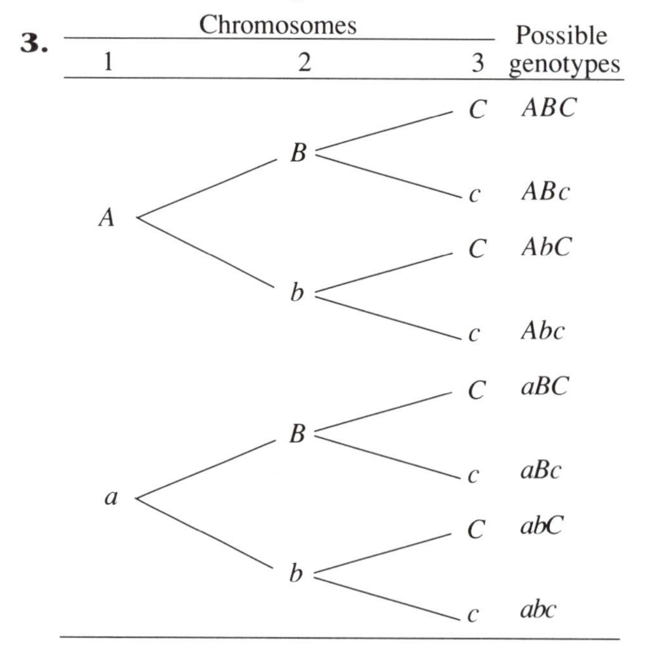

A total of eight different gametes could be produced.

If there are three linkage groups, the species must have a chromosome number of $2n = 6$. The number of different types of chromosomes is the same as the haploid number ($n = 3$ in this case). The haploid number and the number of linkage groups are the same. Let us identify homologous chromosomes in this individual by upper- and lowercase letters. Since each chromosome is individually identifiable, the genotype would be *AaBbCc*.

Each gamete must contain one of each kind of chromosome. Thus the individual could produce two types of gametes with respect to the first chromosome (an *A*-containing gamete and an *a*-containing gamete). Two types of gametes could be produced with respect to each of the three chromosomes, and they could assort in all possible combinations.

4. **(a)** At metaphase of mitosis, the chromosomes have duplicated, and there are therefore 2 chromatids present in each chromosome. In a species with a haploid number of 7, the diploid number is 14. Thus there are 28 chromatids present at metaphase of mitosis.

(b) The chromosome makeup of the first meiotic metaphase is the same as that of the mitotic metaphase, although in meiosis the chromosomes are arranged with homologous sets attached to opposing spindle fibers. Thus there is a total of 28 chromatids at metaphase I of meiosis.

(c) In metaphase II, reduction of chromosome number has occurred. Only 14 chromatids remain in the cell.

5. Eight linkage groups. The number of linkage groups is the same as the number of different types of chromosomes, which is the same as the haploid number of the species. If $2n = 16$, the haploid number (n) is 8.

6. Recombination is visualized as chiasmata between synapsed homologues in prophase I of meiosis.

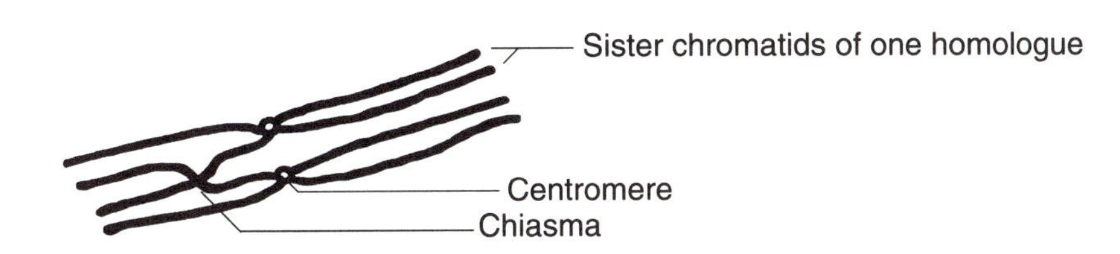

7. Meiosis is basically a reduction division from a diploid to a haploid condition. Thus it cannot occur in a haploid cell. Mitosis, on the other hand, is an orderly division of the existing chromosomes to distribute one copy of whatever is present to each daughter cell. Since chromosomes replicate and separate on the spindle independently of each other, mitosis can occur in a haploid cell.

8.

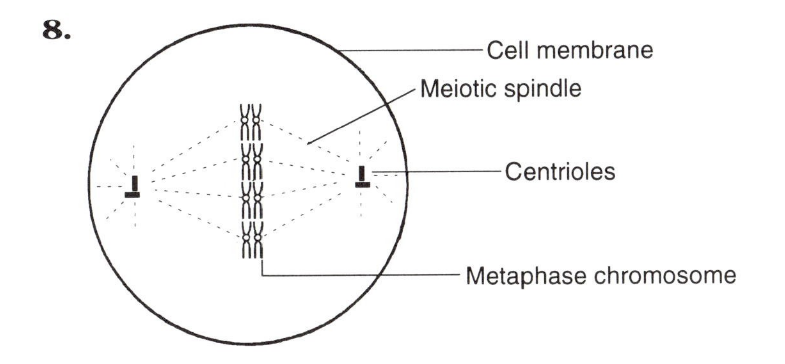

9.

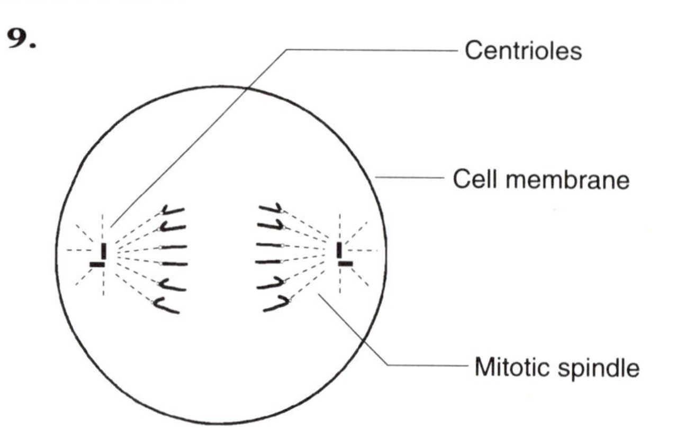

10. Centromeres of the chromosomes are shown as knobs for easy identification (though they are really constrictions). Nuclear membranes are re-forming. Cytokinesis is taking place. Four cells result from the meiotic division of one primary spermatocyte.

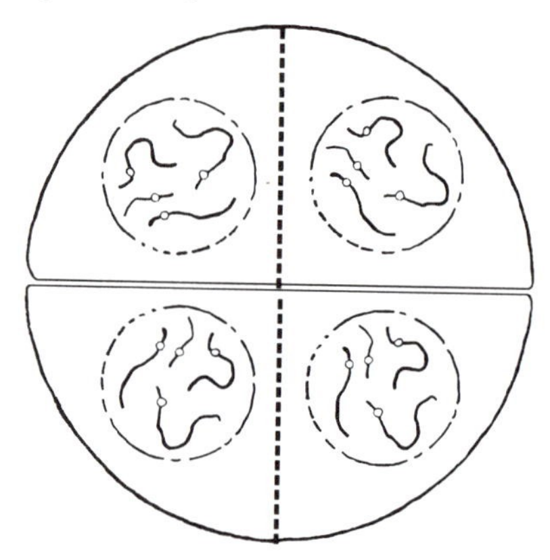

11. Prophase I of meiosis is characterized by synapsis of homologous chromosomes. These homologous chromosomes segregate during the first meiotic division, although the centromeres of the chromosomes do not divide until the second meiotic division. Thus in prophase II chromosome number has been reduced to the haploid number ($n = 3$, in this instance). The dissolution of the nuclear membrane and formation of the meiotic spindle occur in the same way in each.

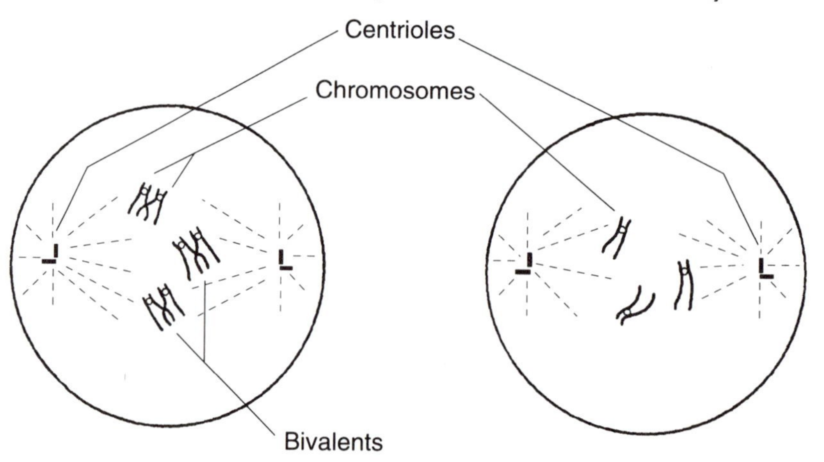

12. In anaphase of mitosis (left figure), homologous chromosomes attach to spindle fibers independently of each other, and chromatids separate. In anaphase I of meiosis (right figure), segregation of homologous chromosomes occurs, and centromeres do not divide. Division of centromeres in meiosis occurs during the second meiotic division. Also note that the first meiotic division (right figure) reduces the chromosome number by half (in this example, from four chromosomes to two in each resulting cell).

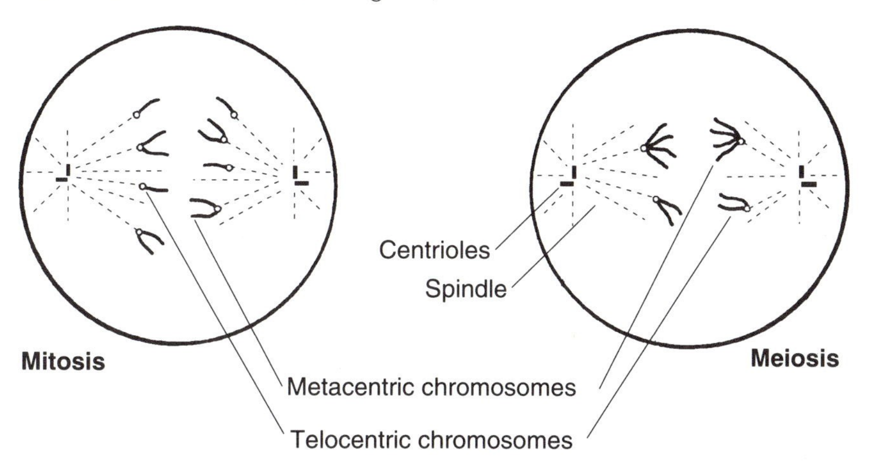

CROSSWORD PUZZLE

MITOSIS AND MEIOSIS

ACROSS

4. Separation of the homologous chromosomes during meiosis I
6. Interphase DNA-protein complex
7. Irregular tightly coiled regions of the chromosome, seen as the chromatin begins condensing during prophase I
13. Stage of mitosis when chromosomes are moving toward the poles
14. Region on chromosome where spindle fibers attach during cell reproduction
16. End of a chromosome
18. Substage of prophase I when the nuclear membrane breaks down and chromosomes move toward the equatorial plate
21. Substage of prophase I when recombination occurs
22. Four associated chromatids of a bivalent
23. One of two replicated chromosome strands held by one centromere
24. Term for a chromosome with the centromere on one end

DOWN

1. Substage of prophase I when the homologous chromosomes pair
2. Random combination of the genes of nonhomologous chromosomes because one pair's orientation at the equatorial plate does not affect the orientation of another pair during metaphase I
3. Cell with one set of chromosomes
5. Term for a chromosome that has the centromere located near one end

8. Dark staining chromatin that is inactivated or structural regions of the chromosome
9. Lightly staining chromatin where genes are located
10. Stage of mitosis when the chromosomes are lined up in the center of the cell
11. Movement of the chiasmata toward the ends of the bivalents during prophase I of meiosis
12. Term for a chromosome with the centromere located near the center
15. Process takes a diploid nucleus and makes four haploid nuclei
17. Genes that are on the same chromosome
18. Having two sets of chromosomes
19. Pair of homologous chromosomes during synapsis
20. Asexual cloning of the nucleus of a cell to produce two identical nuclei

NUCLEIC ACIDS: DNA AND RNA

Nucleic acids play a central role in both the coding and the decoding of genetic information. For this reason it is important to have a fairly detailed understanding of their structure and replication, and of the differences between deoxyribonucleic acid (DNA) and ribonucleic acid (RNA). If you spend a little time looking closely at the types and arrangement of atoms in each molecular component of a nucleotide, you will probably find that the structures are not as complex as they may first appear (for example, $T = 5$ methyl uracil). Indeed, some of the most important are quite similar and are therefore fairly easy to learn and compare. The diagrams that follow can be used to prepare flash cards or other study aids. You can then arrange them to make your own nucleotides or chains.

Comparison of DNA and RNA

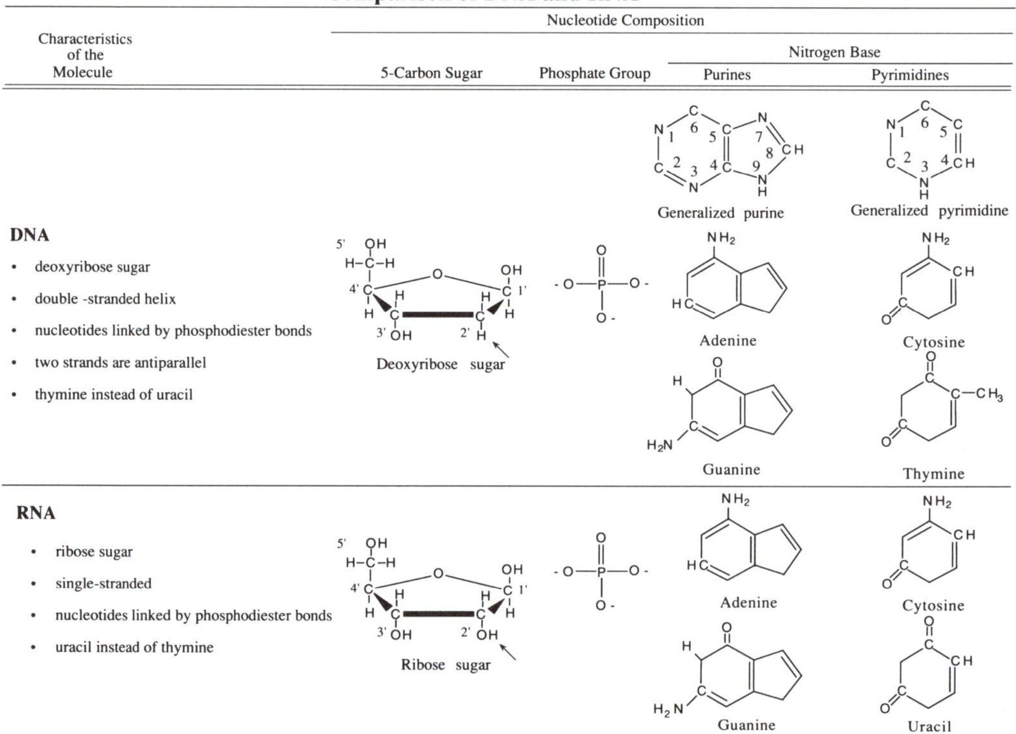

Characteristics of the Molecule	Nucleotide Composition: 5-Carbon Sugar	Nucleotide Composition: Phosphate Group	Nitrogen Base: Purines	Nitrogen Base: Pyrimidines
			Generalized purine	Generalized pyrimidine
DNA • deoxyribose sugar • double -stranded helix • nucleotides linked by phosphodiester bonds • two strands are antiparallel • thymine instead of uracil	Deoxyribose sugar		Adenine Guanine	Cytosine Thymine
RNA • ribose sugar • single-stranded • nucleotides linked by phosphodiester bonds • uracil instead of thymine	Ribose sugar		Adenine Guanine	Cytosine Uracil

Deoxyribonucleic acid (DNA) and ribonucleic acid (RNA) are both composed of chains of subunits called *nucleotides*. Each nucleotide is, in turn, made up of a 5-carbon sugar, a phosphoric acid group, and a nucleotide base. DNA and RNA differ structurally in three main ways, as shown in the table of comparisons. First, the 5-carbon sugar in

DNA is deoxyribose and is missing an oxygen (noted by the arrow) that is present in ribose, the 5-carbon sugar in RNA. Second, the nucleotide bases in DNA are of four types, two purines and two pyrimidines. The same is true of RNA, but there is a difference in one of the pyrimidines. DNA contains thymine, whereas RNA contains uracil. The third difference is not shown in this table. DNA is a double-stranded molecule containing two paired nucleotide chains. RNA, on the other hand, is typically a single-stranded molecule.

As we mentioned earlier, finding an appropriate mnemonic may help you remember important relationships. For instance, you might try "Agents are pure in heart" to help you remember the nucleotide base compositions. Agents will remind you of A (adenine) and G (guanine). *Pure* is for purines, and *heart* reminds you that purines have two attached rings (heartlike, if you have a good imagination). All other nucleotide bases are single-ring pyrimidines. In order to test your understanding of these genetic building blocks, you might set a hypothetical sequence to diagram. Use simplified representations of nucleotide components. Several of the problems in the following Problem Set are designed in this way. An almost endless variety of similar problems can also be written. Just remember, if you can set such a problem for yourself and solve it correctly, you can do the same thing on an examination.

Three classical experiments played a central role in confirming the role of DNA as the genetic material. These experiments were (1) the demonstration of a transforming factor by Griffith (1928); (2) the establishment that this transforming factor was DNA, as shown by Avery, MacLeod, and McCarty (1944); and (3) the confirmation that DNA was not only necessary but sufficient to code for the growth of a new organism, as reported by Hershey and Chase (1952). This sequence of experiments provides an outstanding example of the development and confirmation of an idea and the application of experimental methods. You should look at their designs and conclusions carefully.

In 1953, Watson and Crick proposed a model of DNA as a pair of helically coiled nucleotide chains or strands held together by hydrogen bonding between the nucleotide bases. Within a strand, one nucleotide is bound to the next by a phosphodiester bond between the 3′ carbon of one nucleotide and the 5′ carbon of the next. This creates an important asymmetry in the molecule, with a 3′ end and a 5′ end to each strand.

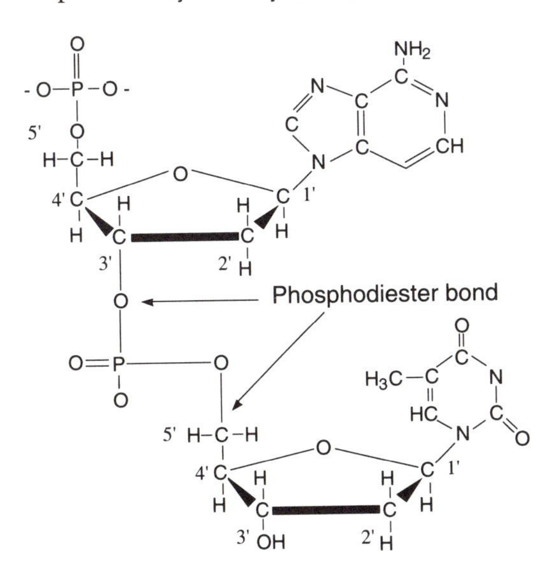

Two nucleotide strands are linked by hydrogen bonds between specific pairs of nucleotide bases. The purine adenine pairs with the pyrimidine thymine by two hydrogen bonds, and the purine guanine pairs with the pyrimidine cytosine by three hydrogen bonds. Furthermore, the orientation of the nucleotide bases in one chain is inverted, when compared to the other chain, so one side has a 3′ to 5′ orientation and the other has a 5′ to 3′ orientation. This is called *antiparallel* since the two strands are parallel but "backward" to each other. In the following diagram, we have exploded a small segment of DNA to show the components and their relationships.

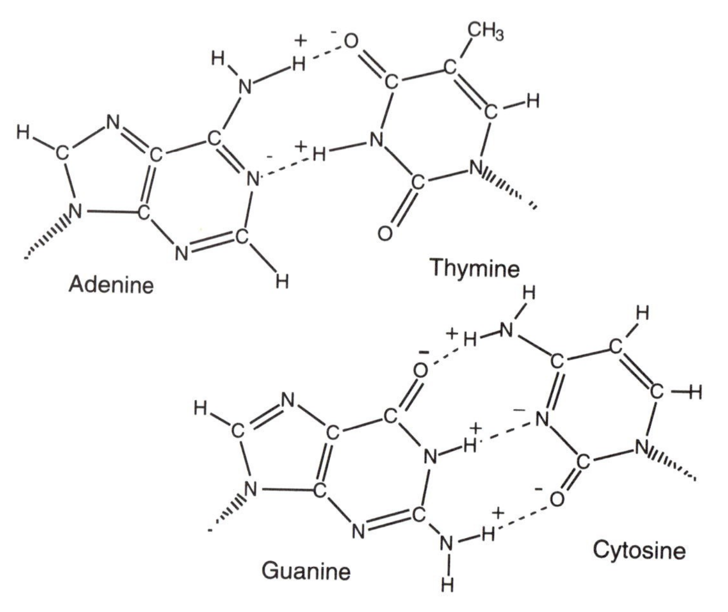

DNA replication is summarized in the diagram. There are several key points to remember. First, the pairing of A and T, and of G and C, gives rise to Chargaff's rule equating the members of each pair [A = T and G = C; A + G = C + T; or (A + G)/(C + T) = 1]. If Chargaff's rule holds for a given set of nucleotides, the molecules must be double-stranded.

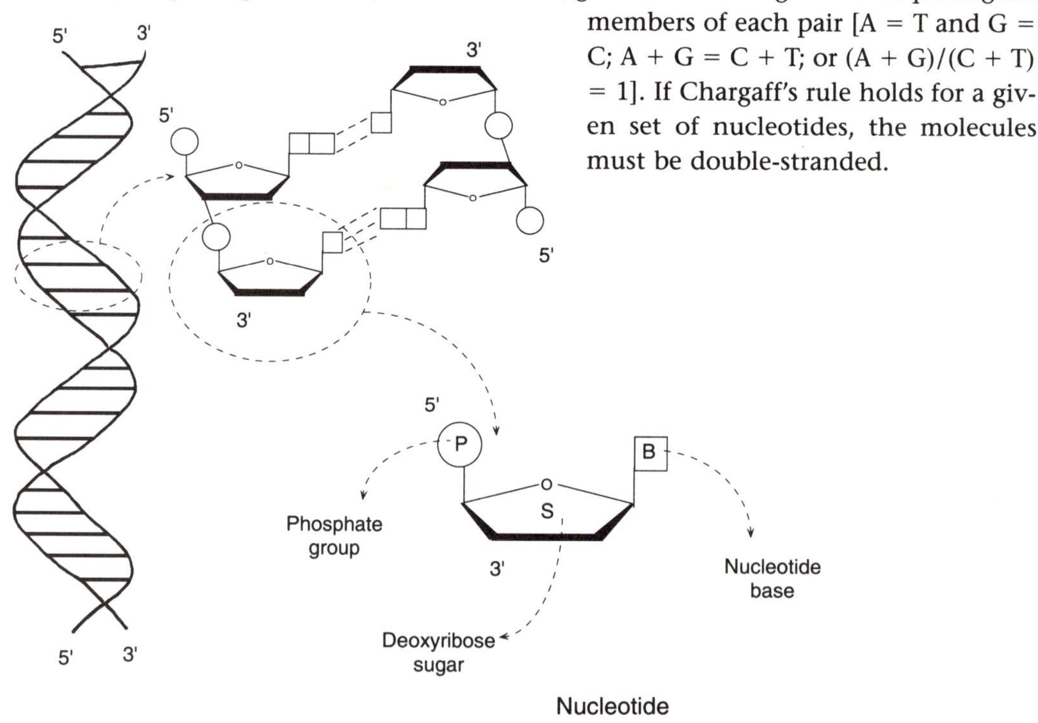

Nucleotide

Second, remember that DNA polymerases synthesize new DNA strands by adding new nucleotides at the 3′ end of a chain. In other words, DNA elongates in the 5′ to 3′ direction. Since the two strands of the double helix are antiparallel, one new strand will be formed as a continuous chain, and the other is formed in short sections as unwinding proceeds. An RNA primer is first laid down, as indicated in the figure by the circles. The short segments, called *Okazaki fragments,* are then joined when the RNA is replaced by DNA nucleotides and the DNA is linked by ligase (see problem 15).

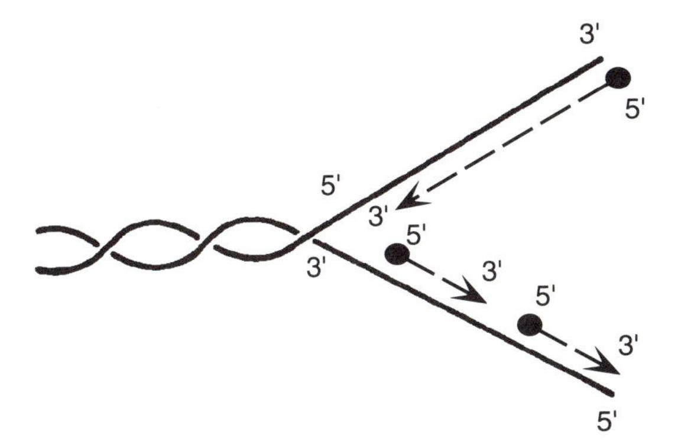

IMPORTANT TERMS

Endonuclease
Eukaryote
Exonuclease
Gyrase
Helicase
Ligase
Nucleoside
Nucleotide
Phosphodiester bond
Polymerase
Primase
Primer
Primosome
Prokaryote
Purine
Pyrimidine
Repetitive DNA
Replication
Semiconservative replication
Single-strand binding proteins
Template DNA
Topoisomerase
Transcription
Translation

PROBLEM SET 3

1. A single strand of DNA contains the base sequence 5′AGCTG3′.

(a) What is the complementary strand sequence?

(b) If the complementary strand is used during transcription, what is the base sequence of the resulting RNA?

2. What kind of evidence indicates that nucleotides of most organisms occur in matched pairs in the DNA molecules?

3. If it is known that about 22 percent of the double-stranded DNA of an organism consists of thymine, can the other base percentages be determined? If so, what are they?

4. The gene for a polypeptide (double-stranded DNA with 300 nucleotide pairs) has a base composition of A = 0.32, G = 0.18, C = 0.18, and T = 0.32. Assume that a single 300-base strand of RNA is transcribed from this gene. Can you determine, from the information given, the base composition of the RNA? If so, what is it?

5. A certain DNA virus has a base ratio of $(A + G)/(C + T) = 0.85$. Is this single- or double-stranded DNA? Please explain your answer.

6. Label the nucleosides in the following figure as RNA or DNA components, and give the reason for your answers.

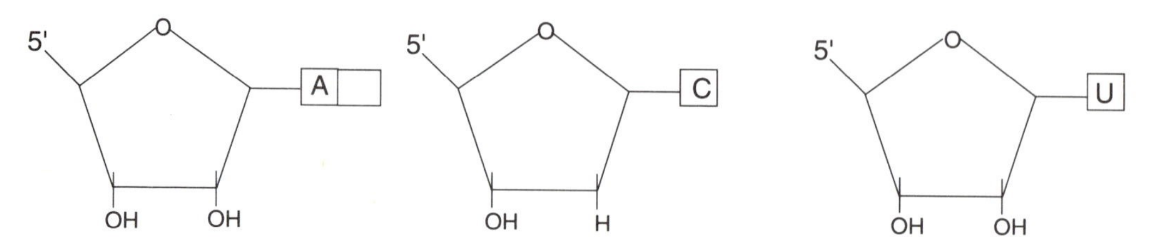

7. Multiple choice (in this and later multiple choice questions): Select the best answer and be certain you can explain the reason for your choice. There is a reasonably good correlation between the amount of DNA and the degree of complexity of the organism in which of the following? **(a)** viruses and bacteria; **(b)** viruses, bacteria, and eukaryotes; **(c)** only eukaryotes.

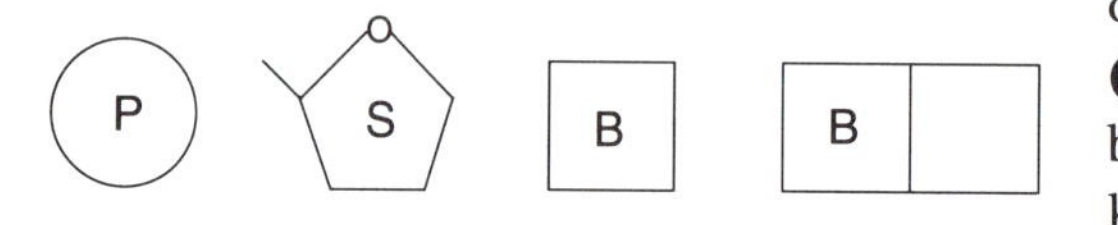

8. Using the following symbols for nucleotide components, construct a single strand of DNA with this sequence: 5′ATGCGT 3′. Now, using your newly constructed DNA strand as a template, construct the complementary strand as if this DNA were replicating. Indicate the direction of growth.

9. Why does DNA polymerase III require an RNA primer?

10. In double-stranded nucleic acid, the proper alignment of bases needed for the hydrogen bonding of complementary bases is possible only when the two polynucleotide chains are oriented in which of the following: **(a)** the same direction? **(b)** opposite directions?

11. The Hershey and Chase experiment used bacteriophage and a chemical difference between proteins and nucleic acids to demonstrate the role of proteins and DNA in genetic transmission. Explain the chemical differences the researchers utilized, and describe their experiment.

12. The genetic material in a hypothetical organism has the base ratio (U + C)/(A + G) of approximately 1.02 as determined experimentally by chemical analysis. This organism is also known to contain 20 percent adenine. Can the other base percentages be determined? If so, what are they?

13. From the following figure, indicate whether transcription is occurring in a prokaryotic or a eukaryotic cell. Please explain.

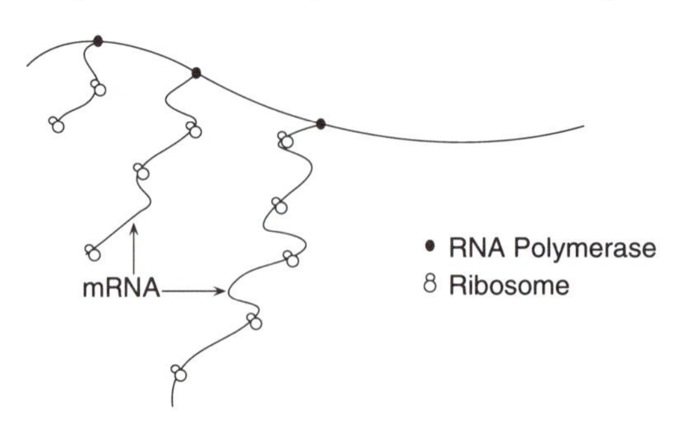

14. The Meselson and Stahl experiment used *Escherichia coli* with ^{15}N-labeled DNA transferred to a ^{14}N medium. The isotope constitution of its replicated DNA was studied by density-gradient separation. Since DNA replication is semiconservative, the Meselson–Stahl experiment would be expected to show which of the following density patterns, after the third replication in ^{14}N?

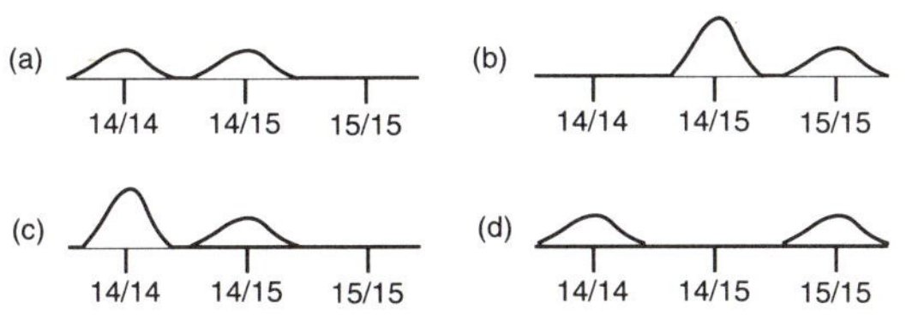

15. The following figure shows a replication fork during the replication of double-stranded DNA. Briefly identify the numbered proteins and describe their functions in DNA replication.

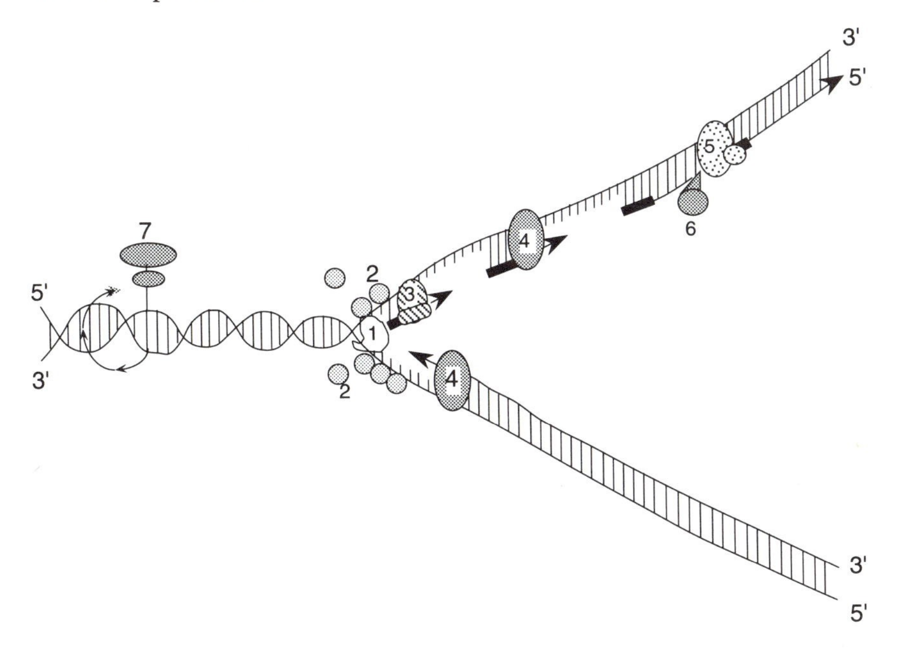

ANSWERS TO PROBLEM SET 3

1. Since adenine pairs with thymine and cytosine pairs with guanine,

(a) RNA polymerase transcribes 5′ to 3′; therefore, the complementary strand is TCGAC.

(b) The base sequence of the resulting RNA is the same as the original base sequence, except that thymine is replaced by uracil: AGCUG.

2. The ratios of adenine to thymine and of cytosine to guanine are 1:1. This relationship can be represented in many different forms (some of which are illustrated in the following answers) and is commonly known as *Chargaff's rule.*

3. Yes, if 22 percent of the molecule is thymine, then 22 percent is also adenine. The remaining 56 percent is divided equally between cytosine and guanine (28 percent each).

4. No, the base composition of the RNA molecule would be determined by the base sequence of only one of the DNA strands. There is no way to determine the base ratios from the data given.

5. The DNA in this virus is single-stranded, because the 1:1 ratio of A to T and of G to C typical of a double-stranded molecule is not found here. Note also that the equality pairs one purine (e.g., adenine) with one pyrimidine (e.g., thymine). In the formula given in the problem, (A + G)/(C + T) should equal 1 in a double-stranded molecule, because the number of purines equals the number of pyrimidines in such a molecule. Be sure to look at a given ratio carefully, however, because a test maker could easily make alterations that would change the expectation; for instance, (A + T)/(C + G) would not be predictable in an organism.

6. **(a)** RNA nucleoside, because the sugar is ribose.
(b) DNA nucleoside, because the sugar is deoxyribose (literally, "missing an oxygen").
(c) RNA nucleoside, because the sugar is ribose and the base is uracil.

7. **(a)** Viruses and bacteria. There is a tendency for an increase in the amount of DNA in eukaryotes. Some protozoa, club mosses, and algae, however, have more DNA per cell than do mammals. There are numerous examples among the lower vertebrates in which the amount of DNA is larger than in more advanced vertebrates. In fact, animals of similar morphologic characteristics may also have quite different DNA content. Depending upon the organism, as much as 90 percent of the DNA in eukaryotes may not code for unique polypeptides and therefore will not reflect morphological or physiological complexity.

8. The direction of growth is the 5′ end of one nucleotide attaching to the 3′ end of the growing strand.

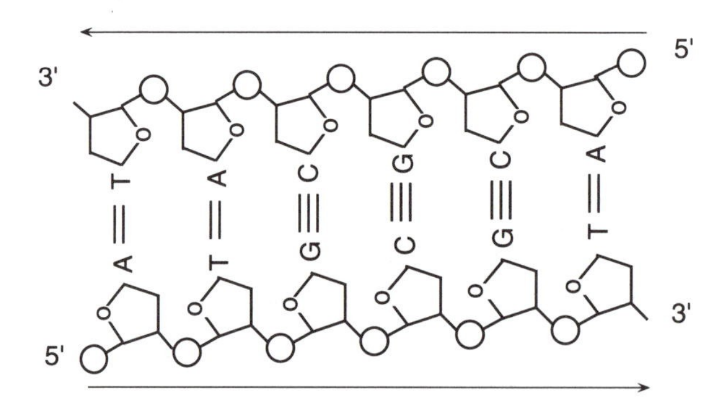

9. DNA polymerase can add nucleotides only to an open 3′ OH end, whereas RNA polymerase does not require such a preexisting "primer." The RNA polymerase produces a 3′ OH end for the DNA polymerase III to begin replication.

10. **(b)** Opposite directions, an arrangement called *antiparallel*.

11. They used the T2 virus, which is composed of a strand of DNA contained within a protein capsule. The protein capsule has some sulfur-containing amino acids but no phosphorus, whereas DNA contains phosphorus but no sulfur. They marked two cultures of T2 viruses by growing one on *E. coli* containing radioactive sulfur and the other on *E. coli* containing radioactive phosphorus. Thus they obtained ^{32}P-labeled DNA in one culture and ^{35}S-labeled protein in the other. These were then used to infect nonradioactive *E. coli*. It was found that 30 percent of the radioactive phosphorus was transmitted to the next generation, but none of the radioactive sulfur was transmitted. This experiment, in agreement with the study by Avery, MacLeod, and McCarty, showed DNA to be the genetic material.

12. Yes, 1.02 is close enough to 1.00 to be the result of a minor experimental error in measuring the equal base ratios of a double-stranded RNA, and the other base percentages would be 20 percent uracil, 30 percent cytosine, and 30 percent guanine.

13. Transcription is occurring in a prokaryote, since the ribosomes have attached to the messenger RNA (mRNA) while it is still being transcribed by RNA polymerase. If this were a eukaryote, the mRNA would have to be processed and leave the nucleus before ribosomes could complex with it.

14. **(c)** After each replication, the proportion of 14/14 will increase. At the first replication, all of the new DNA will be 14/15 hybrid molecules. At the next replication, the ratio of 14/14 to 14/15 will be 1:1. At the third replication, the ratio will be 3:1.

15. **(1)** Helicase or helix unwinding protein, which separates the parental strands.
(2) Single-strand binding protein or helix destabilizing protein, which prevents separated strands from rejoining.
(3) RNA primase or primosome complex, which synthesizes the RNA primer.
(4) DNA polymerase III, which synthesizes DNA using the RNA primer.
(5) DNA polymerase I, which extends the DNA synthesized by DNA polymerase III and removes the RNA primer of the Okazaki fragment in front of it.
(6) Ligase, which joins the last nucleotide of an Okazaki fragment to the adjacent fragment, forming a phosphodiester bond.
(7) Gyrase, which relaxes the supercoils that are extra coils of the coiled molecule caused by unwinding DNA.

CROSSWORD PUZZLE

NUCLEIC ACIDS: DNA AND RNA

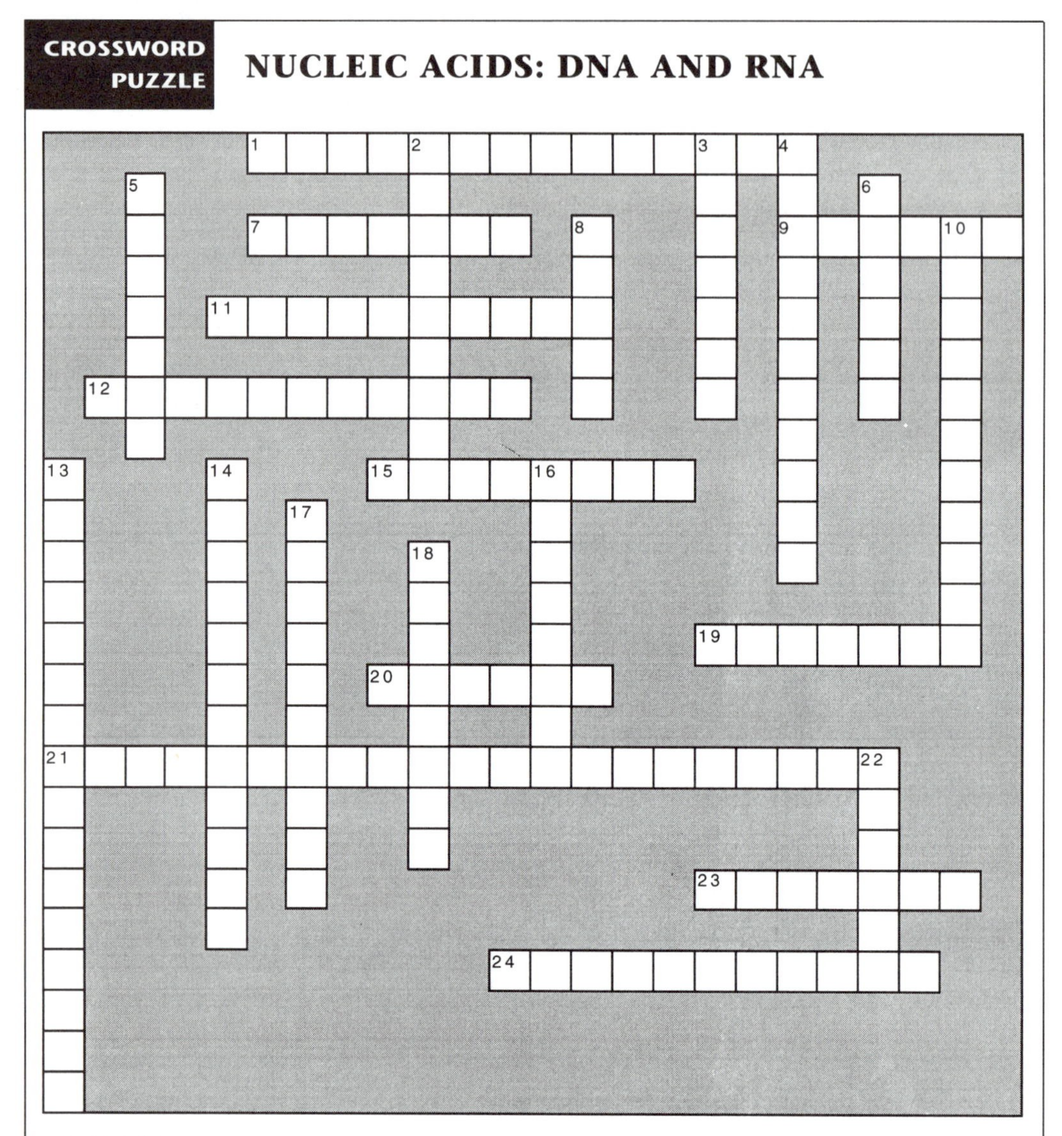

ACROSS

1. Bond between the 3′ end of one nucleotide and the 5′ end of another nucleotide
7. Purine base that does not contain a double-bond oxygen
9. RNA sequence that is required for DNA replication
11. Molecule that contains a nucleoside and a phosphate group
12. DNA's 5-carbon sugar
15. DNA-unwinding enzyme
19. Enzyme responsible for producing the RNA primer of a replicating DNA molecule
20. Pyrimidine base found in RNA but not DNA
21. Protein that prevents the reannealing of DNA during replication
23. Term for the replicating fragments on the so-called lagging strand of DNA
24. Term for the single-ringed nitrogen-containing bases

DOWN

2. Complex of proteins involved in the primer formation
3. Pyrimidine that is found in DNA, but not found in RNA
4. Term for the synthesis of DNA
5. Double-ringed nitrogen-containing bases
6. Enzyme that joins nucleotides together
8. One of the men that demonstrated DNA was the genetic material

10. Enzyme that removes the terminal nucleotide of nucleic acids
13. Describes the way DNA replication occurs
14. Enzyme that cuts the sugar-phosphate backbone of DNA
16. Pyrimidine with only one double-bond oxygen
17. Term for a molecule containing a 5-carbon sugar and a nitrogen containing base
18. Man who first indicated the 1:1 ratio of A to T and G to C.
22. Enzyme that removes supercoils from DNA helix as it unwinds during replication

BASIC MENDELIAN GENETICS

STUDY HINTS

Although Mendel did not know about the events that occur in a dividing cell, his laws of segregation and independent assortment are really descriptions of chromosome behavior during meiosis. *Segregation* simply refers to the fact that the two homologous chromosomes separate to opposite sides of the cell (that is, they segregate) after synapsis in the first meiotic division. *Independent assortment* reflects the fact that the orientation of one pair of homologous chromosomes at the equatorial plate in cell division is independent of the orientation of other pairs; that is, alignment of chromosomes occurs randomly relative to one another.

The secret to solving Mendelian genetics questions is to recognize that there is a pattern hidden within the superficial confusion of facts with which you are presented. The first important step is to have a clear understanding of mitosis and meiosis. Then you must have a thorough working knowledge of the terminology. Beginning students commonly confuse *phenotype* and *genotype,* or *gene* and *allele. Genotype* is the genetic makeup (such as *Aa*), whereas *phenotype* is the expression of the genotype interacting with its environment (such as brown versus black fur). Different *genes* code for totally different protein products (such as a pigment enzyme versus blood membrane protein), whereas *alleles* are the different forms that a specific gene can have (such as normal enzyme *A* versus the lower-activity form of the same enzyme *a*). A list of Important Terms is given at the end of this section, and definitions can be found in the Glossary. We recommend that you use this list of terms as a study aid to test your own ability to define terms. Set up flash cards for difficult terms. Remember that beginning to study a new science is not very different from beginning to learn a foreign language.

Perhaps the most frequent problem that students have in Mendelian genetics is understanding the implications of segregation. As can be seen from meiosis, each parent transmits only one allele of each gene to an offspring. If the parent is homozygous, each gamete will be identical for that gene, but if the parent is heterozygous, two different gamete types will be produced, each of equal frequency. For a dihybrid (for example, *AaBb*), a single allele of each gene must be included in each gamete. If you use a Punnett square to summarize the outcomes of a cross, be sure to double-check that the gametes you list are the correct products of segregation.

The other common difficulty is extracting the appropriate information from a written problem. This takes practice, but there is a logical sequence to follow that might help make the process more routine. First, determine the mode of inheritance. Which trait is dominant, and which is recessive? This may be given, but if not, simply look to see which is expressed in the phenotype of an individual that has inherited both alleles.

Second, define symbols, using a capital letter or letters for the dominant trait and a lowercase letter for the recessive. Then assign genotypes based upon the phenotypes for each individual. Here it is extremely important to distinguish between what you "know" and what you "hypothesize" to be the genotype of an individual. If the individual is homozygous or described as true-breeding or inbred, the genotype can be assigned unambiguously. On the other hand, if the ancestry of offspring phenotypes is unavailable or inconclusive, an individual with a dominant phenotype may only be "known" to carry at least one dominant allele. Do not exceed what you are certain of. Simply represent the second allele by a dash to indicate the fact that it might be either dominant or recessive (i.e., *A* – could be either *AA* or *Aa*).

Next, ask what types of gametes each parent can produce, and summarize the combinations in a Punnett square. Once you have gained some practice, you should then begin to interpret the outcomes on the basis of probabilities.

Finally, reread the question to be certain that you are answering what it has asked. It is often at this point that the correct work can yield the wrong answer if you give genotypic ratios, for example, and the problem asked for phenotypic ratios.

Always remember that solving Mendelian genetics problems is just like working a series of puzzles, using a very precise vocabulary. Characteristics, such as the shape of watermelons and the hair color of guinea pigs, can make questions look very different, but their inheritance patterns generally follow the same set of fairly simple rules.

IMPORTANT TERMS

Allele
Continuous variation
Dihybrid
Discontinuous characters
Dominant
Gene
Genome
Genotype
Heterozygous
Homozygous
Independent assortment
Monohybrid
Pedigree
Phenotype
Recessive
Segregation
Testcross

PROBLEM SET 4

1. In humans, assume that the presence of dimples is a dominant trait. Suppose that in one family both parents have dimples, but their daughter does not. Their son does have dimples. What are the genotypes of the four people in this family?

2. In humans, assume that having attached earlobes is recessive to having free earlobes. A woman and her husband both have free earlobes, although each happens to have a father with attached earlobes. Is it possible for this woman and man to have a child with attached earlobes? Please explain by indicating the genotypes of the parents and child.

3. In the garden pea, tall is dominant to short and red flower color is dominant to white. These genes assort independently. Pure-breeding tall red plants are crossed with pure-breeding short white plants, and the F_1 are crossed with each other to produce an F_2. What proportion of F_2 plants have the same genotype as the F_1 plants?

4. John bought a pair of black mice from his friend Steve. Steve did not know anything about their pedigree, except that at least one great-grandparent mouse had

had brown fur. John discovered that in his first litter of 10 baby mice, 2 had brown fur. What can be concluded about the inheritance of hair pigmentation and the genotypes of his 2 black parent mice?

5. In humans and chimpanzees, the ability to taste phenylthiocarbamide (PTC) is dominant to inability to taste it. The cell of a heterozygote at anaphase I of meiosis would have how many *T* and how many *t* alleles? at metaphase of mitosis? at metaphase II of meiosis? It will help you answer questions like this if you sketch the cells yourself.

6. The characters dumpy wing (*dp*), brown eye color (*bw*), hairy body (*h*), ebony body color (*e*), shortened wing veins (*ve*), and eyeless (*ey*) in *Drosophila* are all produced by recessive alleles. Assuming that these loci are not closely linked, how many different kinds of gametes will be produced by a fly having the genotype *Dp dp bw bw Hh EE Ve ve ey ey*?

7. A normal woman whose father is an albino wishes to marry a phenotypically normal man whose mother was albino. She would like to know her chance of having an albino child. Knowing that in this case albinism is caused by a recessive autosomal gene, what should she be told?

8. Chinchilla coat color in mice is due to a recessive allele (c^{ch}) and the normally pigmented coat, called agouti, is due to the dominant allele. Mice heterozygous for chinchilla coat color are mated to homozygous chinchilla (c^{ch} / c^{ch}) mice, and the resulting progeny include both agouti and chinchilla phenotypes. If one takes one of the agouti progeny and mates it to one of its chinchilla sibs, the expected genotypic ratio among their offspring is **(a)** 3:1, **(b)** 1:1, **(c)** 9:3:3:1, **(d)** 1:2:1, **(e)** none of the above.

9. In a cross of *AaBbccDd* to *AabbCCDD*, the frequency with which we would expect to obtain progeny of the genotype *AABbCcDd* is **(a)** 1/128, **(b)** 1/64, **(c)** 1/32, **(d)** 1/16, **(e)** 1/8, **(f)** none of the above.

10. In a cross of *AaBbCcDd* to *AabbCcDD*, the proportion of progeny that show all four dominant traits is **(a)** 81/256, **(b)** 27/64, **(c)** 54/64, **(d)** 27/256, **(e)** 27/32, **(f)** 9/32, **(g)** 3/32, **(h)** none of the above.

11. If the hypothetical autosomal dominant gene *N* is necessary for the production of pigment required for normal vision and the autosomal dominant gene *S* results in blindness because of the disorganization of neuronal synapses in the eye, what proportion of the progeny produced by the cross *Nnss* × *nnSs* will be blind?

12. Assume that straight hair (*S*), golden brown fur (*G*), and hairy ears (*H*) are dominant to curly hair, dark brown fur, and hairless ears, respectively. All three loci assort independently. In a cross of *Ss Gg HH* with *SS Gg Hh*, how many different phenotypes will be found among the progeny?

13. Referring again to the cross involving straight hair, golden brown fur, and hairy ears, how many different genotypes will be found among the progeny?

14. Hygienic behavior in honeybees is determined by two recessive alleles. Those homozygous for one of these recessives will uncap the brood cells of the developing bees killed by an infection of American foulbrood bacteria. Those homozygous for the other recessive will remove the dead bee and clean the cell, thus removing the source of further infection. Both of these independently assorting loci must be homozygous recessive for hygienic behavior to occur. A testcross of dihybrid non-hygienic honeybees will produce what proportion of colonies that can successfully resist an infection of American foulbrood? Although it doesn't really affect the answer of this problem, you should be aware that the male bee is haploid.

15. Which of the following alternatives is/are correct? For two pairs of alleles, *Aa* and *Bb*, it is theoretically possible to have **(a)** segregation without independent assortment, **(b)** independent assortment without segregation.

16. Assume that curly hair (*s*) in rabbits is recessive to straight hair (*S*). In a cross of *Ss* to *Ss*, the probability of obtaining exactly 3,000 straight-haired and 1,000 curly-haired progeny in a sample of 4,000 offspring is **(a)** 100 percent, **(b)** very high but not quite 100 percent, **(c)** small.

17. Assume that a diploid organism has seven pairs of chromosomes. If the organism is heterozygous at only one locus on each of these seven pairs of chromosomes (*Aa*, *Bb*, . . . , *Gg*), independent assortment will permit the production of how many genetically different gametes? **(a)** 7, **(b)** 14, **(c)** 16, **(d)** 28, **(e)** 49, **(f)** 128.

18. If the organism in problem 17 reproduces by self-fertilization, the number of different genotypes expected among the progeny of this septa hybrid is **(a)** 14, **(b)** 21, **(c)** 49, **(d)** 128, **(e)** 256, **(f)** 512, **(g)** 2,187, **(h)** 4,374, **(i)** none of the above.

19. Cystic fibrosis is inherited as an autosomal recessive trait. Assume that it is possible to identify a heterozygote through a molecular genetic test (we will explore ways of doing this in a later chapter). Suppose that a man whose brother had cystic fibrosis is married and wishes to have children. The result of a skin test indicates that he is a carrier. Although there is no history of cystic fibrosis in his wife's family, her skin test result also shows that she is a carrier. This couple then has three children who do not show any of the characteristics of this condition. What are the chances that their fourth child will have cystic fibrosis?

20. A testcross of a trihybrid (*AaBbCc*) with complete dominance, independent assortment, and no gene interactions will give which genotypic ratio? **(a)** 1:1:1:1, **(b)** 9:3:3:1, **(c)** 1:1:1:1:1:1:1:1, **(d)** 27:9:9:9:3:3:3:1. Which will be its phenotypic ratio?

21. A cross is made of *AaBb* × *AaBb*. The *A* and *B* loci assort independently. The progeny of this dihybrid cross are then allowed to fertilize themselves. The proportion of the progeny that show segregation for the *A* locus (that is, produce *A*– and *aa* progeny) is **(a)** 1/2, **(b)** 9/16, **(c)** 3/16, **(d)** 1/16, **(e)** none of the above.

22. In a cross of *AaBbccDd* × *AaBbCcDD*, with independent assortment for all loci, the fraction of progeny that are homozygous for all four loci is **(a)** 3/4, **(b)** 1/3, **(c)** 1/2, **(d)**1/4, **(e)** 1/8, **(f)** 1/16, **(g)** 1/32.

23. If an individual with the genotype *AaBbCcDDEe* is mated to an individual with the genotype *aaBbCcddEe*, what proportion of the offspring would be *aabbccDdee*? What proportion would be *aabbccddee*?

24. Two guinea pigs are known to be heterozygous for a dominant mutation that causes them to have kinked tails. The breeder would like to establish a true-

breeding kinked-tail line. She first gives away all of the normal-tailed guinea pigs that are born from that pair. What proportion of the remaining pups are homozygous for the kinked-tail allele?

25. Let us assume that two genes affect pod development in strains of beans you are studying. Long pods (*L*) is dominant to short pods (*l*), and round seeds (*R*) is dominant to wrinkled seeds (*r*). In the progeny produced from one cross, you find the following types and proportions of plants: 3/8 long and round, 3/8 long and wrinkled, 1/8 short and round, 1/8 short and wrinkled. What are the phenotypes and genotypes of the parent plants?

ANSWERS TO PROBLEM SET 4

1. The two parents both have dimples and must therefore have at least one dominant allele for dimples. Let us define the dominant as *D* and the recessive as *d*. Their daughter is homozygous recessive (*dd*) for this trait. This means that she received a recessive allele (*d*) from each parent, and the parents must therefore both be heterozygotes (*Dd*). The son has inherited at least one dominant allele from a parent, but it is not possible to tell whether he is homozygous or heterozygous for this trait. We shall simply represent his known genotype as *D*– to indicate that we do not have the evidence to establish the second allele.

2. Since both the woman and her husband had a parent with the recessive trait, they must both be heterozygotes. If we denote the traits as *F* for free earlobes and *f* for attached earlobes, the parents are *Ff* and the child with attached earlobes would be *ff*. The probability of having such a child is 1/4.

3. The F_1 plants of a cross between two homozygotes will be heterozygous for both genes. Let's define *T* as the dominant allele for tall plants, *t* as the recessive allele for short plants, *R* as the dominant allele for red flowers, and *r* as the recessive allele for white flowers. When these *Tt Rr* dihybrid F_1 plants are crossed, nine different genotypes will be produced. Since the genes segregate independently, we can calculate the probability of a dihybrid being formed in the F_2 by looking at the genes one at a time. When *Tt* is crossed to *Tt*, there is 1/2 chance that the progeny will be *Tt*. Similarly, there is a 1/2 chance that the progeny will be *Rr*. The probability of being both *Tt* and *Rr* is the product of the independent probabilities. Thus, 1/4 of the F_2 offspring will be dihybrid like the F_1.

4. Although the mice John bought from Steve showed the dominant trait, they must have both been heterozygous for coat color (which is consistent with there being a homozygous recessive individual in an earlier generation). Two out of 10 is near 1/4, the theoretical expectation from a monohybrid cross. You can therefore conclude that one segregating locus affects coat color in this pedigree and that black is dominant to brown. If we let *B* be the black allele and *b* be the brown allele, the parents John bought were both *Bb*. It is useful to note that the ratio of different kinds of offspring can be compared to theoretical ratios and then predict (hypothesize) what the parent genotypes must have been. That is simply going the opposite direction from the typical thought pathway of using parent genotypes to predict offspring types.

5. At anaphase I of meiosis, segregation would be under way but not yet complete. Thus there would be a total of two *T* and two *t* alleles (remember that DNA replication has duplicated each chromosome in the preceding interphase). At metaphase of mitosis, the same would be true. There would be a total of two *T* and two *t* alleles. At metaphase II of meiosis, however, either two *T* or two *t* would be present, since segregation has occurred but cell division has not yet been completed. The resulting cells would include only a single *T* or a single *t* allele each.

6. If the loci are not closely linked, then all combinations of alleles are possible. All gametes will carry *bw, E,* and *ey,* since the flies are homozygous for these alleles. The gametes will be segregating for all the others. The number of different kinds of gametes = 2^n, where *n* is the number of segregating loci. In this example, $2^3 = 8$. The gametes are shown in the accompanying figure.

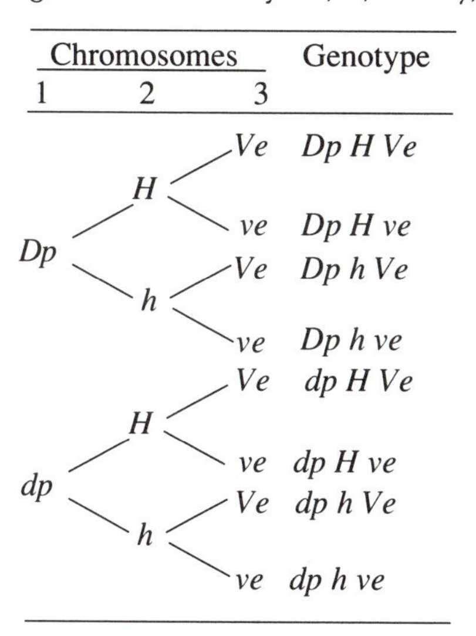

7. The man and the woman are both heterozygous for the recessive albino allele. Since albinism is relatively rare in the population, we can assume that their phenotypically normal parents are homozygous *AA*. There is, therefore, a probability that 1/4 of their children will be homozygous (phenotypically albino). In other words, there is a 1/4 probability that any one of their children will be albino.

8. **(b)** The parents are Cc^{ch} and $c^{ch}c^{ch}$. Since a heterozygous chinchilla mouse carries the dominant *C* allele, in addition to c^{ch}, the progeny consist of half with normal pigmentation (Cc^{ch}) and half with chinchilla color. The cross of the normal F_1 mice to chinchilla mice is therefore Cc^{ch} to $c^{ch}c^{ch}$ and the expected genotypic ratio is half Cc^{ch} and half $c^{ch}c^{ch}$ or 1:1. If wording in a problem like this confuses you, it is useful to write out the symbol and make your own pedigree:

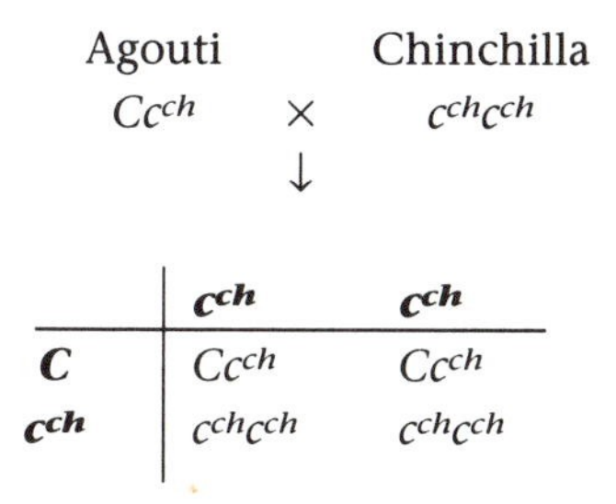
Agouti × Chinchilla

Cc^{ch} × $c^{ch}c^{ch}$

↓

	c^{ch}	c^{ch}
C	Cc^{ch}	Cc^{ch}
c^{ch}	$c^{ch}c^{ch}$	$c^{ch}c^{ch}$

9. **(d)** Taking each gene separately, the probability of getting *AA* from a cross of *Aa* to *Aa* is 1/4; the probability of getting *Bb* from a cross of *Bb* to *bb* is 1/2; the probability of getting *Cc* from a cross of *cc* to *CC* is 1; and the probability of getting *Dd* from a cross of *Dd* to *DD* is 1/2. Assuming independent assortment for all four loci, the probability of getting *AABbCcDd* is the product of the independent probabilities, or 1/4 • 1/2 • 1 • 1/2 = 1/16. The same answer could be obtained in a more visual way by identifying the combinations as in the answer to problem 5.

10. **(f)** Taking each gene separately, the probability of getting *A* – (that is, at least one *A* allele) from a cross of *Aa* to *Aa* is 3/4. the probability of getting *B* – from a cross of *Bb* to *bb* is 1/2; the probability of getting *C* – from a cross of *Cc* to *Cc* is 3/4, the probability of getting *D* – from a cross of *Dd* to *DD* is 1. Assuming independent assortment for all four loci, the probability of inheriting all four dominants is 3/4 • 1/2 • 3/4 • 1 = 9/32. Again, the combinations could be identified and counted as in problem 5.

11. Assuming independent assortment, the gametes formed by the two parents and the progeny produced are

	nS	***ns***
Ns	*NnSs*	*Nnss*
ns	*nnSs*	*nnss*

If the absence of pigment (*nn*) results in blindness and if neural problems result from the *S* allele, then any individual that is *nn* and/or *S* – will be blind. From the Punnett square, we can see that only 1/4 will have vision (*Nnss*); therefore, the answer is that 3/4 will be blind.

12. Since all genes are segregating independently, we can consider them one at a time. From the *Ss* × *SS* part of the cross, all offspring will inherit and *S* allele from the second parent and, thus, have straight hair. From *Gg* × *Gg* there will be a 3:1 ratio of golden brown fur to dark brown fur. Finally, from *HH* × *Hh,* all will inherit the *H* allele from the first parent and have hairy ears. There are, therefore, only two different phenotypes: straight, golden, hairy and straight, dark, hairy.

13. Again, taking the genes one at a time, *Ss* × *SS* yields two genotypes, *Gg* × *Gg* yields three (*GG, Gg,* and *gg*), and *HH* × *hh* yields two. In total there will be 2 × 3 × 2 = 12 different genotypes. You can use a branching diagram like the one we used earlier to show all possible combinations of genotypes.

14. Let us designate the uncapping genotype as *uu* and the removing genotype as *rr.* The heterozygous nonhygienic worker bees are then *UuRr,* and the hygienic males to which they are crossed are *ur* (honeybee males are haploid, but the answer would be the same if you treated them as typical diploids). The gametes and progeny phenotypes are

	Sperm			
	ur	**Uncap**	**Remove**	**"Hygienic"**
Eggs				
UR	*Uu Rr*	No	No	No
Ur	*Uu rr*	No	Yes	No
uR	*uu Rr*	Yes	No	No
ur	*uu rr*	Yes	Yes	Yes

Thus 1/4 of the colonies will be able to resist the infection.

15. By definition, alleles segregate from each other at meiosis. *Independent assortment* is the name given to the situation in which the segregation of one pair of alleles is independent of the segregation of a second pair of alleles. If the two genes are linked and thus not assorting independently, they will each still be segregating. Since independent assortment is really independent segregation, answer **(a)** is correct.

16. The answer is **(c),** small. Theoretical ratios are rarely obtained exactly, even with large numbers. If you flip a coin two times, for example, the theoretical ratio of one head and one tail is expected only half of the time (you expect two heads one-fourth of the time and two tails one-fourth of the time). The results from coin tossing, or rabbit breeding, usually assume a normal distribution with the theoretical ratios as the mean and with deviations in both directions. The exact theoretical ratios here (3,000:1,000) would occur only rarely. It would be analogous to someone's telling you that he or she had flipped a coin 4,000 times and had come up with exactly 2,000 heads and 2,000 tails. Although this is the theoretical value, you would be very dubious of the claim, and rightly so.

17. **(f)** With one locus or one pair of chromosomes (e.g., *Aa*), two genetically different gametes (*A* and *a*) are expected. With two loci or two pairs of chromosomes (e.g., *Aa* and *Bb*), four genetically different gametes (*AB, Ab, aB,* and *ab*) are expected. With three loci or three pairs of chromosomes, eight genetically different gametes are possible. This series can be represented by the expression 2^n, where the number 2 stands for the two alternatives at each locus and *n* is the number of loci or pairs of independently segregating chromosomes. In this case, the number of different gametes is 2^7, or 128. Be careful in applying a formula like this. Don't just count genes; look at the genotype carefully and count only the heterozygous (segregating) genes.

18. **(g)** This is a variant of problem 17. Instead of two possible gametic alternatives at each locus, we are dealing with possible *genotypes* at each locus. In all cases, the number is 3 (e.g., *AA, Aa,* and *aa*), so the number of different genotypes is 3^7 or 2,187.

19. Let us define the normal condition as *C* and the recessive cystic fibrosis as *c*. Both parents are carriers, *Cc*. Each of their children therefore has a 1/4 chance of having cystic fibrosis. It makes no difference how many previous children have been normal. The probabilities remain the same, since no previous events affect which eggs and sperm will fuse at fertilization.

20. **(c)** The trihybrid will produce $2^n = 2^3 = 8$ different kinds of gametes, where *n* is the number of heterozygous loci. A *testcross* is a cross to a homozygous recessive, in which each type of gamete can be identified in the phenotype of the progeny. The genotypic ratio will be 1:1:1:1:1:1:1:1. The phenotypic ratio will be the same, since here each genotype produces a different phenotype.

21. **(a)** Since the question asks only for the *A* locus, we can disregard the *B* locus. The cross is then, essentially, $Aa \times Aa$, and the progeny are 1/4 *AA,* 1/2 *Aa,* and 1/4 *aa.* If these are allowed to self-fertilize, only the *Aa* (1/2 of the progeny) will *show* segregation, producing 3/4 *A*– and 1/4 *aa* individuals. The *AA* will produce only *AA,* and *aa* will produce only *aa.* They will, therefore, *not show* segregation, although of course segregation of chromosomes will still be occurring in these individuals.

22. **(f)** Homozygosity can be due to having either two dominant or two recessive alleles (e.g., *Aa* × *Aa* will produce 1/4 *AA,* 1/2 *Aa,* and 1/4 *aa* or 1/2 homozygous and 1/2 heterozygous individuals). In this problem, we determine the fraction of progeny that are homozygous for each locus. Then, since they are assorting independently, we multiply the individual probabilities to get the probability for homozygosity at all four loci. As in the preceding problems, one could also work out the combinations and tally them, but applying probability expectations is more direct.

For the *A* locus, *Aa* × *Aa,* the probability is 1/2, as just stated.
For the *B* locus, *Bb* × *Bb,* the probability is 1/2.
For the *C* locus, *cc* × *Cc,* the probability is 1/2 (1/2 *Cc*:1/2 *cc*).
For the *D* locus, *Dd* × *DD,* the probability is 1/2 (1/2 *DD:* 1/2 *Dd*).
So, the overall probability is 1/2 • 1/2 • 1/2 • 1/2 = 1/16.

23. Taking the loci one at a time and assuming independent assortment, we have (1/2)(1/4)(1/4)(1)(1/4) = 1/128 and (1/2)(1/4)(1/4)(0)(1/4) = 0.

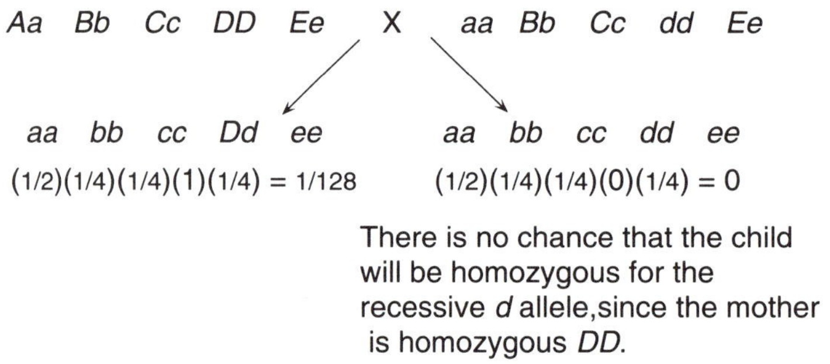

Aa Bb Cc DD Ee X *aa Bb Cc dd Ee*

aa bb cc Dd ee (1/2)(1/4)(1/4)(1)(1/4) = 1/128

aa bb cc dd ee (1/2)(1/4)(1/4)(0)(1/4) = 0

There is no chance that the child will be homozygous for the recessive *d* allele,since the mother is homozygous *DD.*

24. If we let *K* represent the dominant kinked-tail allele and *k* the recessive normal-tail allele, the cross of two heterozygous guinea pigs will be *Kk* × *Kk*. This will initially yield a 1:2:1 ratio of *KK:Kk:kk* progeny. The breeder, however, gives away all of the normal-tail (*kk*) guinea pigs. This leaves a 1:2 ratio of homozygous to heterozygous kinked-tail babies. Thus 1/3 will be homozygous for kinked-tails.

25. The parent strains have long and round seeds (*Ll Rr*) and long and wrinkled seeds (*Ll rr*). The first thing you realize from the offspring is that they do not fit a 9:3:3:1 ratio, so it is not a simple dihybrid cross. Probably the most straightforward approach is to analyze each trait separately. Looking only at pod size, there is a 3:1 ratio of long to short. A 3:1 ratio is the expected outcome of a monohybrid, so you can presume that the parents are *Ll* × *Ll*. Next, looking only at the seed shape, there are equal numbers of round and wrinkled seeds. A 1:1 ratio is the expected outcome of a cross of a heterozygote to a homozygous recessive, *Rr* × *rr.*

CROSSWORD PUZZLE

BASIC MENDELIAN GENETICS

ACROSS

1. Allele that expresses itself in the heterozygote
8. Nucleotide sequence that codes for an RNA sequence
9. Probability of two independent events occurring together is simply the ___ of their individual probabilities
10. Total genetic makeup of an organism
12. Alternative form of a gene
13. Number of genotypes produced in a simple monohybrid cross
15. Number of phenotypes expected in a dihybrid cross
17. Individual has the genotype *AaBbCcDDGg*, with no linkage. How many different kinds of gametes can this individual produce?
19. How many genotypes will be produced by the cross *AABb* × *aabb*?
20. Man responsible for the law of segregation
21. Zygote that has the same allele on the two homologous chromosomes
22. Genetic cross between two individuals that are heterozygous for two different genes
23. Describes an allele that does not express itself in the heterozygote
24. Cross between a homozygous recessive individual and an individual showing a dominant trait
25. Variation that cannot be divided into a few distinct categories

DOWN

1. Phenotypic variation that can be divided into distinct categories
2. Number of genotypes expected in a dihybrid cross

(continued on p. 36)

3. Number of phenotypes produced in a simple monohybrid cross
4. Random combination of alleles of two or more genes caused by being unlinked (on different chromosomes)
5. Term for alleles separating from each other during formation of gametes
6. Genetic cross between two heterozygous individuals in which only one locus is examined
7. Cross between the individuals *AABb* and *aabb* will produce a phenotypic ratio in which ___ of the offspring will be *Aabb*
11. Actual genetic makeup of the organism at a specific locus
14. What is expressed; the result of the genotype interacting with the environment
16. Describes an individual who has two alleles at a given locus
18. Diagram that shows relationships among members of a family over one or more generations

PROBABILITY AND CHI-SQUARE

STUDY HINTS

The ability to determine the probability of an event or series of events is fundamental to many applications of genetic principles. Thinking in terms of probability is not easy at first, but with a set of guidelines and some practice, you should find that you soon have no difficulty. In the following section we will discuss some important terms and then summarize the general types of probability problems you might expect to encounter in genetics.

First, let us contrast two important phrases. *Independent events* are events that have no causal interrelationship. The conception of the first child in a family, for example, cannot biologically influence the fusion of sperm and egg at fertilization for a second child. Each fertilization is an independent event, and each probability for segregation or sex determination must be assessed independently. *Mutually exclusive events,* on the other hand, are related in that the occurrence of one eliminates the possibility that the other will occur. A normal child cannot be both a boy child *and* a girl child. Sex determination yields either of two mutually exclusive events.

Both of these ideas play a role in solving a probability problem in genetics. Depending upon the genotype of the parents, the probabilities of mutually exclusive events, for example, the birth of "normal" progeny as opposed to the birth of "affected" progeny, may be different. In a similar way, the number and combinations of independent events considered in a problem influence not only the answer but also the way in which you find it.

There are perhaps three main levels of complexity in common probability problems: (1) determining the likelihood of a single independent event; (2) determining the likelihood of a sequence of events in which the order either is set by the problem or is not important; and (3) determining the likelihood of a sequence of events in which several different orders must be considered and accounted for. The ways of calculating each type of probability are summarized in the following outline.

I. *Individual independent events*
   A. *Examples:* A gamete carrying a dominant allele being formed; one child being a boy; a heterozygote being produced in a monohybrid cross.
   B. *Calculation:* Determine the proportion of times that such an event is expected to occur in repeated trials (e.g., 1/2 for the probability of a gamete carrying a dominant allele being formed in a heterozygote or 1 in a homozygote).

II. *Sequence of independent events where order is set or irrelevant*
   A. *Examples:* a family of three children, boy–boy–boy (all one class, so order irrelevant); a family of four children, boy–girl–girl–boy (order is set, or specified in the problem).
   B. *Calculation:* Multiply the individual probabilities (e.g., for a family of three boys, $1/2 \cdot 1/2 \cdot 1/2 = 1/8$).

III. *Sequence of events in which different orders must be pooled*

A. *Examples:* a six-child family composed of four girls and two boys in any order; a seven-child family composed of at least two affected children.

B. *Calculation:* Use the probability formula or some expansion of it.

1. For a single sequence,

$$\text{Probability} = \frac{n!}{s!t!}(p)^s(q)^t$$

where n is the number of individuals in the sequence (number of children in the family), p is the probability of the first event (e.g., being normal), q is the probability of the second event (e.g., being affected), s is the number of cases of the first event (e.g., number of normal children in a family), and t is the number of cases of the second event (e.g., number of affected children in the family). The value $n!$ (read "n factorial," that is, the product of all integers from 1 to n) divided by $(s!)(t!)$ gives the number of different ways in which the sequence of events can occur. By definition, $0! = 1$. For example, for a three-child family composed of one boy and two girls, $n!/s!\,t!$ is $3!/1!2!$, and, using the formula

$$\frac{(3 \cdot 2 \cdot 1)}{(1)(2 \cdot 1)} = 3$$

That is, the single boy could be the first, or the second, or the third child.

2. For combining several sequences: Note that one of the examples in Section III.A was a seven-child family composed of at least two affected children. Any family with two affected and five normal children, or three affected and four normal children, or four affected and three normal children, or any of the other possible combinations would fulfill this requirement. The probability can be calculated in either of two ways.

   a. Using the probability formula, calculate the likelihood for each family, and add the figures together. This is an accurate but time-consuming method.

   b. Expand the binomial expression for $(p + q)^n$, where p is the probability of the first event, q is the probability of the second, and n is the size of the sequence (e.g., size of the family). This method is faster and simpler than method (a). For example, for a seven-child family,

$$(p+q)^7 = p^7 + 7p^6q + \underline{21p^5q^2 + 35p^4q^3 + 35p^3q^4 + 21p^2q^5 + 7pq^6 + q^7}$$

   If we let p be the probability of being normal and q be the probability of being affected, the six family makeups underlined in the equation have at least two affected children. Probabilities must include all possiblilities and add up to 1. Note that p (at least two affected) $= 1 - (p^7 + 7p^6q)$, where the terms in parentheses refer to seven-child families with 0 or 1 affected.

Once the probability of an event or set of events has been determined, the next step is often to test the fit of observed data to these expectations. Chi-square (χ^2) tests are frequently used in genetics to test the significance of the deviation between observed and expected numbers.

We shall consider the use of χ^2 for testing two slightly different types of statistical fits. First, let us assume that you have counted the number of "normal" and "mutant" progeny that are produced by a cross between two heterozygous parents and found 137 normal and 63 mutant progeny. Your expectation is that these progeny will be produced in a ratio of 3 "normal" to 1 "mutant." If you have counted, say, 200 progeny, then the expected number of normal ones is 3/4 of these, or 150 progeny. You would expect 50 to have the mutant phenotype.

		Normal	**Mutant**	**Total**
Observed	(O)	137	63	200
Expected	(E)	150	50	200

A χ^2 value is calculated by measuring the deviation between observed and expected numbers ($O - E$), by summing the square of this deviation, and then dividing it by the expected number for each class. That is,

$$\chi^2 = \sum \frac{(O - E)^2}{E}$$

For the data just given, this calculation has the following results:

	Normal	**Mutant**	**Total**
Observed (O)	137	63	200
Expected (E)	150	50	200
$O - E$	-13	13	0
$(O - E)^2$	169	169	
$\frac{(O - E)^2}{E}$	$\frac{169}{150} = 1.127$	$\frac{169}{50} = 3.380$	
$\chi^2 = 1.127 + 3.380 = 4.507$			

For this type of χ^2, the number of degrees of freedom (d.f.) is 1 less than the number of classes. Here there are two classes (normal and mutant), so d.f. = 1. Looking at the table of critical χ^2 values at the end of this book (Table R.1), you see that for 1 degree of freedom you would expect to find a deviation of 3.84 or larger by chance alone only 1 time in 20 (a probability of .05). Since the deviation in our data is greater than this but not as great as the deviation one might find once in 100 tests (.01), we can conclude that there is a significant difference between the observed and expected, and $.01 < p < .05$.

Thus we should reject our initial hypothesis that these progeny would be produced in simple Mendelian proportions. The actual outcome could have been due to differences in viability between classes, errors in classification, or the fact that the trait might have a complex genetic basis. The initial hypothesis, or *null hypothesis,* can only be rejected by our statistical test; it cannot be proved. (The null hypothesis is discussed in more detail in Chapter 9.)

A χ^2 test can also be used to test whether two sets of data are independent. This is called a *contingency* χ^2 *test* and is easily calculated by substituting the values from a table of data into the formula given in the following table.

	Class A	Class B	Totals
Group 1	a	b	$a + b$
Group 2	c	d	$c + d$
	$a + c$	$b + d$	$a + b + c + d = n$

$$\chi^2 = \frac{[|ad - bc| - 1/2(n)]^2 n}{(a + b)(a + c)(c + d)(b + d)}$$

where the vertical bars around the term $|ad - bc|$ signify the absolute value (the positive magnitude) of the difference between *ad* and *bc*. For example, consider the calculations for the following set of data from two replicate experiments.

	Normal	Mutant	Totals
Replicate 1	78	72	150
Replicate 2	44	41	85
	122	113	235

$$\chi^2 = \frac{[|(78)(41) - (72)(44)| - (1/2)(235)]^2(235)}{(150)(122)(85)(113)} = .0102$$

Thus we have no statistical basis for rejecting the null hypothesis that the two replicates are producing normal and mutant progeny in the same proportions.

The 1/2 in the formula just given is a correction factor (Yates's correction factor) that is typically included in χ^2 tests in which the number of observations in any of the categories is less than five. It is also frequently used when the number of classes is small, as in our contingency χ^2. We could have also used it in the earlier χ^2, in which case $\chi^2 =$ 4.167, and p is still less than .05.

One key point to remember is that χ^2 tests take the sample size into consideration. Thus one cannot do a χ^2 analysis of percentages. An illustration of this is included in problems 14 and 15 at the end of the chapter.

Finally, remember that statistical tests evaluate only the likelihood that a given set of data is inconsistent with a stated hypothesis. One cannot prove that a particular hypothesis is correct, since proof would require observing all possible examples. One can only reject a hypothesis (that is, the null hypothesis) by marshaling evidence against it. If the null hypothesis is rejected, then by elimination we have supported the alternative.

IMPORTANT TERMS

Chi-square (χ^2)
Contingency test
Degrees of freedom
Hypothesis
Independent events
Mutually exclusive events
Null hypothesis
Yates's correction factor

PROBLEM SET 5

This problem set concentrates on applying the basic rules of probability to genetics questions and on testing hypotheses by using the χ^2 test of goodness of fit. The first 12 questions involve probability, and the last 5 are χ^2 problems.

1. The sex-linked trait hemophilia is found in the history of a particular family. A phenotypically normal couple produces a son who is a "bleeder." The probability that the next child will be a bleeder is **(a)** 0, **(b)** 3/4, **(c)** 1/2, **(d)** 1/4, **(e)** none of the above.

2. A married student, talking to the teacher after a lecture on probability, asks the following question. "My husband and I have been married for almost five years, and we plan to have a family of three children. If I tell you that at least two of them are girls, what is the probability that all of our children are girls?" What should the teacher answer?

3. The gene pool of a human population has five different alleles at an autosomal locus. The number of different genotypes that are possible at this locus is **(a)** 5, **(b)** 10, **(c)** 15, **(d)** 20, **(e)** 25, **(f)** none of the above.

4. A trihybrid pea plant, having the genotype $AaBbC^1C^2$, is self-fertilized. All loci are unlinked. There is complete dominance at the *A* and *B* loci but incomplete dominance at the *C* locus. The fraction of progeny that will be phenotypically different from the parent is **(a)** 1/8, **(b)** 7/8, **(c)** 37/64, **(d)** 23/32, **(e)** none of the above.

5. Red flower color is dominant (*R* –) to white (*rr*) in garden peas. In a cross of *Rr* to *rr,* 4,400 progeny are recovered. The probability that these 4,400 plants will consist of 2,200 red- and 2,200 white-flowered plants is **(a)** 100 percent, **(b)** 50 percent, **(c)** 25 percent, **(d)** much less than 25 percent.

6. Phenylketonuria, a metabolic disorder in humans, is inherited as an autosomal recessive trait. A husband and wife, both heterozygous for this gene, plan to have six children. What is the probability that four of the offspring will be normal and two will have phenylketonuria?

7. Heterozygotes (carriers) for the autosomal sickle cell anemia gene occurr in the U.S. black population with a frequency of about 1 in 10. If two phenotypically normal people from the general black population marry, what is the probability that their first child will have sickle cell anemia? **(a)** 1/10, **(b)** 1/40, **(c)** 1/100, **(d)** 1/400.

8. A deleterious trait, inherited as an autosomal recessive (*b*), has a penetrance of 60 percent. In a cross of *Bb* and *Bb,* the expected frequency of individuals showing the deleterious trait is **(a)** .15, **(b)** .25, **(c)** .40, **(d)** .45, **(e)** .60, **(f)** none of the above.

9. A woman whose mother was heterozygous for the retinal-cancer mutation retinoblastoma (*R*), a dominant allele with 90 percent penetrance, marries a man who is heterozygous for the mutation. Assuming that all other people in the pedigree are homozygous normal, what is the probability that their first child will suffer from retinal cancer?

10. A thirty-year-old woman with four children starts showing the unmistakable symptoms of Huntington's chorea, the rare autosomal dominant trait that killed the American folk singer Woody Guthrie.

(a) What is the probability that two of her four children carry the gene for Huntington's chorea?
(b) What is the probability that the woman's father carries the gene?
(c) What is the probability that the woman's mother carries the gene?
(d) What is the probability that the woman's cousin on her father's side of the family carries the gene?

11. What proportion of all five-child families from a cross of *Ss* and *Ss* will be expected to include at least one child who is homozygous *ss*? **(a)** 1/32, **(b)** 781/1,024, **(c)** 243/1,024, **(d)** 31/32, **(e)** none of the above.

12. Consider all six-child families produced by parents that include one heterozygous for a common recessive autosomal condition and one homozygous for the condition ($Rr \times rr$). What proportion of all such six-child families will include at least three affected children?

13. A testcross of a monohybrid gray mouse to an albino strain results in 64 gray and 48 albino progeny. Test the goodness of fit of these data to a 1:1 ratio, using the χ^2 test.

14. Self-fertilization of several phenotypically tall pea plants results in the production of 100 progeny: 79 tall and 21 short plants. Test the goodness of fit of these data to the hypothesis that the cross should yield progeny in the expected ratio of 3 tall:1 short.

15. In the same experiment as in problem 14, assume that 1,000 plants are produced, including 790 tall and 210 short (the same frequencies). Test the goodness of fit to the same 3:1 ratio, and compare your results to those you obtained in problem 14. Although in the same proportions is chi square the same?

16. A pure-breeding, tall pea plant with white flowers is crossed with a pure-breeding, short plant with red flowers. The F_1 plants are tall, with red flowers. When allowed to fertilize themselves, these produce the following F_2: 326 tall, red; 104 tall, white; 117 short, red; and 29 short, white. Explain these data, indicating genotypes of the parents, and the F_1, and the different F_2 phenotypes. What are the expected numbers of the various F_2 classes? Test the goodness of fit between the data and your hypothesis, using χ^2.

17. A pure-breeding plant with colored flowers is crossed with a pure-breeding plant with colorless flowers. The F_1 are all colored, and when self-fertilized these produce an F_2 consisting of 196 colored and 92 with colorless blooms. Two alternative hypotheses might be proposed. First, this could be a monohybrid cross, with colored flowers dominant to colorless flowers. A second explanation is that this is a dihybrid cross, with flower pigment dependent on two dominant complementary genes that are not linked. In this second alternative, only the $A-B-$ genotype would produce colored flowers. Test the goodness of fit of the data to the two expected ratios. Which hypothesis fits the data better?

ANSWERS TO PROBLEM SET 5

1. **(d)** 1/4. Since hemophilia is sex-linked, the father of this family must be genotypically normal, or he would have hemophilia. A sex-linked recessive trait is passed from the mother to her sons. Thus the mother must be a heterozygote. If she were homozygous, she too would have hemophilia. The mating and all possible offspring are summarized as follows. (Let *H* represent the normal allele and *h* represent the hemophilia-causing allele. The father's Y chromosome does not carry an allele for this locus.)

$Hh \times Hy$

↓

All female children will inherit the *H* allele from the father and will be either *HH* or *Hh* and are phenotypically normal. Females will comprise 1/2 of all children, on the average. Half of the male offspring will inherit the *h* allele from their mothers and will have hemophilia. Since being male and inheriting the *h* allele are independent events, the probabilities are multiplied: $1/2 \cdot 1/2 = 1/4$.

2. 1/4.The purpose of this problem is to help you see that it is critically important to determine which of all possible events (three-child families in this case) are consistent with the available data (two girls already born in this case). Families with two or more boys are not relevant to the problem. A three-child family is the outcome of three independent events in which there is a 50 percent probability of the child's being a girl and a 50 percent probability of its being a boy. The possible makeups of three-child families are as follows:

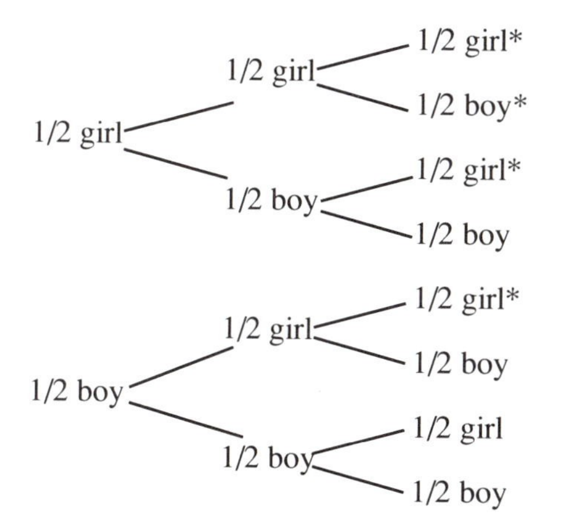

The four families that include at least two girls (marked with an asterisk) constitute the base sample that must be considered. In only one are all three children girls.

3. **(c)** 15. Each individual genotype must contain only 2 of these alleles. The homozygotes total 5. In addition, there are 10 heterozygotes. Remember that A^1A^2 is genotypically the same as A^2A^1. Numbering the alleles from 1 to 5, we get the following genotypes:

	A^1	A^2	A^3	A^4	A^5
A^1	A^1A^1	A^1A^2	A^1A^3	A^1A^4	A^1A^5
A^2		A^2A^2	A^2A^3	A^2A^4	A^2A^5
A^3			A^3A^3	A^3A^4	A^3A^5
A^4				A^4A^4	A^4A^5
A^5					A^5A^5

4. **(d)** 23/32. The three loci will assort independently. To answer this question, consider each locus separately. First, in a cross of $Aa \times Aa$, 3/4 will show the dominant phenotype like the parents. The same is true of the *B* locus. Since the *C* locus displays incomplete dominance, however, only 1/2 of the offspring will be heterozygotes like the parent. The fraction of all progeny that will be phenotypically the same as the parent will be the product of the individual probabilities: $3/4 \cdot 3/4 \cdot 1/2 = 9/32$. The proportion that will be phenotypically different is the remainder of all combinations: $1 - 9/32 = 23/32$. You can also calculate the probability of being different, but if you choose to approach the problem this way, remember that one can be different in some of the genes but like the parents for others. Thus, there are many phenotype combinations that are different in some way. All of these must be counted.

5. **(d)** Much less than 25 percent. The 1:1 ratio is only the expectation. Equal numbers of dominant and recessive phenotypes from this cross would not occur very often, however, because of the sampling variation that always exists in a chance event. Chi-square tests are used to compare the observed and expected data.

6. The probability is .297. For this type of problem, one should use the probability formula. Since the parents are both heterozygotes, the probability of the child's having a normal phenotype is 3/4, and the probability of having the mutant phenotype is 1/4. In a family of six children, $n = 6$. The stipulated makeup of four normal and two affected children means that $s = 4$ and $t = 2$. Substituting into the probability formula, we have

$$\text{Probability} = \frac{n!}{s!t!}(p)^s(q)^t$$

$$\text{Probability} = \frac{6!}{4!2!}\left(\frac{3}{4}\right)^4\left(\frac{1}{4}\right)^2 = \left(\frac{6 \cdot 5 \cdot 4 \cdot 3 \cdot 2 \cdot 1}{4 \cdot 3 \cdot 2 \cdot 1 \cdot 2 \cdot 1}\right)\left(\frac{81}{4{,}096}\right)$$

$$= \frac{1{,}215}{4{,}096} = .297$$

7. **(d)** 1/400. For a child to have sickle cell anemia, both normal parents must be heterozygotes. The probability that the mother is a carrier is 1/10. The same is true for the father. If both are carriers, the probability that the child is a homozygote for sickle cell anemia is 1/4. These are independent events, and all probabilities should be multiplied: $1/10 \cdot 1/10 \cdot 1/4 = 1/400$.

8. **(a)**.15. There is a 1/4 probability of a *bb* individual being produced by the cross of $Bb \times Bb$. If the individuals are *bb*, there is only a 60 percent probability that they will express the trait. These are independent events, and the frequencies are therefore multiplied: $1/4 \cdot 6/10 = 6/40 = .15$.

9. The answer is 45/80. The woman has a 1/2 chance of having inherited the retinoblastoma mutation from her mother. Since the man she marries is known to be a heterozygote, there is a 3/4 probability that their child will have at least one dominant mutant allele (*R* –), assuming that both parents are carriers. If the child inherits the mutation, there is only a 9/10 chance that it will express the trait. All three of these are independent events, and their probabilities are multiplied:

1/2 chance that the woman is a heterozygote
× 3/4 chance that the child will inherit the *R* allele if both parents are carriers
× 9/10 chance that the child will express the trait
+ 1/2 chance that the woman is *rr*
× 1/2 chance that the child will inherit the *R* allele from its *Rr* father
× 9/10 chance that the child will express the trait
= (1/2 × 3/4 • 9/10) + (1/2 • 1/2 • 9/10) = 45/80

10. **(a)** 3/8. The mother is a heterozygote for this dominant trait, since it is rare, and it is therefore unlikely that both of her parents were carriers (the only way in which she could be a homozygote). There is, therefore, a 1/2 chance that a child will be a carrier and a 1/2 chance that it will be normal. Finding the proper solution involves substituting these probabilities and family composition into the probability formula:

$$\text{Probability} = \frac{n!}{s!t!}(p)^s(q)^t$$

$$\text{Probability} = \frac{4!}{2!2!}\left(\frac{1}{2}\right)^2\left(\frac{1}{2}\right)^2 = \left(\frac{4 \cdot 3 \cdot 2 \cdot 1}{2 \cdot 1 \cdot 2 \cdot 1}\right)\left(\frac{1}{16}\right) = \frac{3}{8}$$

(b) 1/2.

(c) 1/2. The answer to both **(b)** and **(c)** is the same. The woman in this problem carries a dominant trait that she inherited from either her mother or her father. Since it is an autosomal trait, there is an equal probability that it was her mother or her father who carried it.

(d) 1/8. There is a 50 percent probability that the mutant allele will be transmitted in each generation. There is a 50 percent chance that one of the father's parents carries the mutant, which means that his parents have a 50 percent chance of being *Rr* × *rr*. The probability that the *R* allele will be passed on to the cousin's parent is 50 percent. The probability that the *R* allele will be passed on to the cousin is also 50 percent. Thus the probability that the cousin carries the gene is 1/2 • 1/2 • 1/2 = 1/8.

11. 781/1,024. All of the five-child families from such a mating will include at least one homozygous *ss* child except the family composed only of children with the dominant phenotype. Since the cross is between two heterozygotes, there is a 3/4 chance that a child will have the dominant phenotype. Because the birth of each child is an independent event and order is irrelevant (since they are all the same), you would simply multiply probabilities to determine the proportion of completely normal families for this mating: $(3/4)^5$ = 243/1,024. All of the rest have at least one affected (*ss*) individual: 1 – (243/1,024) = 781/1,024.

12. 21/32. This problem requires that the proportion of several different family make-ups be calculated. The most efficient way to do this is to use the expanded binomial for six-child families.

$$(p + q)^6 = p^6 + 6p^5q + 15p^4q^2 + \underline{20p^3q^3 + 15p^2q^4 + 6pq^5 + q^6}$$

The four families that have three or more affected individuals are underlined. In order to determine what proportion of all six-child families have at least three affected individuals, one must substitute the probability that the child will be normal (p) from the type of mating. Here the mating is between an affected individual and a heterozygote, so $p = 1/2$ and $q = 1/2$. Substituting in the underlined portion of the binomial gives

$$\begin{aligned}\text{Probability} &= 20\left(\frac{1}{2}\right)^3\left(\frac{1}{2}\right)^3 + 15\left(\frac{1}{2}\right)^2\left(\frac{1}{2}\right)^4 + 6\left(\frac{1}{2}\right)\left(\frac{1}{2}\right)^5 + \left(\frac{1}{2}\right)^6 \\ &= \frac{20}{64} + \frac{15}{64} + \frac{6}{64} + \frac{1}{64} \\ &= \frac{42}{64} = \frac{21}{32}\end{aligned}$$

13. Set up your χ^2 test as shown in the following table. If you pay attention to the sums of the first two rows, you will be less likely to make a mathematical error, and if you do all the intermediate steps it will be easier to detect a mistake.

	Gray	Albino	Total	Notes
Observed	64	48	112	should be the same
Expected (1:1)	56	56	112	
Deviation (Observed-Expected)	+8	−8	0	should always be zero
$(\text{Deviation})^2$	64	64		
$(\text{Deviation})^2/\text{Expected}$	1.14	1.14		

$$\chi^2 = \sum \frac{(\text{Deviation})^2}{\text{Expected}} = 1.14 + 1.14 = 2.28$$

With two classes, gray and albino, there is $2 - 1 = 1$ degree of freedom. The χ^2 table at the end of this book (Table R.1) shows that the value 2.28 lies above the .05 cutoff. You would expect, on the basis of chance alone, deviations as large or larger from 10 percent to 50 percent of the time. This is not sufficiently rare to require us to reject the null hypothesis that the data sets reflect a 1:1 ratio.

14.

	Tall	Short	Σ	Tall	Short	Σ
Observed	79	21	100	790	210	1,000
Expected (3:1)	75	25	100	750	250	1,000
Deviation (Observed-Expected)	+4	−4	0	+40	−40	0
$(\text{Deviation})^2$	16	16		1,600	1,600	

$$\chi^2 = \sum \frac{(\text{Deviation})^2}{\text{Expected}} = 0.21 + 0.64 = 0.85 \qquad 2.13 + 6.40 = 8.53$$

The χ^2 calculations in the left-hand side of the table are done for a 3:1 expected ratio, with a sample size of 100. For a better contrast, we have also put the calculations required for problem 15 on the right-hand side of the table. For problem 14, however, the χ^2 value of .85 with 1 degree of freedom gives a p value of between .5 and .1. We can conclude that the data fit the expectation, in that the deviation is not significantly large enough to compel us to reject the null hypothesis.

15. The calculations for this χ^2 test are given in the answer for problem 14 (right-hand table), so that you can contrast them readily with those for the same proportions but different sample size. In this case, the χ^2 value of 8.53 gives a p value of much less than .01. In other words, you would expect deviations this large or larger by chance in less than 1 in 100 such data sets. We can therefore reject the null hypothesis that the observed and expected frequencies are the same. Thus one clearly cannot use percentages in calculating χ^2; sample size is very important.

16. Tall and red seem to be dominant to short and white, and the cross is a dihybrid, with each parent contributing one dominant and one recessive trait. The F_2 shows segregation and independent assortment generating a 9:3:3:1 expected ratio, as indicated in the following table.

	9 Tall, red	**3 Tall, white**	**3 Short, red**	**1 Short, white**	**Total**
Observed	326	104	117	29	576
Expected	324	108	108	36	576
Deviation (Observed-Expected)	+2	−4	+9	−7	0
(Deviation)^2	4	16	81	49	

$$\chi^2 = \sum \frac{(\text{Deviation})^2}{\text{Expected}} = 0.01 + 0.15 + 0.75 + 1.36$$

Chi-square is therefore 2.27. The number of degrees of freedom is 4 (the number of classes) minus 1, or 3. The probability is therefore between .9 and .5, and we conclude that the data are consistent with the hypothesis.

17. If this is a monohybrid cross, the 288 F_2 plants are expected to include $3/4 \cdot 288 = 216$ colored and $1/4 \cdot 288 = 72$ colorless plants. For the second explanation, only $A-B-$ plants are colored. This gives $9/16 \cdot 288 = 162$ colored. All others are colorless; $7/16 \cdot 288 = 126$.

	Colored	**Colorless**	**Total**
Observed	196	92	288
Expected (3:1)	216	72	288
Deviation (Observed-Expected)	−20	+20	0
(Deviation)^2	400	400	

$$\chi^2 = \sum \frac{(\text{Deviation})^2}{\text{Expected}} = 1.85 + 5.56 = 7.41$$

	Colored	Colorless	Total
Observed	196	92	288
Expected (9:7)	162	126	288
Deviation (Observed-Expected)	+34	−34	0
$(\text{Deviation})^2$	1,156	1,156	

$$\chi^2 = \sum \frac{(\text{Deviation})^2}{\text{Expected}} = 7.14 + 9.17 = 16.31$$

The smaller χ^2 value occurs with the 3:1 ratio, so this hypothesis gives a better fit between the observed and the expected. Still, both are significantly different, so in practice you would want to repeat the experiment and consider additional alternative hypotheses.

SEX-LINKAGE AND GENE INTERACTIONS

STUDY HINTS

Regular segregation and the assortment of alleles in a heterozygote produce the familiar genotypic and phenotypic ratios that you investigated in Problem Set 4. In a real sense these ratios are also "hypotheses," in that they are the expectations appropriate to a particular genetic situation. For example, a cross between two heterozygotes, $Aa \times Aa$, yields a 3:1 phenotypic ratio among the offspring. Turning this around, if one finds a 3:1 phenotypic ratio in a family, it is reasonable to hypothesize that the parents were both heterozygotes. Ratios are therefore an important key to establishing the genetic basis of an unfamiliar trait.

Sex-linkage, multiple alleles, gene interactions, and maternal and cytoplasmic effects are natural complications that can modify these underlying patterns and ratios. The secret to solving these types of problems is to be familiar with the clues that are often embedded in modified ratios. A distinct difference between male and female phenotypic ratios leads one to consider the possibility of sex-linkage and/or a maternal or cytoplasmic involvement. Lethality would lead to truncated ratios (e.g., 1:2), whereas the gene interactions such as epistasis would merge certain genotypic classes into the same phenotypic class. Textbooks often describe a large number of these modified ratios, but for convenience we have summarized some of the most commonly encountered ones in Table 6.1

TABLE 6.1 SUMMARY OF MODIFIED RATIOS

Ratio	Possible Interpretations
1:1	Monohybrid testcross: genotype or phenotype
1:1:1:1	Dihybrid testcross: genotype or phenotype
1:2:1	Monohybrid genotypic ratio
	Monohybrid phenotypic ratio for an incompletely dominant or codominant trait
1:2	Monohybrid phenotypic or genotypic ratio in which the dominant or recessive homozygote dies (a recessive lethal)
3:1	Monohybrid phenotypic ratio
9:3:3:1	Dihybrid phenotypic ratio
12:3:1	Dominant epistasis, in which one dominant locus masks the expression of the second locus
9:3:4	Recessive epistatis, in which one recessive homozygous locus masks the expression of a second locus
15:1	Either dominant trait sufficient to produce the phenotype
9:7	Both dominant traits required to produce the phenotype

In addition to traits that can be traced to nuclear genes and their interactions, phenotypes are often dependent upon cytoplasmic interactions. The behavior of nuclear genes is most familiar, because they follow the patterns summarized by Mendelian rules. Cytoplasmic genes, such as those in mitochondria or chloroplasts, on the other hand, are inherited in each generation through the maternal cytoplasm. Thus, although cytoplasmic genes persist within a family they do not segregate in a Mendelian fashion.

In contrast, maternal effects are often transient. Maternal effects involve genes that affect the cytoplasm, such as by coding for some material that is deposited in the egg and that functions until it is replaced by products of the offspring's nucleus, early in development. For example, the pigment precursor kynurenine, produced by a dominant gene, *A,* causes brown-pigmented areas in larval and adult flour moths, *Ephestia kuhniella.* In a cross between *Aa* females and *aa* males, all progeny are pigmented at first, but as they mature only the *Aa* larvae retain their pigmentation, because the kynurenine deposited in the egg cytoplasm by the mother cannot be replenished by the larval genome in *aa* individuals.

Another example, the direction of snail shell coiling in *Limnaea,* demonstrates that maternal effects can sometimes persist in an individual. Maternal effects and cytoplasmic inheritance can generally be distinguished by the fact that the extrachromosomal factors in cytoplasmic inheritance can persist for many generations.

IMPORTANT TERMS

Antibody
Antigen
Attached X
Autosome
Balanced lethal
Codominance
Complementary gene action
Cytoplasmic inheritance
Epistasis
Expressivity
Hemizygous
Heterogametic sex
Histocompatibility
Homogametic sex
Incomplete dominance
Isoallele
Lethal
Maternal effects
Nondisjunction
Overdominance
Parthenogenesis
Penetrance
Petite mutations
Phenocopy
Pleiotropism
Segregation distortion
Sex-influenced
Sex-linked

PROBLEM SET 6

1. In a family, the father and mother have normal red–green color vision. Red–green color blindness is a sex-linked recessive trait in humans. If their son is color-blind and their daughter has normal color vision, give the genotypes of all members of the family.

2. If a normal man marries a woman who is heterozygous for a sex-linked recessive lethal mutation, what would be the expected sex ratio of their offspring?

3. In a paternity suit, it is discovered that the mother and the child are both heterozygous for the MN blood group and are both blood type B. The suspected father has blood type A, group MN, and, in addition, is an albino and is heterozygous for the dominant nervous degenerative condition Huntington's chorea. The accused man **(a)** can, **(b)** cannot be the father of the child.

4. A woman with blood type O has a child with blood type O. She claims that a friend of hers is the child's father.

(a) His blood type is A. Can he be excluded as the father on this evidence alone?
(b) Does the fact that the accused man's mother is type A and his father is type AB permit him to be excluded?
(c) Does the additional information that his mother's parents are both AB permit him to be excluded?

5. Assume that two different albino strains of mice give pigmented progeny when crossed with pigmented strains. In both cases, the F_2 segregates three pigmented to one albino. When albino strain I is crossed to albino strain II, however, all progeny are pigmented. How could you explain these results, and what further cross could be made to test your hypothesis?

6. Here are the results of two crosses of a short-bristle mutant in *Drosophila.*

(1) short-bristle × wild-type:

36 short-bristle females	32 short-bristle males	
42 wild-type females	34 wild-type males	$n = 144$

(2) short-bristle × short-bristle:

39 short-bristle females	33 short-bristle males	
21 wild-type females	18 wild-type males	$n = 111$

Crosses of type **(2)** invariably have fewer progeny than those of type **(1).** How would you explain these results?

7. Assume in humans a sex-linked locus with three alleles, A^1, A^2, and A^3, and two independently assorting autosomal loci: B, with four alleles (B^1, B^2, B^3, and B^4), and C, with three alleles (C^1, C^2, and C^3). Given this potential genetic variability, the number of genetically different males that are possible is **(a)** 600, **(b)** 11, **(c)** 48, **(d)** 180, **(e)** 72, **(f)** none of the above.

8. Assume that a gene for baldness (B) is dominant to its allele for hair production (b) and that a gene for curly hair (Cy) is dominant to its allele for straight hair (cy). The two loci are not linked, and baldness shows dominant epistasis over hair form. In a cross of $Bb\ Cycy \times bb\ cycy$, the expected progeny will be **(a)** 12 bald: 3 curly: 1 straight, **(b)** 9 bald: 4 curly: 3 straight, **(c)** 4 bald: 3 curly: 1 straight, **(d)** 2 bald: 1 curly: 1 straight, **(e)** none of the above.

9. There are at least 4 closely linked genes in the histocompatibility complex in humans. These genes are multiallelic, having as many as 35 codominant alleles. They are so closely linked that recombination among them can be disregarded. An individual is tested and found to have the genotype

$A^1A^{34}\ B^3B^{24}\ C^1C^{23}\ D^{10}D^{11}$

His wife has the genotype

$A^1A^{20}\ B^{10}B^{11}\ C^3C^{12}\ D^9D^{13}$

Their son is

$A^1A^{34}\ B^{11}B^{24}\ C^1C^{12}\ D^9D^{10}$

Diagram the cross that produced this son, and show the allele linkages (the haplotype) on the chromosomes of the parents and their son.

10. Assume that in rabbits there are three different independently assorting autosomal loci that affect coat color. A colorless-pigment precursor *V* is converted to a colorless precursor *W* by action of the *A* allele. *W* is converted to tan pigment by action of the *T* allele, and the tan pigment is converted to black pigment by action of the *B* allele.

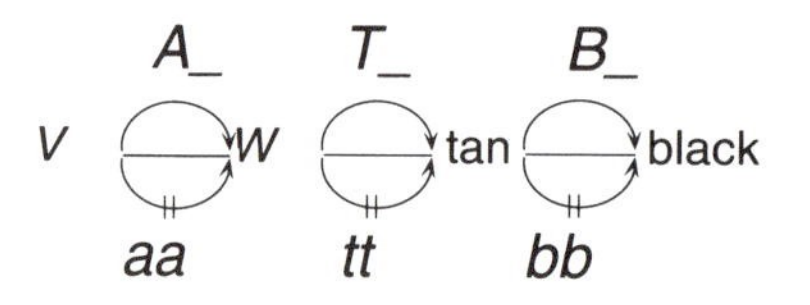

The homozygous recessive condition at each locus results in loss of enzyme activity for the reaction controlled by that gene. A cross of trihybrid bunnies, *AaTtBb* × *AaTtBb*, would be expected to give rise to which of the following progeny? **(a)** 12 black: 3 tan: 1 white, **(b)** 9 black: 3 tan: 4 white, **(c)** 4 black: 3 tan: 9 white, **(d)** 27 black: 21 tan: 16 white, **(e)** 27 black: 9 tan: 28 white, **(f)** 21 black: 16 tan: 27 white, **(g)** none of the above

11. In a certain plant species, an F_2 generation produced 113 round seeds, 75 oval seeds, and 12 curled seeds. One of the original homozygous parent strains had round seeds, and the other had curled seeds. What is the minimum number of genes that would adequately explain the data, and what genotypic ratio and explanation best fit these results?

12. The presence of dominant alleles at two different loci (*A* – *B* –) in poultry results in walnut comb. Pea comb is produced by the genotype *aaB* –, rose comb by *A* – *bb*, and single comb by *aabb*. If you were given a rooster with a walnut comb, known to be heterozygous for both genes, and a rose-comb hen from a true-breeding strain, what cross or crosses, and how many generations, would it take to obtain true-breeding pea-comb poultry?

13. When two genes interact, phenotypic ratios are modified from the 9:3:3:1 independent segregation ratio. What will the phenotypic ratios be when **(a)** one recessive homozygote suppresses the expression of an independently assorting gene? **(b)** either recessive homozygote suppresses the expression of the other locus? **(c)** the presence of one dominant allele suppresses the expression of the other locus?

14. The autosomal genes Curly (*Cy*) and Plum (*Pm*) in *Drosophila* are independently segregating, phenotypically dominant mutations that are recessive lethals. In other words, they have a distinct phenotype as heterozygotes but are lethal as homozygotes. When Curly and Plum are separately crossed to wild flies and the F_1 mutant progeny are intercrossed, what will the resulting phenotypic ratio be?

15. Assume that in mice there are a recessive lethal gene (*d*) and a recessive gene (*f*) that, when homozygous, results in female sterility. The two loci are both on chromosome 3. Assume that you have available one strain carrying the lethal allele and another strain with the *f* allele together with a crossover inhibitor that prevents exchange between the *d* and *f* loci. Describe the balanced lethal stock that can be made using these two chromosomes.

16. In the snail *Limnaea,* dextral (right) coiling is dominant to sinistral (left) coiling and shows a maternal effect. What are the possible genotypes for direction of coiling in a phenotypically dextral snail?

17. When Mendel crossed plants heterozygous for genes determining height ($Tt \times Tt$), he got 3/4 tall (*TT* and *Tt*) and 1/4 short (*tt*) in the next generation. Does a comparable genetic ratio result in *Limnaea* from a cross of $Dd \times Dd$? Please explain.

18. The variegation that is seen in plants such as *Mirabilis,* where leaves or parts of leaves are full green, lighter green, variegated green and white, or completely white, provides evidence that chloroplasts (which are capable of replication) **(a)** are distributed more or less at random to daughter cells during mitosis, **(b)** behave much like chromosomes during mitosis, in that they divide and separate to opposite poles of the cell.

19. In some of the first experiments carried out using *Drosophila,* Calvin Bridges crossed white-eyed *Drosophila* females to red-eyed males. The progeny are red-eyed females and white-eyed males. Which of the following one or more statements is true? **(a)** The male is the heterogametic sex. **(b)** The white-eyed allele is located on the X chromosome. **(c)** Red eyes (the wild-type) is dominant to white eyes.

20. The genetic makeup of a population of mice has four different alleles for a certain sex-linked gene and five different alleles for an autosomal locus. Basing your analysis upon these two loci, what is the number of genetically different males in the population?

21. Prune (*pn*) is a recessive eye-color mutant located on the X chromosome of *Drosophila.* An autosomal dominant gene, Prune-killer (*K-pn*), causes the death of all prune larvae. If a cross is made between a *pn* /+; *K-pn* / + female and a + / Y; *K-pn* / + male, what will be the sex ratio of the progeny, and what will be the frequencies of wild-type and prune phenotypes among the progeny?

22. In a plant species capable of either self-fertilization or outcrossing, two pure-breeding, white flower strains occur. One strain has white flowers because of a dominant cytoplasmic gene. The other has white flowers because of a recessive chromosomal gene. If reciprocal crosses are made between these two strains, what phenotypes and what proportions would be expected in the progeny? If the F_1 were self-fertilized, what ratios and phenotypes would you expect?

23. Some years ago, the well-known movie star Charlie Chaplin was named in a paternity suit by an actress. Her blood type was A, the child's was B, and Chaplin's was O. The first jury was hung; the second found him guilty. What do you think of the verdict, and why?

24. The ABO locus in humans has four common alleles: A^1, A^2, B, and O. These are sometimes called the "public" antigenic variants. There are also several less common alleles at this locus, some so rare that they may very well be found in only one family or a very few families. These are called "private" antigenic variants. A man who is A^9B (where A^9 is a hypothetical "private" allele) has been wrongfully accused of fathering a child with blood type A. Should he look forward to serological tests with trepidation or with confidence?

25. A man with type O blood marries a woman with type A. One of their children has type B blood, and the other is blood type AB. Assuming that the man is indeed the father of these children, what hypothesis or hypotheses might you propose to explain his phenotype and those of his children?

26. A haploid segregational petite in yeast (*p*) is crossed with a haploid wild-type yeast (+). What will be the genotype(s) and phenotype(s) of **(a)** the diploid vegetative progeny? **(b)** the haploid progeny, after meiosis?

27. A large population of haploid yeasts consists of 1/4 segregational petites (*p*) of *a* mating type; 1/4 segregational petites of a^+ mating type; 1/4 normal (+) of *a* mating type; and 1/4 normal of a^+ mating type. If a large number of random matings occur, giving rise to diploid vegetative strains, what will be their genotypes and phenotypes?

28. In a certain bird species, two color morphs are known: gray and red. Red chicks are often produced by gray parents, but gray chicks can also be produced by red parents. There are also pure-breeding birds that only produce offspring the color of the parents. If a female from a pure-breeding red line is mated to a male from a pure-breeding gray line, the F_1 are all red. If the reciprocal cross is made, the F_1 are all gray. If the F_1 female progeny from either cross are mated to males from either the pure-breeding gray strain or the pure-breeding red strain, the offspring are always red. However, if the F_1 males are mated to the pure-breeding lines, the offspring are phenotypically the color of the pure-breeding line. What is the probable mode of inheritance of color in these birds?

29. In cockatiels (medium-sized parrots), gray birds often produce a clutch that has white chicks as well as gray ones. The white chicks from such a cross are always females. White parents produce only white chicks. What is the probable mode of inheritance of color in these birds?

30. In a hypothetical rodent species, a gene that produces floppy ears was found segregating in the population. When two floppy-eared rodents are crossed, a ratio of 2 floppy ears to 3 straight ears is found in the offspring. If a floppy-eared rodent is crossed to a straight-eared individual, a ratio of approximately 1:1 is produced. What is the probable mode of inheritance of ear phenotype in this species?

ANSWERS TO PROBLEM SET 6

1. Sons get their X chromosome from their mothers. The combination of a normal mother with a color-blind son tells us that the mother was heterozygous (*Cc*) and that the son is hemizygous (*c*). The father is phenotypically normal for color vision and is therefore *C*. The daughter is either *CC* or *Cc*, each with a 1/2 probability.

2. The genotypes may be written as *Aa* female × *A* male (where *a* is lethal). The possible offspring will consist of 1/4 *AA* females: 1/4 *Aa* females: 1/4 *A* males: 1/4 *a* males (a ratio of 2 normal females: 1 normal male: 1 lethal-carrying male [dead]). The sex ratio of the children will be 2 females: 1 male.

3. The MN blood-group system shows codominance, where MM = M blood group,

NN = N blood group, and MN = the MN blood group. In the ABO system, I^A and I^B are both dominant to i and are codominant to each other. ABO genotypes are summarized as follows:

I^AI^A and I^Ai	A	blood type
I^BI^B and I^Bi	B	blood type
$I^A\ I^B$	AB	blood type
ii	O	blood type

Albinism is an autosomal recessive condition, and Huntington's chorea is an autosomal dominant condition. The Huntington's and albino conditions are of little or no value here, since albino carriers are relatively infrequent, and, in this case, even if the mother is a carrier (*Aa*), only half the children from the alleged parentage will be albino. The average age for the onset of Huntington's chorea is 35 to 40, so that condition is useless as a marker in this instance. Even if the young child carries the Huntington's allele from the alleged father, the child will not be expected to manifest it at such an early age. So we have to follow the blood types:

$$MN\ I^B - \text{(mother)} \times MN\ I^A - \text{(alleged father)}$$
$$\downarrow$$
$$MN\ I^B - \text{(child)}$$

The man cannot be excluded, since half of this couple's progeny would be expected to be MN, and if he is I^Ai (one of the possible genotypes that will be expressed as an A blood type), then the child could have inherited the i allele from him.

4. **(a)** The mother is *ii*, the child is *ii*, and the man can be I^AI^A or I^Ai. In the latter case, he could be the father, so he cannot be excluded.

(b) His mother is either I^AI^A or I^Ai, and his father is I^AI^B. He could be I^Ai if his mother had transmitted the i allele. So this information will not exclude him from being the child's father.

(c) If his mother's parents are both AB, then his mother must be I^AI^A, and he, in turn, must be I^AI^A. This additional information will permit him to be excluded.

5. The segregation pattern described in the first part of this problem shows us that the albino trait is inherited as an autosomal recessive trait in both strains. The progeny of albino I crossed with albino strain II are all normal. The only way to account for this is for the albino phenotype to be produced by a different genetic locus in each strain. The cross is therefore

$$wwCC \text{ (albino I)} \times WWcc \text{ (albino II)}$$
$$\downarrow$$
$$WwCc \text{ (pigmented offspring)}$$

A test of this hypothesis would involve crossing the F_1 progeny to produce an F_2. The F_2 would include

9/16 *W- C-*	9/16 pigmented
3/16 *W- cc*	7/16 albino
3/16 *wwC-*	
1/16 *wwcc*	

6. In cross **(1),** the short-bristle trait is clearly not recessive to wild-type, but the mutant flies used here are not homozygous for the short-bristle trait, since the F_1 segregates 1 mutant (68 flies) to 1 wild-type (76 flies). In cross **(2),** the short-bristle parents are again heterozygous for the mutant and wild-type alleles. Instead of the expected ratio of 3 short: 1 wild-type, we get 72 short: 39 wild-type, something close to a 2:1 ratio. With the clue of fewer progeny in cross **(2),** it appears that the dominant short-bristle allele is lethal when homozygous. The 2:1 ratio is therefore a modification of the 1:2:1 ratio, with one homozygous class dying.
Cross (1) = $Ss \times ss$, which gives 1/2 *Ss* (short) and 1/2 *ss* (wild). **Cross (2)** = $Ss \times Ss$, which gives 1/4 *SS* (die): 1/2 *Ss* (short): 1/4 *ss* (wild).

7. **(d)** Since males have a single X chromosome, there are three *A* genotypes available to them: A^1 A^2, and A^3. The genotypes available at the *B* locus include all combinations of the four alleles taken two at a time, since it is an autosomal locus. There are 10 such combinations (B^1B^1, B^1B^2, B^1B^3, etc.). The formula to determine this is

$$\text{Number of combinations} = \frac{n(n+1)}{2}$$

where *n* is the number of alleles. The combinations can, of course, be written out and tallied. For the *C* locus there are six combinations:

$$C\text{-locus combinations} = \frac{n(n+1)}{2} = \frac{3(3+1)}{2} = 6$$

Since these are all assorting independently of each other, we multiply the number of possible genotypes at each locus, and the answer is 3 • 10 • 6 =180.

8. **(d)** Dominant epistasis of the *B* (baldness) allele means that when *B* is present, one cannot tell whether the hair is curly (*Cy* –) or straight (*cycy*), since there is no hair at all. The gametes and progeny expected from this cross are

	1/4 *B Cy*	**1/4 *B cy***	**1/4 *b Cy***	**1/4 *b cy***
All *b cy*	*Bb Cycy*	*Bb cycy*	*bb Cycy*	*bb cycy*
Trait	Bald	Bald	Curly	Straight

9. Since we can rule out recombination among the four histocompatibility loci, we can consider each *ABCD* combination as a single segregating unit. The puzzle is, therefore, to arrange the alleles so that one haploid set from the mother and one from the father will account for the genotype of the offspring.

$$\frac{A^{1}B^{11}C^{12}D^{9}}{A^{20}B^{10}C^{3}D^{13}}\text{ Mother} \quad \times \quad \frac{A^{34}B^{24}C^{1}D^{10}}{A^{1}B^{3}C^{23}D^{11}}\text{ Father}$$

↓

$$\frac{A^{1}B^{11}C^{12}D^{9}}{A^{34}B^{24}C^{1}D^{10}}\text{ Son}$$

10. **(e)** This cross consists of three separate monohyhrid crosses:

$Aa \times Aa \quad Tt \times Tt \quad Bb \times Bb$

Each gives a ratio of 3 dominant to 1 recessive, with all assorting independently. In the reaction

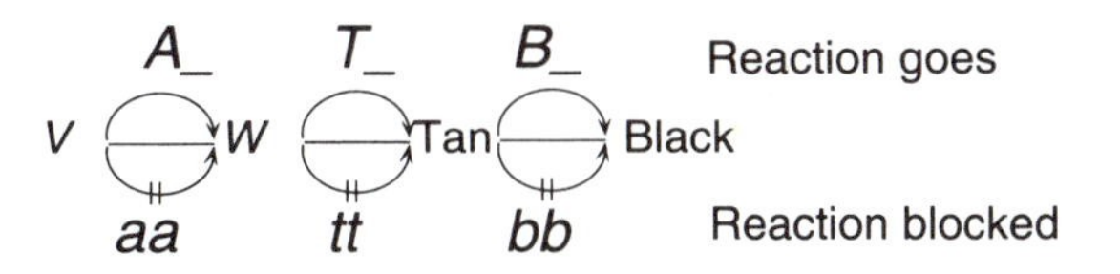

each step requires at least one dominant allele at the appropriate locus. Working back from the end point, black coat color requires the genotype $A-T-B-$ (that is, at least one dominant allele at each locus), which has the probability $3/4 \cdot 3/4 \cdot 3/4 = 27/64$. Tan coat color results from the genotype $A-T-bb$ (that is, the last step is blocked), and occurs with the probability $3/4 \cdot 3/4 \cdot 1/4 = 9/64$. An individual homozygous for the recessive allele at the *A* and/or *T* loci (*aaT* –, *A* – *tt*, or *aatt*) will be white. These make up the remainder of the progeny:

$$1 - 27/64 - 9/24 = 28/64 \text{ white}$$

11. This ratio is approximately 9:6:1, the ratio one would expect of an interaction in which both dominants are needed to produce round seeds (*A* – *B* –). Either dominant alone will produce oval seeds (*A* – *bb* or *aaB* –), and the double homozygous recessive will have curled seeds (*aabb*). The loci are assorting independently.

12. The walnut-comb rooster is heterozygous for both loci (*AaBb*). He is crossed to a true-breeding rose-comb strain, which must have the genotype *AAbb.* The genotype one is trying to obtain is *aaBB,* true-breeding pea comb. The F_1 progeny include four different genotypes:

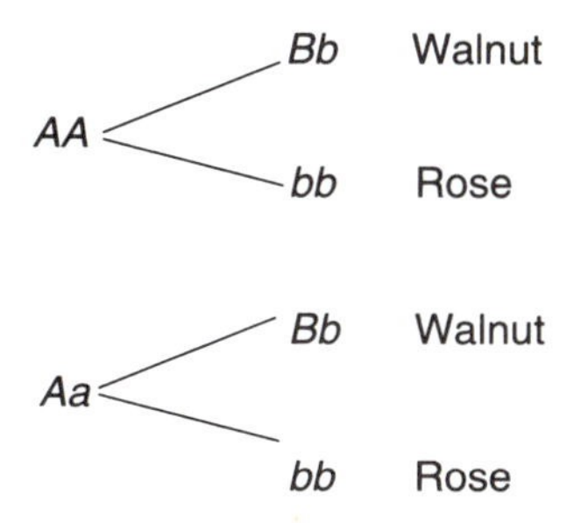

One step would then be to mate the original *AaBb* rooster to walnut-comb F_1 hens. Half of such matings would be to *AaBb* hens. In their offspring, 3/16 will have pea combs, but you still need to take an additional step before you will have a pure-breeding line, since only 1/3 of the pea-comb chickens will be homozygotes (*aaBB*). The other 2/3 are *aaBb.* Here a testcross would be useful. If one mated each of the pea-comb chickens to a single-comb (*aabb*) sibling, one could eliminate from further consideration any chickens that produce single-comb chicks, because these would have had to be heterozygous at the *B* locus. Those that only produce pea-comb chicks could be mated to each other, since they would be homozygous at both loci.

13. **(a)** 9:3:4
(b) 9:7
(c) 12:3:1

14. Since these are dominant, independently assorting genes, the cross is

$$F_1 \quad \frac{Cy}{Cy^+} \; \frac{Pm^+}{Pm^+} \quad X \quad \frac{Cy^+}{Cy^+} \; \frac{Pm}{Pm^+}$$

In the offspring, half will be *Cy*, half will be Cy^+, where the plus (+) refers to the wild-type allele. Independently of these phenotypes, half will be *Pm* and half will be Pm^+. There will therefore be four progeny phenotypes when both loci are considered: 1/4 Curly: 1/4 Curly Plum: 1/4 Plum: 1/4 wild-type.

15. When a female mouse that is *d* + /+ *f* is crossed to the same type of male, the progeny will be 1/4 *d* + /*d* +, which die in utero; 1/2 *d* +/+ *f*, which are viable and fertile as females; and 1/4 + *f* /+ *f*, which are sterile females but fertile males. The only viable and fertile females are the *d* + /+ *f* individuals, which make up 2/3 of the females. There are two types of viable and fertile males: 2/3 are *d* + /+ *f* and 1/3 are + *f* /+ *f*. When the *d* + /+ *f* females are mated to the same type of brother, the progeny will again be 2/3 *d* + /+ *f* and 1/3 + *f* /+ *f*. When the *d* + /+ *f* females are mated with their + *f* /+ *f* brothers, the progeny consist of 1/2 *d* + /+ *f* and 1/2 + *f* /+ *f*. So the only viable and fertile mice will be *d* + /+ *f* females and *d* + /+*f* and + *f* /+ *f* males, generation after generation. The recessive lethal gene (as well as the *f* gene) is maintained in a balanced stock.

16. The ovum is one of the mother's cells, and prior to fertilization it already has its future mitotic spindle orientation (and ultimately the coiling of the adult snail) determined in response to *her* genotype. It is independent of the genotype of the fertilized egg. The only thing that one can say with certainty about a dextral snail is that its mother had at least one *D* allele and that the genotype of the male parent was immaterial. So any of the following crosses will give rise only to dextral snails: *DD* or *Dd* female × *DD, Dd,* or *dd* males. It therefore follows that the genotype of a phenotypically dextral snail can be *DD, Dd,* or *dd*.

17. Yes, but not in the next generation, as it does in peas and other organisms. The progeny will all be coiled dextrally, since the mother had a *D* allele. However, the progeny will be 1/4 *DD:* 1/2 *Dd:* 1/4 *dd*. When these, as females, produce progeny, 3/4 of them (*DD* and *Dd*) will produce only dextral snails, and 1/4 will be *dd* and produce only sinistral snails, for an overall 3:1 ratio.

18. **(a)** If they behaved as chromosomes, daughter cells would get the same proportions of normal and defective chloroplasts, with little or no variegation. The variegation supports the idea of random separation of normal and defective chloroplasts, with some cells (and larger regions of the leaves) getting more (darker green) and some getting fewer (lighter green to white) of the normal chloroplasts.

19. All three statements are correct.

20. For the sex-linked locus, the mice will be hemizygous and will have any of four different X chromosome genotypes. For the autosomes, however, the five alleles of the autosomal locus can occur in all combinations of pairs of alleles (A^1A^1, A^1A^2,

A^1A^3, and so forth). There are 15 such combinations (remember that A^2A^3, for example, is the same as A^3A^2). The total number of genotypes is the product of these independent combinations, giving a total of 60 different male genotypes. Note, too, that the number of female genotypes would be much larger (150), since one would have various combinations of sex-linked alleles to take into consideration.

21. Since *pn* is a sex-linked recessive trait, the flies that are killed by the *K-pn* / *K-pn* or *K-pn*$^+$ genotypes are *pnpn* females and *pn* / *Y* males. To answer the question it is necessary to determine the frequencies of all possible genotypes and then ascertain the phenotypes of these classes of flies. Since the male parent is *pn*$^+$, all his daughters will receive this X-linked gene, so that none of the female progeny will be prune. On the other hand, half of the progeny will be males and of these, half the sons of the heterozygous female will be prune, and half will be wild-type. Since both parents are *K-pn*$^+$, 3/4 of the progeny will carry at least one *K-pn* allele, and 1/4 will be wild-type. Since the two loci assort independently, the progeny will consist of

1/2 (= 8/16) female, all of which are normal
3/16 *pn* / *Y*; *K-pn* / +, which die
1/16 *pn* / Y; + / +, which have the prune phenotypes, and
4/16 that are + / Y; *K-pn* / +, or + / Y; + / +, and are wild-type

So all females (8/16) are recovered as wild-type; 3/16 of the male larvae die, 1/16 are recovered as prune, and 4/16 are recovered as wild-type males. The sex ratio is 8 females to 5 males.

22. Since one of the genes is cytoplasmic, it will have an effect on the phenotype only when transmitted by the female. Let us represent the dominant chromosomal color allele as *A*, with *aa* being white, and represent the cytoplasmic factor as *C*, with *c* being color. In the reciprocal crosses, different F_1 phenotypes would be expected:

CAA female × *aa* male	*caa* female × *AA* male
↓	↓
all *CAa* (white)	all *cAa* (colored)

When the F_1 individuals from the left-hand cross are self-fertilized, all would be white because of the cytoplasmic factor. When the F_1 on the right are self-fertilized, 1/4 would be white because of the segregation of the chromosomal locus.

23. He should have been exonerated. The child could be I^BI^B or I^Bi. Since the mother had no I^B allele, the child must be I^Bi, and the mother must be I^Ai. The father must have either the B or AB blood type to contribute the I^B allele. Chaplin, with the O blood type (*ii*), could not have fathered the child.

24. He should look forward to the tests with confidence, since the child will almost certainly not have the rare A^9 allele. If this man were the father, the child would have received his A^9 allele, since the child's blood type is not B or AB. Note that you could use the symbols I^A, I^B, I^{A9}, and so forth, if you prefer.

25. The best explanation would be that the father carries a gene that masks the expression of his blood-type locus. This occurs in individuals homozygous for the rare Bombay locus (*hh*). These have an O blood type, because the Bombay locus is epistatic to the ABO blood-type antigen alleles. Alternatively, a germ-line mutation could have occurred in the man, changing *i* to I^B.

26. **(a)** All will be $+p$ and normal, since the plus (+) allele restores the activity of the innately normal mitochondria from the segregational petite parent and the normally functioning mitochondria from the (+) parent continue to function normally in the $+p$ diploid.

(b) Half will get the (+) allele and will be normal; half will get the *p* allele and will be phenotypically petite.

27. A total of 1/4 will be + + (normal), 1/2 will be $+p$ (normal), and 1/4 will be *pp* (petite).

28. There is a maternal effect such that the offspring phenotypically express the genotype of the mother, with red dominant to gray.

29. This factor is sex-linked. In birds, the female is the heterogametic sex. The sex chromosomes in the male are denoted WW, while those in the female are WZ. Thus, in a cross between a pair of gray birds in which white female chicks were found, the male must have been heterozygous for the recessive white allele.

30. The floppy-ear phenotype is a recessive lethal, but morphologically dominant condition. Thus a cross between two floppy-eared individuals is always a cross between two heterozygotes. If we define *F* to represent the floppy-ear allele and *f* the straight-ear allele, $Ff \times Ff$ gives an F_1 composed of 1/4 *FF*, which die; 1/2 *Ff*, which have floppy ears; and 1/4 *ff*, which have straight ears. Of the surviving rodents, 2/3 have floppy ears. The cross of floppy to straight ears is a testcross yielding a 1:1 ratio.

CROSSWORD PUZZLE

SEX-LINKAGE AND INTERACTIONS

ACROSS

1. One gene with more than one phenotypic effect
6. Any chromosome other than the sex chromosomes
7. Proportion of phenotypic variation in a given trait that results from genetic segregation
8. Interference of one gene with the expression of another gene
9. Separation of homologous chromosomes during first meiotic division
14. Term for the sex that produces only one kind of gamete relative to the sex chromosomes
17. One gene with more than one phenotypic effect
18. Environmental mimic of a phenotype typically associated with a specific genotype
19. Form of linkage in which alleles of two genes occur together more often than predicted by their frequencies

DOWN

2. Mutation that causes death before reproductive age
3. Frequency with which a gene manifests itself in the phenotype of the heterozygote
4. Effect through which the genotype of mother is expressed in the offspring rather than the offspring expressing the offspring's own genotype
5. Form of dominance in which the phenotype of the heterozygote falls outside the range of either homozyogote

(*continued on p. 62*)

10. Type of dominance in which the phenotype of the heterozygote is intermediate between those of the homozygotes
11. Irregular distribution of chromosome or chromatids due to mishaps in cell reproduction
12. Genes present only once in the genotype, such as the genes of the X chromosome in the human male
13. Genes that are found on the X chromosome and not on the Y chromosome
14. Term for the sex that produces two kinds of gametes
15. Degree of phenotypic expression in those showing a given phenotype
16. Type of dominance in which two alleles express products in the heterozygote independently

PEDIGREE ANALYSIS

A human pedigree is a shorthand way of showing the relationships in a family in which a genetic trait is segregating. The analysis of pedigrees is one of the most direct ways we have to determine the mode of transmission for human inherited conditions. It is not always possible to reach a definitive solution, however, because of small family sizes and the limited information available about most family histories. Sometimes several explanations are possible, but one may be better (that is, more probable) than another. Here we shall show some of the symbols commonly used in pedigrees and outline some simple rules that should help you in analyzing pedigrees. Generations are given Roman numerals; individuals are numbered from left to right within each generation.

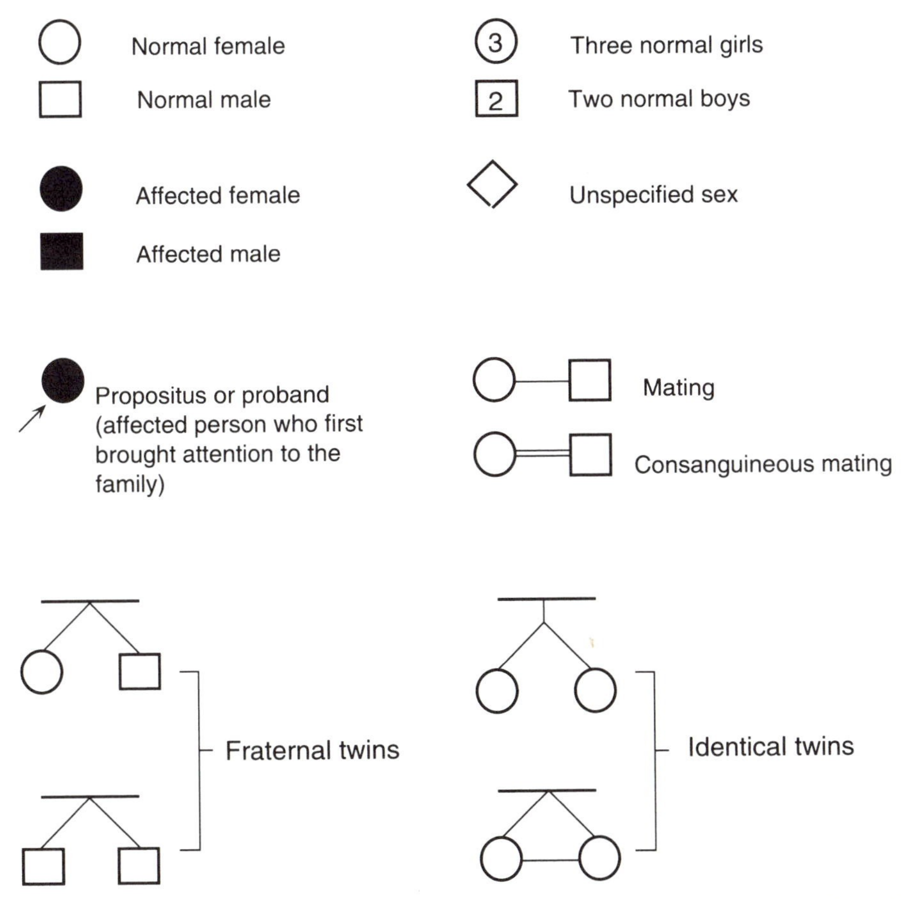

Although one cannot make experimental matings to investigate the inheritance of a human trait, one can use the pattern of its transmission in a pedigree to make predictions about the mode of inheritance. First, it is important to recognize that most traits tend to fall into certain limited categories: dominant or recessive, autosomal or X-linked, and completely penetrant or incompletely penetrant. The initial step is therefore to try to classify an unknown trait in terms of these categories, recognizing that a single pedigree may not provide sufficient information to do this unambiguously.

Start by determining whether the trait is dominant or recessive. The following rules are designed to help you ask some useful questions about each pedigree you analyze. In the absence of evidence to the contrary, when evaluating a pedigree of a rare condition assume that individuals marrying into the pedigree are homozygous normal.

I. *Dominance relationships:* A dominant trait is the easiest to recognize, so begin by evaluating the pedigree for its fit to the predictions for dominant inheritance. If it does not fit these, then it is a recessive trait or shows some complex aspect, such as incomplete penetrance.

   A. *Dominant traits*

      1. Reversing the typical Mendelian logic, remember that a dominant trait will not occur in an individual unless it also appears in at least one of the parents (exceptions include a new mutation or incomplete penetrance).
      2. A fully dominant trait will not skip generations. It will therefore often appear to be relatively common in a pedigree.
      3. Unaffected sibs will have only unaffected offspring.

   B. *Recessive traits*

      1. In a marriage of two affected individuals, all of the offspring will be affected.
      2. Recessive conditions are frequently found in pedigrees that include marriage between close relatives (a consanguineous mating).
      3. A recessive trait commonly skips one or more generations because it is masked in heterozygotes.

II. *Linkage relationships:* Look first for evidence of sex-linkage. Remember that a single pedigree may not provide enough information to make an unambiguous classification.

   A. *Sex-linked inheritance*

      1. A sex-linked trait can never be passed from a father to his son, since the father's X is passed to daughters and the Y does not carry typical coding genes. Therefore a single example of father-to-son transmission is sufficient proof that the trait is *not* sex-linked.
      2. If the trait is recessive, all sons of a female who expresses the trait will also be affected.
      3. If recessive, the trait will occur most frequently in males.
      4. If the trait is dominant, it will be expected to occur slightly more often in females.

   B. *Autosomal inheritance*

      1. An autosomal trait can be passed from a father to his son.
      2. Especially for a recessive autosomal trait, approximately the same number of males and females will be affected.

More complex relationships: Although the rules just listed will help you to analyze simple Mendelian conditions in a human pedigree, you should always remember that many situations can complicate the observed transmission patterns and gene expression. Incomplete penetrance, sex-limited expression (for example, testicular feminization, which can appear only in males), new mutation, two different loci that affect the same trait, adoption, mistaken parentage, and promiscuous mating can confuse the interpretation of a pedigree. In addition, the Y chromosome may be associated with the inheritance of some factors such as molecular polymorphisms. Such markers are called *holandric,* but there are few, if any, morphological traits that have definitely been traced to the Y chromosome.

IMPORTANT TERMS

Consanguinity
Holandric
Pedigree
Penetrance
Proband
Propositus (male)
Proposita (female)

PROBLEM SET 7

In the following pedigree diagrams, the generations and individuals within a generation are not numbered. This test of your ability to identify the appropriate individuals by numbering the pedigree properly is part of each problem. All pedigrees are fictitious. Any resemblance to a family living or dead is purely coincidental.

1. This pedigree is for the autosomal dominant trait achondroplasia, a rare form of dwarfism. Assuming complete penetrance, the probability that III-l and III-2 will have an affected child is **(a)** zero, **(b)** 1/4, **(c)** 1/2, **(d)** 1/16, **(e)** 1/8, **(f)** 1/32, **(g)** none of the above.

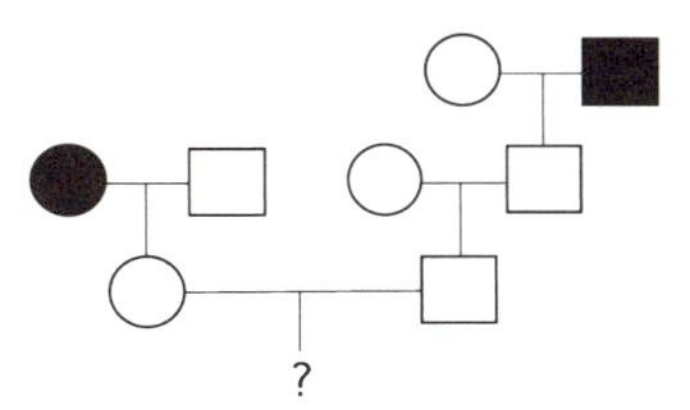

2. Todd and Sara (V-1 and V-2) wish to wed but are unaware that they share a common great-great-grandmother (I-2), who was heterozygous for a very rare autosomal recessive disease. The probability that their first child will be affected is **(a)** 1/16, **(b)** 1/64, **(c)** 1/256, **(d)** 1/1,024, **(e)** 1/4,096, **(f)** none of the above.

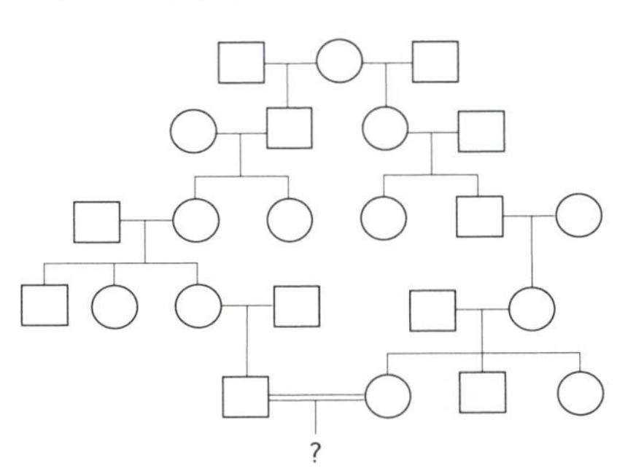

3. Two phenotypically normal people marry. Unknown to them, they have a common great-grandfather who suffered from a rare disorder (1 in 30,000 affected in the general population) that is inherited as an autosomal recessive trait (*dd*). Assume that the people marrying into this pedigree are all *DD*, that there has been no earlier consanguinity, and that no affected (*dd*) individuals have occurred. What is the probability that the first child of this couple (a baby girl) will be affected? Draw the pedigree.

4. Laura and Michael (II-5 and II-6) meet while attending a lecture concerning a rare human metabolic disease that is inherited as an autosomal recessive trait. They both have a sib who has this disease. They fall in love but are apprehensive about having children, since they know that the chance their first child will be affected is **(a)** 1/2, **(b)** 1/4, **(c)** 1/8, **(d)** 1/9, **(e)** 1/10, **(f)** 1/16, **(g)** 1/32, **(h)** none of the above.

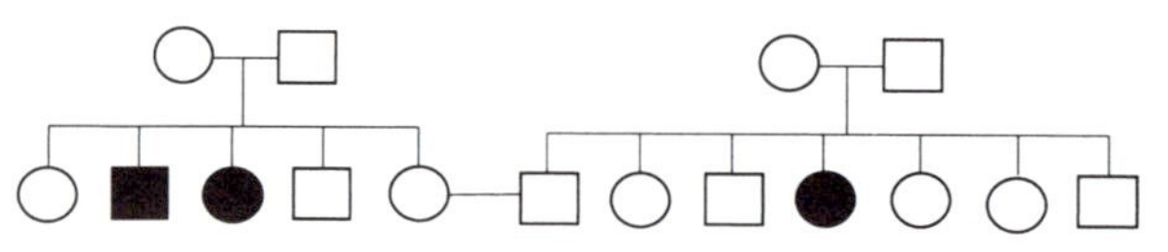

5. Consider the following pedigree.

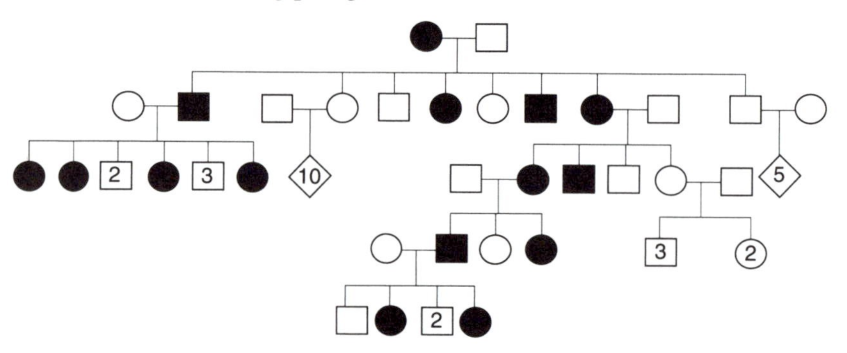

(a) What pattern of transmission is most consistent with this pedigree?
(1) autosomal recessive, (2) autosomal dominant, (3) X-linked recessive, (4) X-linked dominant.

(b) If individual V-2 marries a normal individual, and if the condition has a penetrance of 85 percent, what is the probability that their *second* child will express the trait?

(c) On the third line, what does the diamond with a *10* in the middle mean?

6. The accompanying pedigree is for a trait with 100 percent penetrance. The trait is not necessarily uncommon in the population, but individual I-l is homozygous normal.

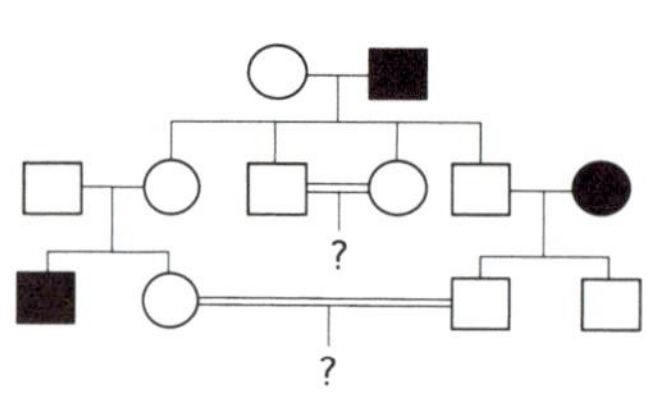

(a) What is the probable mode of inheritance?

(b) What is the probability that the child of the brother–sister mating between II-3 and II-4 will show the trait?

(c) What is the probability that a child of the first-cousin marriage (III-2 and III-3) will have the trait?

7. What kind of inheritance pattern is illustrated in this pedigree? Please name one or more specific traits that might fit this pattern of transmission and expression.

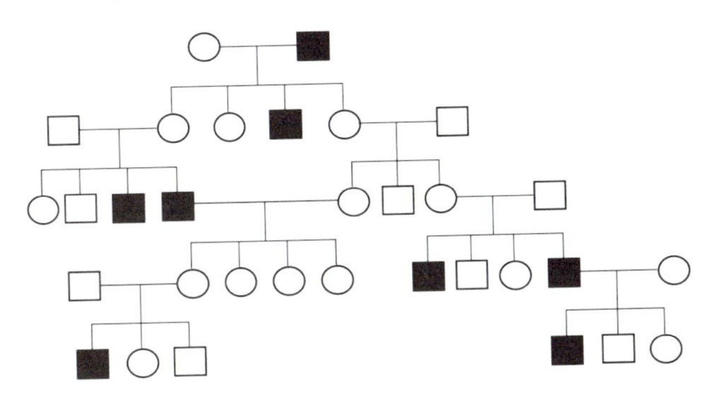

8. Consider the following pedigree of a trait having 100 percent penetrance.

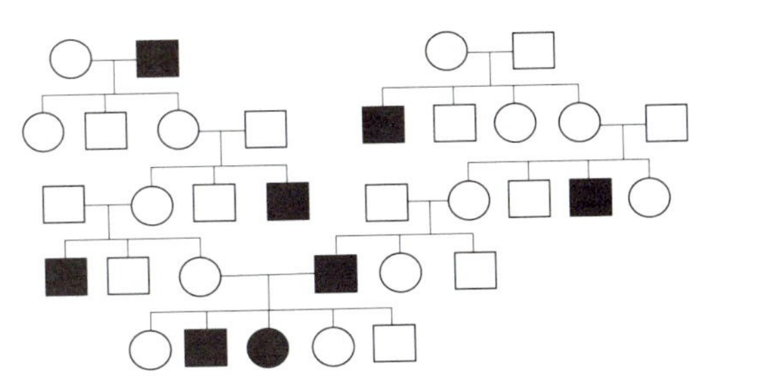

(a) The probable mode of inheritance of this trait is (1) X-linked dominant, (2) X-linked recessive, (3) autosomal recessive, (4) autosomal dominant.

(b) If individual V-2 marries a homozygous normal person, what is the probability that their first child will be a carrier?

9. This pedigree is for a rare autosomal recessive trait. The probability that a child of III-4 and III-5 will show the trait is **(a)** 1/2, **(b)** 1/4, **(c)** 1/8, **(d)** 1/12, **(e)** 1/16, **(f)** 1/32, **(g)** 1/64, **(h)** none of the above.

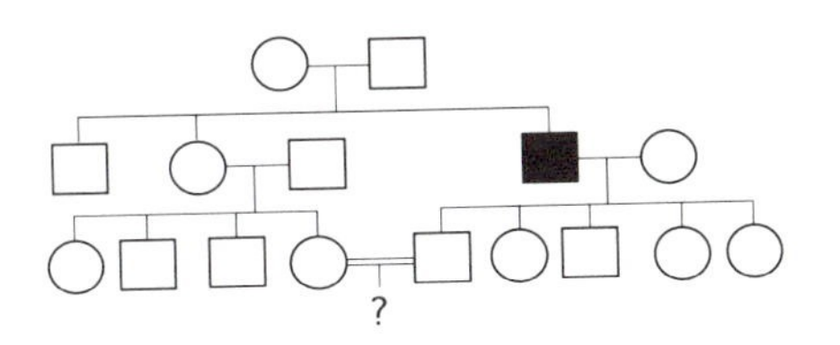

10. What kind of inheritance pattern is illustrated in this hypothetical pedigree? What is the probability that the offspring shown as a question mark will express the trait if it is a male child? What is the probability if it is a female child?

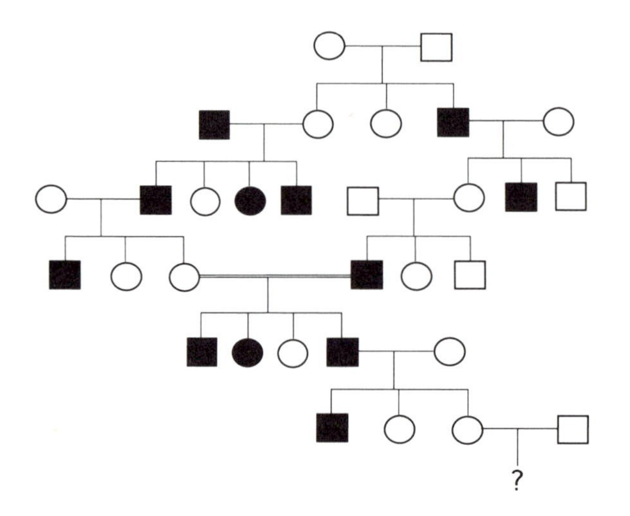

11. What can you tell the propositus about the inheritance of the trait he is exhibiting? How would you explain the entry of the gene into the pedigree?

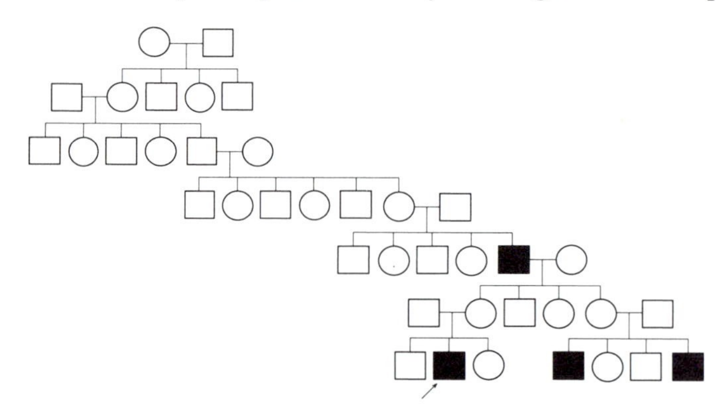

12. Shown is a pedigree for polydactyly in humans. What is the mode of inheritance for polydactyly, and what specific clues does the pedigree provide to support your hypothesis?

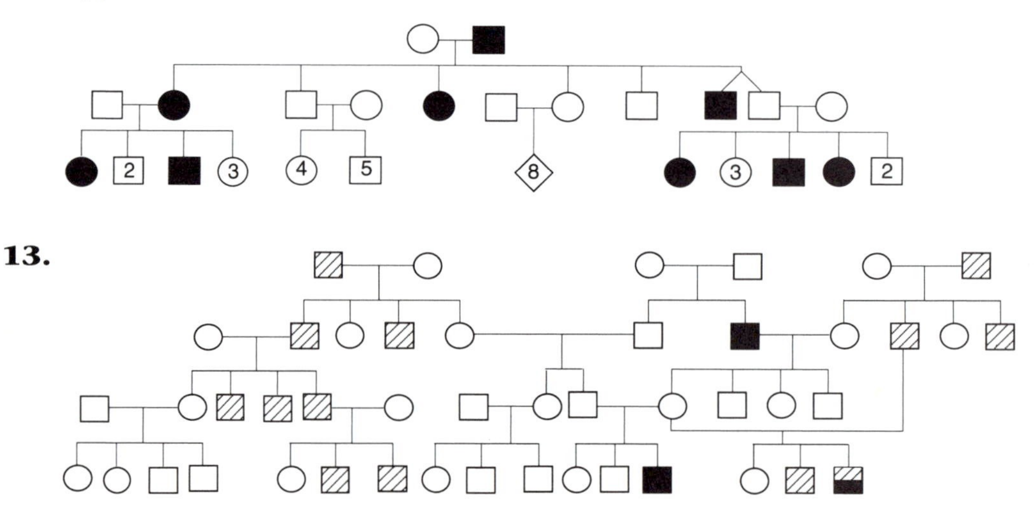

13.

This pedigree shows the inheritance of two traits. One, marked with diagonal lines, is the inheritance of a DNA marker; and the other, marked as a solid black symbol, is for total color blindness where an affected individual sees objects only in shades of white, gray, and black.

(a) Since two traits are segregating in the same pedigree, would we classify the genetic expression as pleiotropic? Why or why not?
(b) What is the inheritance pattern shown by each trait?
(c) If male IV-16 were to marry female IV-5, what is the probability that their first child would have the DNA marker?
(d) What is the probability that the first son will be totally color-blind?
(e) What is the probability that the first child will both be color-blind and carry the DNA marker?

ANSWERS TO PROBLEM SET 7

1. **(a)** Since the trait is inherited as a rare, completely penetrant dominant condition, we can assume that the affected people in this pedigree are heterozygous for the mutant allele. If the children in generation III had inherited the condition, they would express it. It then follows that the probability of an affected child's occurring in generation IV is zero.

2. **(d)** We must first determine the probability that Todd and Sara are both carriers (*Aa*). Because of the rare nature of the disease, we can assume that all people marrying into the pedigree are *AA*. The marriages in generation I are therefore *AA* males with the *Aa* female. Probabilities are traced in the pedigree, where the half-filled circle in generation I indicates heterozygosity. Tracing descendants II-2 and II-3 indicate that each has a 1/2 probability of carrying the recessive allele (1/2 of the offspring from a cross *AA* x *Aa* will be *Aa*). Since at each generation the person marrying into the pedigree will be *AA*, the probability the recessive allele will be retained in the line will be halved in each generation. If Todd and Sara are both *Aa*, the probability of having an *aa* child is 1/4. Thus the total probability of the gene's traversing all generations and the parents' having an *aa* child is

$$(1/2 \cdot 1/2 \cdot 1/2 \cdot 1/2 \cdot 1/2 \cdot 1/2 \cdot 1/2 \cdot 1/2) \cdot 1/4 = 1/1{,}024$$

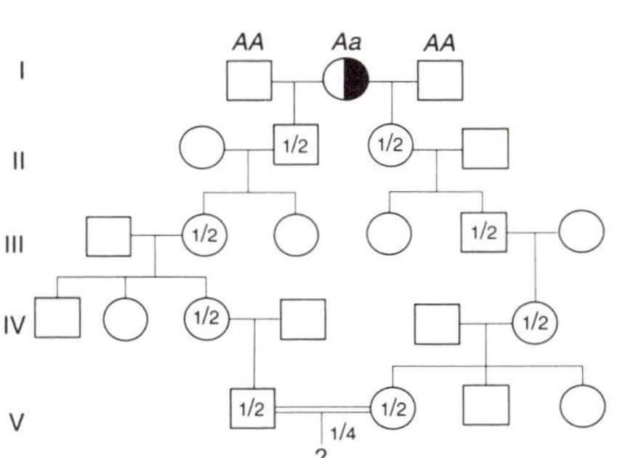

3. For simplicity's sake, only the people in the direct line from the *dd* common great-grandparent are shown in the pedigree we have drawn. Their sexes are arbitrary, since the trait is autosomal, and we are making up the pedigree. Since the great-grandfather is *dd*, all of his children in generation II will be heterozygous. For each subsequent generation, the probability that the *d* allele will be transmitted is 1/2. So the probability that it is transmitted to generation IV is 1 • 1/2 • 1/2 = 1/4 for each of the parents. The probability that *both* of the parents are *Dd* is therefore 1/4 • 1/4 = 1/16. The probability that the first child of IV-1 and IV-2 (regardless of sex) will be *dd* is 1/16 • 1/4 = 1/64.

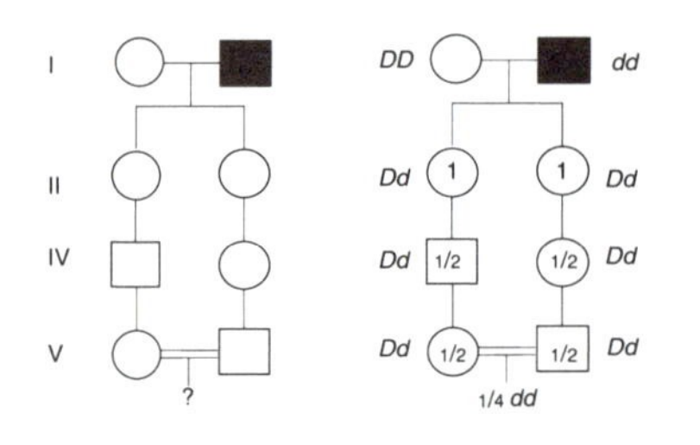

4. (d) Since the trait is an autosomal recessive one, both sets of parents in generation I must be *Aa*. That is the only way they could produce affected children without themselves showing the trait. The expected progeny for such a cross are 1/4 *AA* + 1/2 *Aa* + 1/4 *aa*. That is, 3/4 are phenotypically normal, and 1/4 are affected. Since Laura and Michael are phenotypically normal, we can exclude the probability that they are *aa*. They are either *AA* or *Aa*, with a 1/3 chance of their being *AA* and a 2/3 chance that they are *Aa*. Thus the probability that they *both* are carriers is 2/3 • 2/3 = 4/9. Multiplying this by 1/4 (the probability that the heterozygotes will have an affected child), we get an overall probability of 4/36 = 1/9.

5. (a) The pattern of transmission appears to be X-linked dominant. Every affected individual has an affected parent, and normal parents have only normal progeny. The trait appears to be X-linked, because the female offspring of affected males are all affected, but no sons are affected. Affected females produce both affected sons and affected daughters.

(b) Female V-2 will transmit the mutant allele to both male and female offspring with a probability of .5. Whether it is the first, second, third, or a later child makes no difference. The probability that it will be expressed in a carrier is .85. Thus the overall probability that it will be both inherited and expressed is .50 • .85 = 0.425.

(c) The diamond signifies 10 phenotypically normal children whose sex is either unknown or irrelevant to the interpretation of the pedigree.

6.

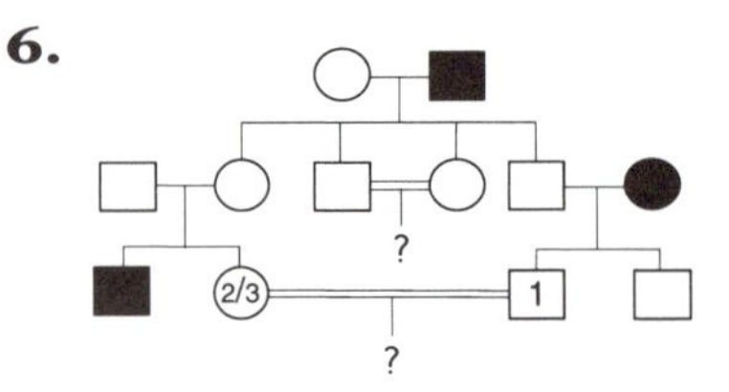

(a) Since the trait is 100 percent penetrant, it cannot be dominant, because III-l is affected but has phenotypically normal parents. If it is a sex-linked recessive trait, then the III-3 and III-4 males should be affected, since their mother is homozygous recessive and would transmit the trait to all her sons. But since they are normal, the trait must be inherited as an autosomal recessive. The parents in generation I are *AA* female and *aa* male.

(b) Both II-3 and II-4 are heterozygous *Aa*. The probability of having an affected child is therefore 1/4.

(c) Since III-l is affected, the parents (II-l and II-2) are both heterozygous. This is a situation in which an individual marrying into the pedigree is a carrier, as often happens when the trait is not particularly rare. Since 3/4 of the offspring from a mating of two heterozygotes would be phenotypically normal (1/3 being *AA* and 2/3 being *Aa*), the probability that III-2 is a carrier is 2/3. Individual III-3 *must* be a carrier. His father was *Aa,* and his mother was *aa,* but he is not *aa* (he had 1/2 chance of being *Aa* and 1/2 of being *aa*). The probability that *both* III-2 and III-3 will be carriers is therefore 2/3 • 1 = 2/3. The probability that an affected child's being produced from a mating of two heterozygotes is 1/4. Thus the answer to this question is 2/3 • 1/4 = 1/6.

7. Since this trait skips generations and is expressed in males, one might jump to the conclusion that it is a sex-linked (X-linked) recessive character. But it is always important to see whether the evidence is consistent with your initial hypothesis. A sex-linked trait will never be passed from a father to his son. But that happens twice in this pedigree: I-2 passing the trait to male II-4, and IV-9 passing it to male V-4. So it must be an autosomal recessive trait. Now, what autosomal recessive traits might have an expression that is limited to males? Examples of sex-limited traits in males include those affecting secondary sexual characters like beard growth, breast size, and voice changes. Indeed, autosomal genes affecting any process or structure that is limited to one or the other sex could potentially fit this pattern. Examples in females include milk quality and quantity in dairy cattle and hydrometrocolpos, a condition in humans in which fluid accumulates in the uterus and vagina (A. P. Mange and E. J. Mange, *Genetics: Human Aspects, second edition* [Sunderland, MA: Sinauer Associates, 1990]).

8.

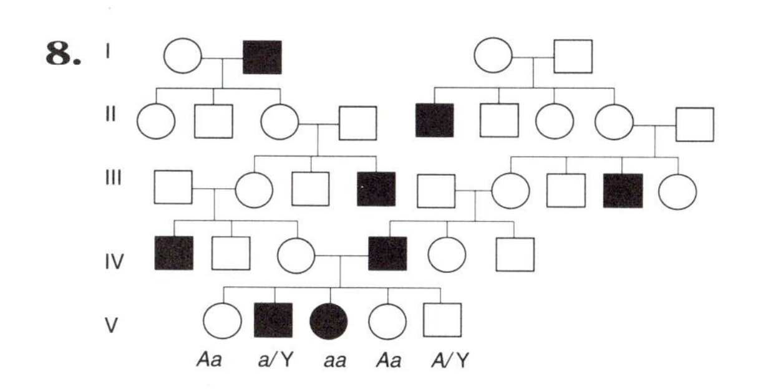

(a) The answer is (2): The mode of inheritance is sex-linked recessive. It is recessive because affected individuals do not necessarily have an affected parent (remember, 100 percent penetrance). It is sex-linked because all but one of the affected individuals is a male, apparently coming from a carrier *Aa* mother. The affected female, V-3, is a homozygote produced from merging two branches of the pedigree. Her mother was *Aa* and her father hemizygous *a*/Y.

(b) A carrier for a sex-linked recessive trait can only be a female (a male would always show such a trait). In this case the mating would be between an affected male (*a*/Y) and a normal female. All the daughters would inherit the father's X chromosome with the *a* allele, but none of the sons would. Thus there is 1/2 chance that a child will be a carrier, since there is 1/2 chance that it will be a girl.

9.

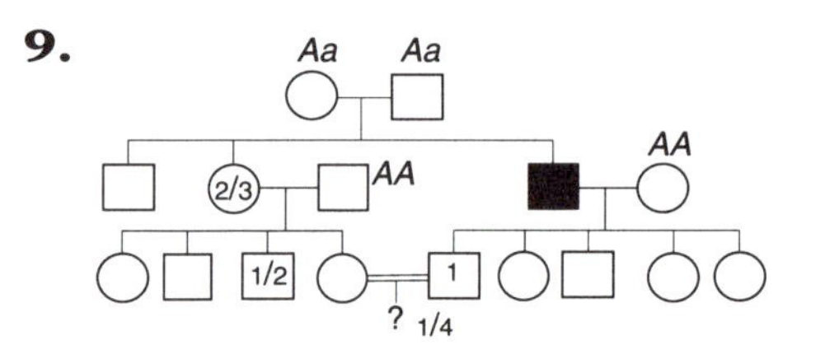

(d) 1/2 Since II-4 is affected (*aa*), we can conclude that both his parents are *Aa*. Since the trait is rare, we can assume that his wife is *AA* and that all his children will therefore be *Aa*. Each of the phenotypically normal children of I-l and I-2 has a 1/3 chance of being *AA* and a 2/3 chance of being *Aa*. If II-2 is a heterozygote, the daughter (III-4) has a 1/2 chance of inheriting the *a* allele. If II-2 has a 2/3 chance of being *Aa* and she is married to an *AA* man, the overall probability that III-4 is heterozygous is 2/3 • 1/2 = 1/3. The overall probability of having an affected child is 1/3 (probability of III-4 being *Aa*) • 1 (the probability of III-5 being *Aa*) • 1/4 (probability of two carriers having an affected child) = 1/12.

10. This trait is autosomal, because it is passed from fathers to sons and so cannot be sex-linked. It is not a sex-limited trait, because females can be affected, although at a much lower rate than males. Furthermore, the females can inherit the trait from fathers and pass it on to sons but not show it themselves, unless they are homozygous (such as females III-4 and V-2). Thus, the trait is dominant in males and recessive in females. Expression is sex-influenced. Examples of human sex-influenced traits that are more commonly expressed in males include male pattern baldness, and pyloric stenosis. Thus, if it is male, the child has (1/2)(1/2) = 1/4 chance of being affected. A daughter could not be homozygous, so the probability is zero. Examples in females include congenital hip dysplasia, osteoporosis, and some autoimmune diseases such as lupus.

11. The trait looks most like a sex-linked recessive. It could have entered the pedigree through a female carrier, III-6. It might also have entered through a I-1 carrier female, but if so, it was passed down to IV-6 without affecting any of the eight males in generation II, III, and IV (including III-5, who would have to be non-penetrant). Since half of them would have been expected to show the trait, this source is less likely. The best hypothesis is that it was brought in by III-6. Another comparatively unlikely explanation is that it originated as a mutation of *A* to *a* in the germ line of IV-6 or of III-5. Since mutation is a rare event, this explanation is used only if all others are less likely or otherwise excluded. One might test our hypothesis (III-6 carrier female) by tracing the pedigree of her family to see whether one can confirm the possibility of her being *Aa*.

12. It looks like an autosomal dominant trait with incomplete penetrance. Dominance is indicated by its frequency and by the fact that in all but one sibship the affected people have an affected parent. In that exception (the offspring of II-10), their father is a monozygotic twin with a polydactylous brother. The twins therefore both carried the dominant gene for polydactyly, but it is nonpenetrant in II-10. Its autosomal nature is indicated by the fact that the affected male, I-2, has an affected son. If the trait were X-linked, you would not expect an affected son, since the father's X chromosome goes only to his daughters. Similarly, you would not expect a normal daughter unless, of course, it were nonpenetrant in her. In summary, the pattern of expression is consistent with the expectations of an autosomal dominant trait. The discordant monozygotic twins and the fact that a normal II-10 male produces affected children indicate less than 100 percent penetrance.

13. **(a)** Pleiotropy is defined as two or more traits that are due to the same genetic change. Here there are clearly two different genetic factors segregating, so genetic expression would not be classified as pleiotropic.

(b) The DNA marker is passed from a father to all of his sons, and never from father to daughter. The only chromosome that shows that pattern of transmission is the Y, so the DNA marker could be considered a holandric trait. Total color blindness, on the other hand, is not passed to sons but can be transmitted through a daughter to grandsons. Total color blindness, like the more familiar "red–green color blindness" is X-linked.

(c) Male IV-16 is totally color-blind and carries the Y chromosome DNA marker. Female IV-5 has normal color vision and, although her father carried the Y chromosome marker, she could not have inherited it (or "she" would be a "he"). The probability that their first child will have the DNA marker is 1/2, since all sons will inherit the Y from the father and there is a 1/2 chance that the child will be a son.

(d) Since color blindness is a sex-linked recessive, it will not be passed to the son from the father. The probability is, therefore, zero. You might also note that the probability that a daughter will be color-blind is also zero, since it is a recessive trait and would be heterozygous, not homozygous, in any daughters. All daughters would be heterozygous carriers.

(e) To be both color-blind and a carrier of the DNA marker is not possible in this marriage. As noted in part d, no children will be color-blind though, from part c, half will have the Y-linked DNA marker.

OVERVIEW OF BASIC STATISTICAL TESTING

In earlier chapters, such as Chapter 5, we discussed probability and statistics as they apply to a specific kind of genetic problem. Here we want to take a more general view of biostatistics and introduce some of the different ways one can describe relationships and test hypotheses.

The term *statistics* refers to the mathematical process of collecting, analyzing, interpreting, and presenting numerical data. Some statistical measures are purely descriptive, such as the sample mean or values of dispersal like the range, standard deviation, and variance. Other statistical measures are designed to evaluate relationships among groups of data or to test hypotheses about them.

Some descriptive statistics important in genetic analyses are discussed in more detail in Chapter 9, which focuses on quantitative genetic traits. Many quantitatively varying traits, such as seed number and tail length, approximate a normal distribution. The mean is the average value for a data set, and the variance is a measure of dispersal around the mean. The standard deviation is the square root of the variance and divides a normal distribution into subgroups of known size (for example, 68 percent of the data points fall within one standard deviation of the mean, 95 percent fall within two standard deviations, and 99 percent fall within three).

In describing relationships within and among the data points, it is useful to distinguish between two general types of statistical tests. *Parametric statistic tests* assume a normal distribution of the data; *nonparametric statistics* do not. Parametric tests can only be used when the following conditions are met: the data fit a statistically normal frequency distribution, individual data points are independent, and all observations are on the same continuous scale of measurement. On the other hand, nonparametric tests can be used to analyze almost any type of data distribution, but they are generally less powerful than the parametric tests. Most of the common statistical tests used in genetics, such as the *t*-test and analysis of variance, are parametric tests.

In the following outline, we have organized a brief description of various statistical tests in terms of the kind of data presented and the type of question or hypothesis being tested. Most genetics textbooks have sections that describe these tests in more detail. *Biometry* by R. R. Sokal and F. J. Rohlf (San Francisco: W. H. Freeman and Company, 1981) and other statistical books are also valuable references.

I. Differences
   A. Differences between distributions of data in categories
      1. Between observed sample and theoretical distribution

         χ^2 test of goodness of fit: This tests the difference between one observed sample and a theoretical distribution (examples: the fit between observed numbers of brown and tan mice produced from a monohybrid cross where the theoretical expectation is a 3:1 ratio; the fit of male and female offspring numbers to a theoretical expectation of a 1:1 ratio). The theoretical distribution is set by the

hypothesis you choose to test (examples: fit to a 1:1 ratio in one test and fit to a 1:2 ratio in another). The value of chi-square is

$$\chi^2 = \text{sum} \frac{(\text{observed} - \text{expected})^2}{\text{expected}}$$

and the number of degrees of freedom (d.f.) is the number of classes minus 1 (example: to test the fit to a 3:1 ratio, there are two classes and d.f. = 1; to test the fit to a 9:3:3:1 ratio, there are four classes and d.f. = 3). If the number in any class is small, Yates's correction factor is used to give a more accurate value of χ^2. The numerator becomes $(|\text{observed} - \text{expected}| - ½)^2$.

For a chi-square test, the *null hypothesis* is that the observed distribution fits the theoretical distribution, which is defined by the hypothesis you have chosen. A table of chi-square values is given in the Appendix.

2. For independence between two distributions:

χ^2 test of independence (*contingency* χ^2): This is used to compare two samples taken under different conditions (examples: a treatment group compared to a control group; one replicate in an experiment compared to another replicate; one sample compared to another sample taken at a different time). There are no specific expected numbers, as there are when comparing to a theoretical ratio.

For the following data

	Classes of observation		
	1	2	Totals
I	a	b	$a + b$
II	c	d	$c + d$
	$a + c$	$b + d$	$a + b + c + d = N$

with Yates's correction factor included, the contingency chi-square value is

$$\chi^2 = \frac{[|ad - bc| - 1/2(n)]^2 n}{(a + b)(a + c)(c + d)(b + d)}$$

A related test, the *homogeneity chi-square test,* can be done on multiple samples.

B. Differences between means

1. Two samples

Student's t-test: This test uses the means, variances, and sample sizes to test whether two samples were drawn from the same underlying distribution. The null hypothesis is that the two distributions are random samples from the same theoretical distribution. The number of degrees of freedom is the sum of the two sample sizes minus 2. A table of t values is given in the Appendix and the formula for a t-test is presented in Chapter 9. A significant value of t indicates that, with the given variances and sample sizes, the difference between the two means is larger than would be expected by chance alone (for example, for a calculated value of t that the statistical table shows has a p value of less than 0.01, a difference in means as large as or larger than that in the data would be expected only once in every 100 times).

2. More than two samples

   Analysis of variance (ANOVA): An ANOVA can be thought of as a series of *t-tests,* and it is a powerful test that can be applied to many different experimental designs. We refer you to a biostatistics book for an introduction to this important type of analysis.

C. Differences between variances

   F-test: The *F*-test or variance ratio tests whether one variance is significantly larger than another and involves two separate degrees of freedom, one for each variance.

II. Correlations and regressions

There is often some confusion between correlation and regression, because the two measures are mathematically and conceptually related. Here we distinguish between them, and we refer you again to a biostatistics reference for their formulae and for worked-out examples.

A. *Correlation* is concerned with whether two variables are dependent on each other, that is, whether they vary together (examples: does the amount of protein intake covary with the number of eggs laid by a fly; to what extent in the tail length in mice correlated with overall body weight). One variable is not expressed in terms of the other. Correlations are quantified in terms of a number, the *correlation coefficient.*

B. *Regression,* on the other hand, is a formula intended to describe the dependence of one variable on another, independent variable. Regression equations can be used to explore hypotheses about causal relationships between the two variables and to predict the value of one variable in terms of the other (examples: from a regression of protein intake versus fly egg number, we can predict the number of eggs that will be laid given a certain level of protein nutrient; from a regression of temperature on pigment deposition, we can predict the color phenotype when the animal is raised at a specified temperature; using a regression of age versus frequency of chromosomal mutations, we can explore the causal relationship between these two variables).

QUANTITATIVE INHERITANCE

STUDY HINTS

A large part of genetic analysis depends upon accurate statistical treatment of data. As we have pointed out earlier, probabilities help form the hypotheses against which patterns of inheritance and gene action can be tested. The same is true of quantitative genetics. In this instance, however, the patterns are complicated by the fact that one is dealing not only with the segregation of several genes but also with environmental factors that can mask the expression of alleles in the genotype. Several important genetic specialties, such as biometrical genetics, are concerned with ways of extracting information about the genotype from quantitative phenotypes.

Perhaps the most commonly used measures of quantitative distributions are the mean, standard deviation, variance, and standard error. These measures are described briefly in the following table (and in the Glossary).

Measure	Symbol	Calculation
Mean	$\bar{x}$	The sum of all individual measurements divided by the sample size (n)
Standard deviation	s, s.d., or σ	$\sqrt{\frac{\Sigma x^2 - \bar{x}^2(n)}{n-1}}$, where $\Sigma \chi^2$ is the sum of each individual measurement squared
Variance	s^2, σ^2, $\hat{s}^2$	$\frac{\Sigma(x - \bar{x})^2}{n-1}$, square of the standard deviation
Standard error	s.e. or $S\bar{x}$	$s/\sqrt{n}$ is standard deviation of the means

Using these measurements of mean and of dispersal around the mean, one can accurately describe many aspects of a quantitative distribution. Of these, perhaps the standard deviation is the most useful statistically. Predictable proportions of the population of values will fall within one, two, or three *standard deviations* of the mean. These are shown precisely in some texts, but for convenience the divisions are often rounded off as illustrated in the following figure.

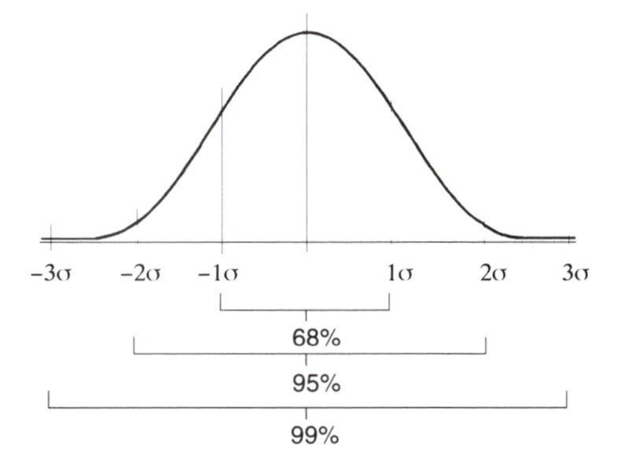

Since quantitative phenotypic distributions, such as those for height or IQ, involve both genetic segregation and environmental influences, it is useful to have ways of predicting the relative importance of each of these factors for a given trait or population. Although gene number can only be estimated crudely, consider the expectations based on simple Mendelian segregation. If one pair of alleles is segregating, 1/4 of the F_2 progeny should be as extreme in phenotype as each of the original parents (e.g., 1/4 should be as short as the shorter parent and 1/4 as tall as the taller parent in an original cross between short and tall). When two pairs of alleles contribute to the trait, 1/16, or $(1/4)^2$, of the F_2 progeny are as extreme as each of the original parental phenotypes. Generalizing from this, $(1/4)^n$ of the F_2 progeny should be in this extreme class, where n is the number of loci segregating in the cross.

In a more general way, we can ask what proportion of the total phenotypic variance is due to genetic causes. This proportion of heritable variance is called *heritability* and is symbolized h^2 (the squared term simply indicates that it is derived from variances). In the most restricted or narrow sense, the estimated genetic effects are limited to additive genetic segregation. In the broad sense, heritability reflects gene interactions, variations in dominance, and other genetic complications. The heritability resulting from additive segregation is the most important type for predicting the response of a population to selection or patterns of transmission. We will therefore limit our discussion to heritability in the narrow sense.

Calculations of heritability can be done in a number of ways, such as by measuring the regression of offspring phenotypes on parent phenotypes or by comparing the variances in generations bred from crosses between homozygous strains. We can illustrate the logic of such measurements by considering the sources of variation in the F_1 and F_2 generations produced by crossing inbred parental lines.

All variation is environmental.

The individuals in the distribution are genetically the same; all variation (V) is environmental (V_E), since all individuals are homozygous for alleles they share, and all are heterozygous for alleles that differ between the parental strains.

Variation is the result of both environmental factors (V_E) and genetic segregation (V_G).

F2 VF2

Thus one way to calculate heritability is as follows:

$$h^2 = \frac{V_{F_2} - V_{F_1}}{V_{F_2}} = \frac{V_G + V_E - V_E}{V_G + V_E} = \frac{V_G}{V_G + V_E}$$

When evaluating distributions, we frequently would like to know how different they are from each other. The *t*-test is a statistical test that allows us to calculate the significance of the difference between two means. It takes into account the variances (s_1^2 and s_2^2) and the sample sizes (n_1 and n_2) of each distribution.

$$t = \frac{\bar{x}_1 - \bar{x}_2}{\sqrt{s_3^2(1/n_1 + 1/n_2)}} \quad \text{where} \quad s_3^2 = \frac{(n_1 - 1)s_1^2 + (n_2 - 1)s_2^2}{n_1 + n_2 - 2}$$

In a *t*-test, the null hypothesis is that the two distributions are samples from the same population of measurements; that is, the two distributions are not significantly different. The use of a null hypothesis comes from a consideration of the limitations of information and the structure of logic. It is a sort of default position. Consider, for example, the statement "All genetics students know a person named Richard." This statement is difficult (or even impossible) to prove correct, since it would involve surveying every known student of genetics, but it is easy to disprove. All one needs to do is find one example of a student who does not know someone named Richard. The null hypothesis in a statistical test works the same way. One cannot readily prove an idea, but one can disprove it, or at least establish strong evidence against it. In a *t*-test and similar tests used in quantitative genetics, the formulation of the appropriate null hypothesis is very important. Usually, the null hypothesis is that the two things or sets being compared are the same. One can readily marshal evidence to support the idea that they are different, so one ends up approaching the problem backward. By failing to disprove the null hypothesis, one indirectly supports the alternative.

IMPORTANT TERMS

Analysis of variance
Class intervals
Coefficient of variation
Correlation
Covariance
Frequency distribution
Heritability
Multiple-factor inheritance
Null hypothesis
Polygene
Population
Quantitative characters
Regression
Standard deviation
Standard error
Statistics
t-test
Variance

PROBLEM SET 9

1. In a hypothetical experiment, we cross two strains of chickens that are genetically different with respect to several gene loci that affect egg weight. We discover that the F_2 variance in egg weight is .195, but the F_1 variance is only .130. From this we can conclude that the heritability for egg weight in these chickens is about **(a)** .66, **(b)** .20, **(c)** .33, **(d)** .13, **(e)** none of the above.

2. In the F_2 generation of a cross between tall snapdragons and short snapdragons, a total of 30 plants (of 7,678 plants measured) were as short as the original short snapdragon parental strain. From these data, please estimate the number of segregating loci affecting plant height in this cross.

3. In a random sample of 100,000 fifth-graders, the average IQ was found to be 100, and the standard deviation was 15 IQ points. **(a)** What number of students would be expected to have an IQ over 130? **(b)** How many would have an IQ between 70 and 100? **(c)** How many would have an IQ within one standard deviation of the mean?

3 1303 00195 0980

4. Assume an animal population has just one segregating locus on each of its 20 pairs of chromosomes, and assume that each locus has 3 alleles (chromosome 1 has alleles A^1, A^2, or A^3; chromosome 2 has alleles B^1, B^2, or B^3, and so forth). What is the number of genetically different gametes that such an animal can produce?

5. For the animal population described in problem 4, how many different genotypes are possible?

6. Assume that three strains of garden peas (A, B, and C) produce the samples of peas indicated in the accompanying figure. All three samples contain the same number of individuals and come from plants raised in the same environment. The strain that is the most inbred is probably **(a)** A, **(b)** B, **(c)** C.

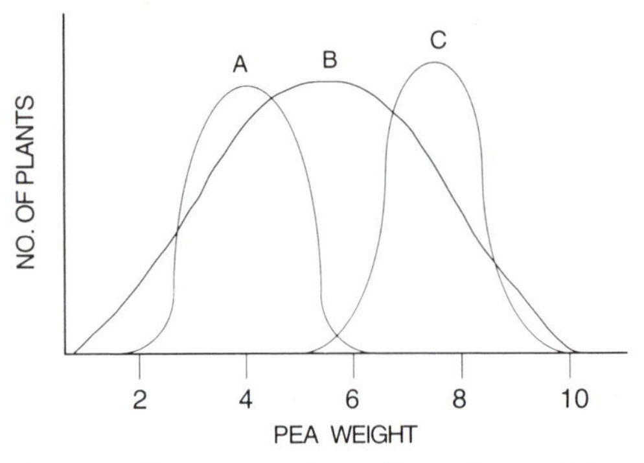

7. Assume that a pure-breeding plant with red flowers is crossed with a pure-breeding plant with white flowers and that flower color is a polygenic trait, with the two parent strains differing at seven loci. The F_1 will all be intermediate in flower color. In the F_2, about one plant in how many will be expected to have white flowers? **(a)** 8, **(b)** 16, **(c)** 32, **(d)** 64, **(e)** 4,000, **(f)** 16,000, **(g)** 64,000, **(h)** 256,000.

8. Plant populations A and B each have a mean height of 1.8 meters, with a standard deviation of 0.3 meter in each case. Population A contains 50 plants, and population B consists of 100 plants. The standard error (standard deviation of the means) of population A will be **(a)** the same as that of population B, **(b)** larger than that of population B, **(c)** smaller than that of population B.

9. A sample of animals has a mean weight of 250 kilograms (kg) and a variance of 100. With this information we can predict that 34 percent of the animals will have weights ranging from **(a)** 150 to 350 kg, **(b)** 240 to 250 kg, **(c)** 150 to 250 kg, **(d)** 240 to 260 kg.

10. The idealized, or classical, model of polygenic inheritance predicts a normal F_2 distribution for various traits (such as height). On the other hand, if some of the loci had a dominant allele for increased height, what would the distribution look like?

11. Calculate the mean, variance, and standard deviation for each of the following sets of data obtained by measuring bristle numbers on the fruit fly *Drosophila melanogaster.*

Data set 1: 12, 10, 5, 21, 8, 14, 18, 9, 18, 7
Data set 2: 19, 21, 16, 19, 20, 17, 22, 14, 18, 15

12. Using a *t*-test, determine whether the two sets of data given in problem 11 are consistent with the hypothesis that they came from the same population of *Drosophila.*

13. Which of the following sets of data has the largest variance?

	Mean	Sample size	Standard error
1	19.348	25	0.9282
2	0.400	25	0.0494
3	113.375	16	0.9215

14. The following data were obtained by measuring the lengths of new growth in two plots of a plant species maintained in different light–dark cycles. Is there a significant difference in the response to lighting conditions?

Sample 1: 25.5, 19.7, 16.5, 23.6, 21.3, 12.2, 16.8, 14.5, 27.4, 16.0
Sample 2: 16.7, 22.1, 28.3, 12.1, 24.8, 25.4, 16.8, 15.5, 22.0, 18.9

15. In which one or more of the following traits is genetic variance more influential than environmental factors in determining the phenotype? (Heritability estimates were obtained from D. S. Falconer, *Introduction to Quantitative Genetics* [London: Oliver & Boyd, 1960]).

Trait	Heritability
Milk production in cattle	0.60
Litter size in pigs	0.30
Egg production in poultry	0.20
Puberty age in rats	0.15
Wool length in sheep	0.55
Amount of spotting in Friesian cattle	0.95
Body weight in sheep	0.35
Abdominal bristle number of *Drosophila*	0.50

ANSWERS TO PROBLEM SET 9

1. **(c)** *Heritability* is defined as the proportion of phenotypic variance that is the result of genetic segregation. It can be calculated from the F_1 variance (which is totally environmental, V_E) and the F_2 variance (which is composed of both environmental and genetic components, V_G).

$$h^2 = \frac{V_{F_2} - V_{F_1}}{V_{F_2}} \text{ where } V_{F_2} = V_E + V_G \text{ and } V_{F_1} = V_E$$

$$= \frac{V_E + V_G - V_E}{V_E + V_G}$$

$$= \frac{0.195 - 0.130}{0.195} = \frac{0.065}{0.195} = 0.33$$

2. The number of segregating loci is estimated from the proportion of F_2 offspring that are phenotypically like one of the original P generation parents (that is, are as extreme as one of the original extreme parental strains). For example, if there is only one locus segregating (*A* and *a*), then in the F_2 a total of 1/4 of the progeny will be *aa,* like one (phenotypically extreme) original parent. Thus gene number can be estimated by reducing the observed data to the form $(1/4)^n$ where n is the number of segregating loci. In this problem 30/7,678 is almost exactly $(1/4)^4$. There are therefore four segregating loci estimated from this evidence. Two important assumptions of this model are that **(a)** there are only two alleles per locus and **(b)** they are assorting independently. Unless there is evidence to the contrary, these assumptions should be made when evaluating one of our problems.

3.

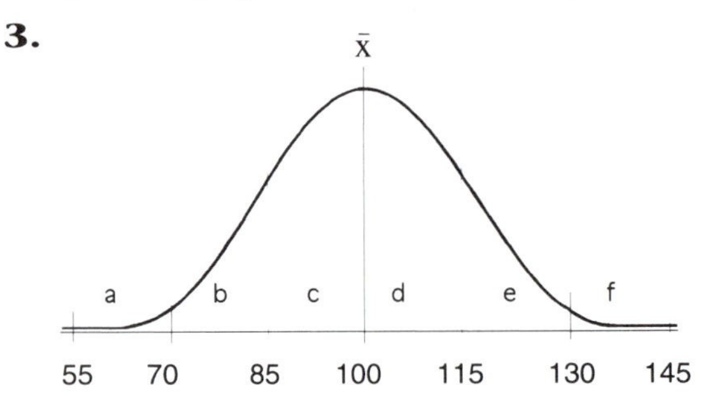

This figure shows the IQ distribution, with a mean of 100 and a standard deviation of 15. Standard deviations are marked to the left and right of the mean . Each region under the curve has been given a letter for convenience in discussing this problem.

(a) What proportion of students would be expected to have an IQ of over 130? Since 130 is 2 standard deviations above the mean, 95 percent of the population would have an IQ between 70 and 130 (sections b, c, d, e). Half of the remainder would have an IQ above 130 (section f). This is half of 5 percent, or 2.5 percent, which is 2,500 individuals from our sample.
(b) How many would have an IQ between 70 and 100? A total of 95 percent of the population falls within 2 standard deviations of the mean. Thus, half of 95 percent would have an IQ in the range 70–100 (sections b + c). This is 47,500 individuals.
(c) How many would have an IQ within 1 standard deviation of the mean? A total of 68 percent of the population has a phenotype within 1 standard deviation of the mean (an IQ in the range 85–115; sections c + d). This would involve 68,000 fifth-graders in our sample population.

4. For each of the 20 chromosomes there are three possibilities (allele 1, 2, or 3), independent of the genotypes of the other 19 chromosomes. The answer is therefore $(3)^{20} = 3,486,784,401$.

5. For each of the 20 loci there are $n(n + 1)/2$ combinations $=3(3 + 1)/2 = 6$ different genotypes possible (A^1A^1, A^2A^2, A^3A^3, A^1A^2, A^1A^3, and A^2A^3). The answer is thus $(6)^{20}$, which is 3,656,158,440,062,976. This large number is achieved with only 20 independently assorting loci, each with only 3 alleles, a striking demonstration of the ability of sexual reproduction, with its accompanying segregation and independent assortment, to generate hereditary variability.

6. **(c)** Inbreeding increases the proportion of homozygotes and thus decreases the variance of the phenotype. The strain with the smallest variance, and therefore probably the most inbreeding, is C.

7. **(f)** The proportion that will be as extreme as one original parental line is $(1/4)^n$ where n is the number of segregating loci. In this situation, $(1/4)^7 = 1/16{,}384$. The closest answer is 16,000.

8. **(b)** The standard error is defined as the standard deviation divided by the square root of the sample size (n). As n increases, the standard error decreases. Thus population A (where $n = 50$) will have a larger standard error than will population B ($n = 100$).

9. **(b)** Since 34 percent is 1 standard deviation either above or below the mean, the only range that fits is 240–250 kg. Remember that the standard deviation is the square root of the variance. If you chose an answer such as **(c),** 150–250 kg, you may have done so because you did not read the question carefully enough to evaluate which statistics you were given to work with. This is a common error in working problems in genetics, but it is also an easy error to avoid. Remember, it is less painful to learn the lesson in a problem set like this than on a graded examination.

10. If polygenic alleles showed dominance, rather than being additive as they are commonly modeled, the distribution would be skewed toward increased height, as in the following figure.

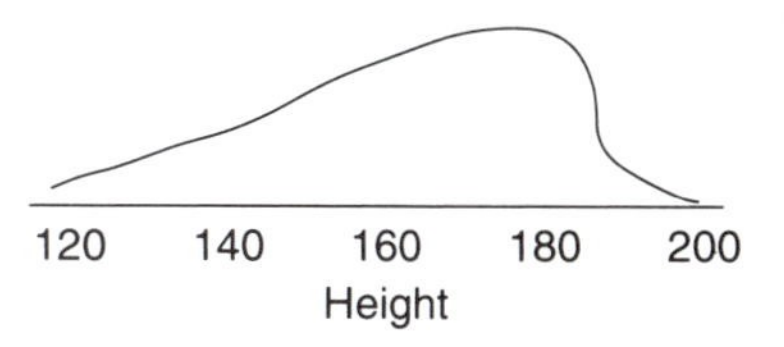

11. The equations for these parameters were given in the Study Hints for this chapter. For data set 1, the mean is 12.2, the variance is 28.844, and the standard deviation is 5.371. For data set 2, the mean is 18.1, the variance is 6.767, and the standard deviation is 2.601. If your calculations are similar to, but not exactly the same as, the ones we have given, it might simply be due to differences in the way in which the calculations are rounded at each step.

12. The null hypothesis for a *t*-test is that the two samples come from the same population. A significantly large difference (i.e., large *t*-value) between the two sets of samples is evidence against the null hypothesis. To evaluate the similarity of this pair of samples, we substitute the following values into the equation for *t* given in the Study Hints:

$\bar{x}_1 = 12.2 \quad S_1^2 = 28.844 \quad n_1 = 10$

$\bar{x}_2 = 18.1 \quad S_2^2 = 6.767 \quad n_2 = 10$

From this we find that

$$t = \frac{|12.2 - 18.1|}{\sqrt{(17.806)(1/10 + 1/10)}} = \frac{5.9}{\sqrt{3/5{,}612}} = \frac{5.9}{1.887} = 3.127$$

The number of degrees of freedom is $n_1 + n_2 - 2 = 18$. Using the reference table of *t*-values (Table R.2), we find that for this magnitude of *t* and 18 degrees of freedom, p is between .01 and .001. This means that if one took two samples from the same population, one would expect to observe deviations this large or larger by chance alone with a frequency of between once in 100 and once in 1,000 times. This is clearly not likely, and one can interpret this as evidence against the hypothesis that the two data sets are from the same population.

13. The purpose of this question is to test your understanding of the relationships among the various measurements of a distribution. The standard error (*s.e.*) is the standard deviation (*s.d.* or *s*) divided by the square root of the sample size (*n*). This can be rearranged as

$$\frac{s.d.}{\sqrt{n}} = s.e.$$

$$s.d. = s.e.\ (\sqrt{n})$$

Substituting from the first set of data,

$$s.d. = 0.9282(5) = 4.641$$

The variance is the square of the standard deviation, so in this case the variance equals 21.539. Your answer might be slightly different if you did not round off in the same way. Variances for the two remaining data sets are 0.061 and 13.587, respectively. Therefore, sample 1 has the largest variance.

14. The mean and variance for sample 1 are 19.35 and 25.012, whereas for sample 2 they are 20.26 and 25.980; $n = 10$ in both samples. Substituting into the formula for *t*,

$$t = \frac{|19.35 - 20.26|}{\sqrt{25.496(1/10 + 1/10)}} \qquad \text{where } s_3^2 = \frac{(9)(25.012) + (9)(25.980)}{18} = 25.496$$

Thus,

$$t = \frac{0.91}{\sqrt{5.09}} = 0.403$$

The number of degrees of freedom is $n_1 + n_2 - 2 = 18$ in this problem. From Table R.2 it can be seen that a *t*-value this large or larger would occur frequently by chance alone. There is therefore no statistically significant difference between these samples.

15. Heritability is the proportion of total phenotypic variance that can be accounted for by genetic factors. Thus genetic influences are greater than environmental ones for traits in which h^2 is greater than 0.5. This is true of milk production in cattle, length of wool in sheep, and amount of spotting in Friesian cattle. Abdominal bristle number in *Drosophila* apparently has equal genetic and environmental influences in this sample of heritabilities.

CROSSWORD PUZZLE

QUANTITATIVE INHERITANCE

ACROSS

1. Inheritance in which several genes influence the phenotype of one quantitative character
6. Proportion of phenotypic variation in a given trait in a specified population that is accounted for by genetic segregation
8. Hypothesis that assumes that two samples are drawn from the same population or that a sample fits a specific model
9. Traits in which the phenotype is measured on a continuous scale
10. When two or more characteristics are dependent on, and vary with, each other
12. Measure of the extent to which values within a distribution depart from the mean
14. Values computed from sample data
15. Gene with a small effect that is involved in determination of a quantitative trait's phenotype
16. Sum of all values, divided by the total number of data points

DOWN

2. Subdivisions made in frequency distributions to facilitate presentation and analysis
3. Group of individuals of the same species, in a defined area
4. Square root of the variance of the mean.
5. Distribution that arranges statistical data by demonstrating the number of occurrences of each value in a data set
7. Square root of the variance
11. Statistical method that will predict one character by using another that is correlated
13. Statistical test used to compare the differences between two distributions

10 OVERVIEW OF GENETIC MAPPING

Genetic mapping is the process of determining where a given gene is located in the genome. There are several different levels, or degrees of resolution, and a wide variety of techniques one can use. For example, one can map a gene to a major cell region, such as nuclear DNA versus mitochondrial or cytoplasmic DNA. Within the nucleus, a gene can be mapped to the X chromosome or to one of the autosomes. Within a chromosome, techniques like the frequency of recombination between pairs of heterozygous loci, the expression of recessive alleles in deletion heterozygotes, and biochemical markers detected in nuclei of somatic cell hybrids can give geneticists evidence about the linkage map order and relative spacing of genes.

A basic question behind genetic mapping is, What is a gene? A gene can be defined in molecular terms as a locus or DNA sequence on a chromosome. It can also be defined in terms of function through the phenotype it produces. Before mapping a gene, one generally has a phenotype to work with, but not the associated DNA sequence. That is what mapping is trying to identify. But there are pitfalls in working simply with phenotypes; one of the major ones is that different genetic loci acting on the same biochemical process can have the same phenotype. The fine structure mapping done by S. Benzer (1955) showed that recombination can occur both within and between genes and led to the concept of *complementation.* To be complementary is to complete or to supply another's lack. What one cannot do, the other can, and vice versa. A *complementation test* is essentially designed to test whether two mutants are allelic or not. As seen in the following diagram, if two mutations are in different genes (that is, they are not allelic), they complement each other since the normal *A* and *B* alleles in the hybrid can provide what was missing in the recessive homozygous parents. If two mutations are allelic, on the other hand, the progeny of a cross between them are still homozygous and the two mutant strains do not complement each other.

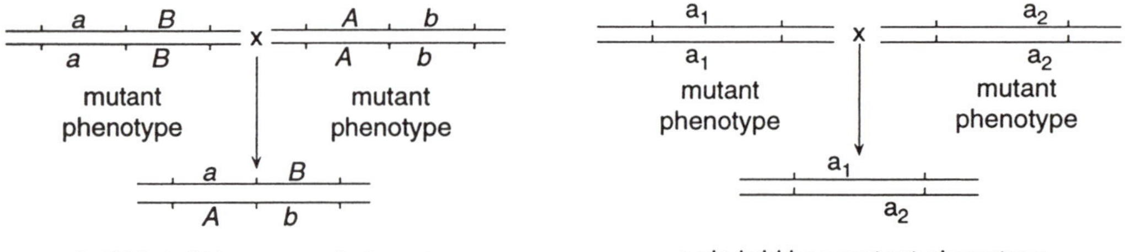

AaBb hybrid has normal phenotype

∴ The two mutants complement each other and are in <u>different</u> complementation groups.

aa hybrid has mutant phenotype

∴ The two mutants do not complement each other and must be in the same complementation group.

Mapping to a chromosome can be done by tracing the inheritance of a new mutation with respect to genes of known location. For example, X-linkage can be distinguished from autosomal linkage by crossing females homozygous for the new mutant to homozygous normal males. If the gene is X-linked, all the F_1 males will have the mutant trait since they inherit their X from their mother. If the males are phenotypically normal, they are heterozygous and the gene is on an autosome. Similarly, a gene can be mapped to an autosomal linkage group by the way it segregates together with, or independently of, genes of known map position.

The first genetic map was produced by A. H. Sturtevant (1913) in *Drosophila* using crossing over between pairs of heterozygous loci to assess gene order and relative distance between gene loci. In 1931, C. Stern and, in an independent study, H. B. Creighton and B. McClintock provided cytological proof that crossing over involves a physical exchange between homologous chromosomes during meiosis. This is a normal phenomenon that can occur at any point along the chromosome. Sturtevant used the frequency of crossover between two points as a measure of relative distance; this topic is discussed in more detail in Chapter 12.

In *Drosophila* and most other organisms, however, the four strands produced from each meiosis are mixed with other meiotic products. In this strand analysis, it is impossible to identify the individuals produced from the gametes of a single meiosis. One of the fascinating points about genetics is that unusual aspects of the development of some organisms can provide unique insight into general genetic principles. This is well illustrated by the Ascomycetes (including the yeasts and *Neurospora*), in which all the spores from a single meiosis are packaged in a sac (or ascus).

In yeasts the four ascospores are found in no particular order, but in *Neurospora* the meiotic products are linearly arranged in an "ordered tetrad," because the developing ascus is narrow and elongated in one plane (see the following figure).

Tetrad analysis, in which we are able to study all four products of meiosis, is a powerful tool in the analysis of crossing over.

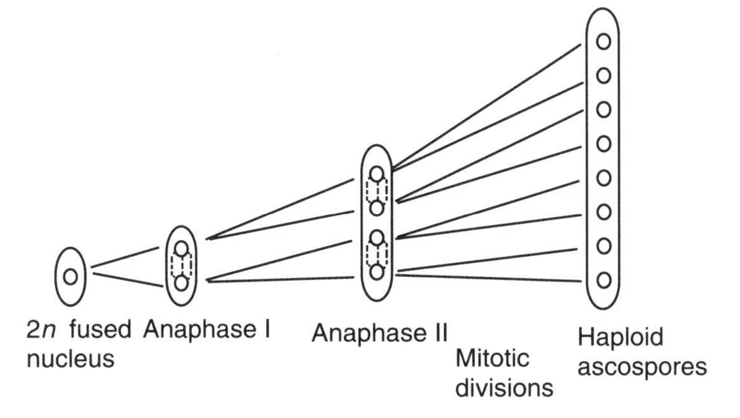

An interesting consequence of ordered tetrads is illustrated in the following figure. On the left is an ascus in which there is no crossover between the *a* locus and the centromere. The alleles segregate at the first meiotic division (division I), as do their centromeres. On the right, an exchange between the *a* locus and the centromere results in the segregation of the *a* from the plus (+) allele at the second meiotic division (division II).

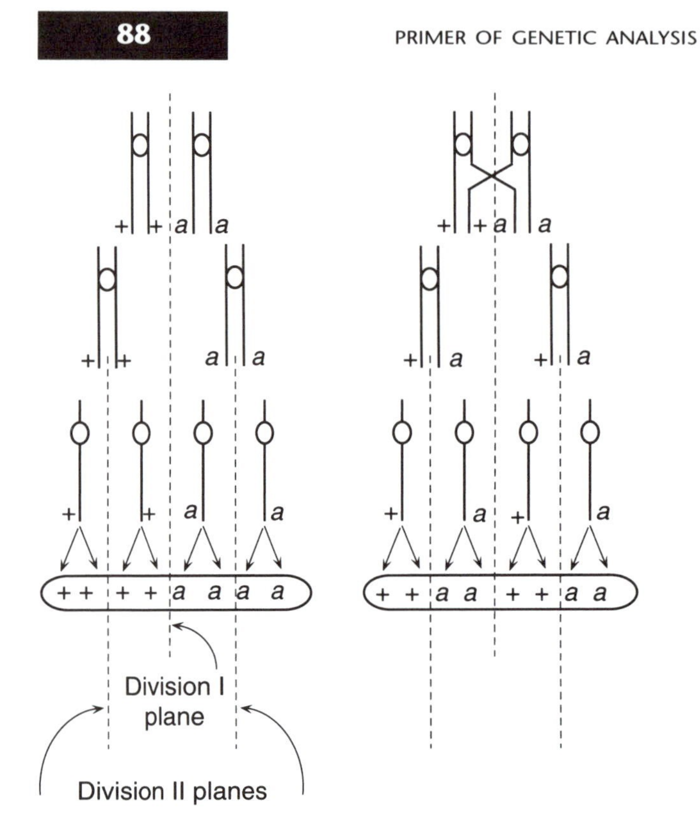

The frequency of second-division segregation for a locus is a measure of the linkage of the locus to its centromere, a measure impossible to obtain in organisms that provide only strand data. A single crossover exchange involves only two of the four strands and results in two crossover and two noncrossover products. A tetrad with a crossover is indicated by an ascus showing second-division segregation, so the number of crossover strands equals half the number of asci showing second-division segregation. For example, a gene that shows 9 percent second-division segregation is $9/2 = 4.5$ map units from its centromere and shows 4.5 percent crossing over.

When two loci are considered, they either assort independently or are linked. Assuming a cross of + + and *ab*, with independent assortment (and omitting, for simplicity, the mitotic duplication of the four meiotic products), we have

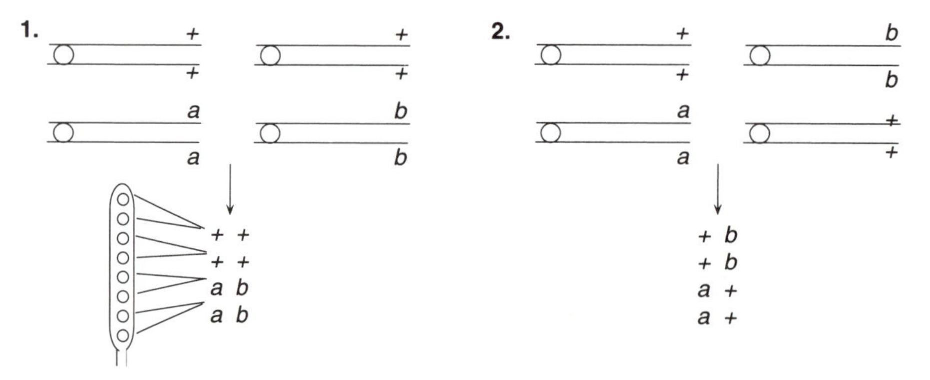

In ascus 1 (left) we have only two types of ascospores. Since these are the same as the

parents, this is termed a *parental ditype* (PD). In ascus 2 (right) we again have only two types of ascospores, but since they are both unlike the original parents this is termed a *nonparental ditype* (NPD).

If the *a* and *b* loci were not linked, we would expect PD and NPD classes with equal frequency, as a result of independent assortment. A complicating factor in *Neurospora* is that a crossover proximal to either locus can generate a third ascus class called the *tetratype* (T) containing all four possible combinations of alleles.

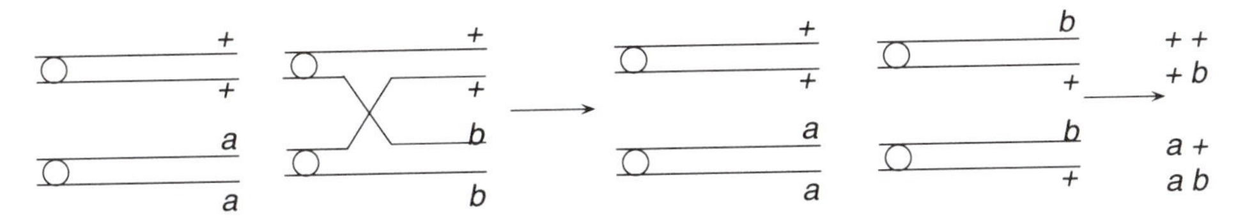

If the numbers of PD and NPD asci are approximately equal, we can assume that the loci are unlinked. This can be tested by chi-square (χ^2). If the two genes are linked, however, they can also produce PD, NPD, and T ascus classes, but the relative proportions of PD and NPD will favor the PD class. (Indeed, a NPD could only arise through two crossovers involving all four strands.) For example, consider two linked genes, *b* and *f*, and the three classes of tetrads that would result from noncrossover and single-crossover events. Note that the sequence of ascospore genotypes would differ, depending upon the orientation (that is, toward top or bottom of the ascus) of the centromeres at anaphase I and at anaphase II.

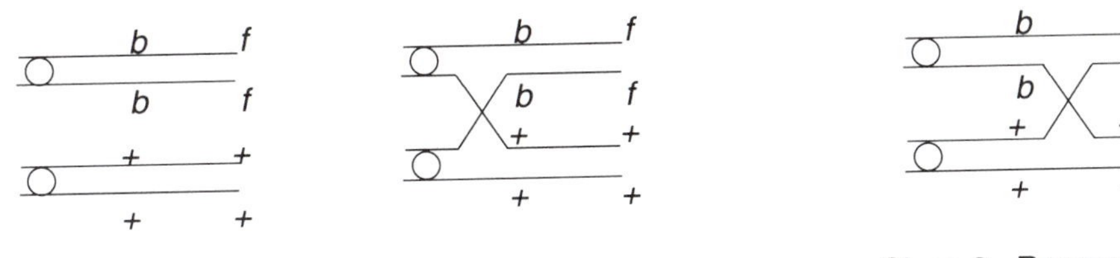

Class 1. Noncrossover

Class 2. Representative single crossover between centromere and both loci

Class 3. Representative single crossover between *b* and *f*

Class 1	Class 2		Class 3			
b f	b f	b f	b f	b f	b +	b +
b f	+ +	+ +	b +	b +	b f	b f
+ +	b f	+ +	+ f	+ +	+ f	+ +
+ +	+ +	b f	+ +	+ f	+ +	+ f

- *Class 1:* All are PD and show first-division segregation for *b* and *f*.
- *Class 2:* All are PD but show second-division segregation for both loci, indicating that the crossover took place between the centromere and both the *b* and *f* loci. Only two of the four possible ascus types are indicated.
- *Class 3:* All are T, indicating exchange between the *b* and *f* loci. Half of the ascospores are parental (*bf* and + +), and half are crossovers (*b* + and + *f*). In addition, they are all first-division segregants for *b* and second-division segregants for *f*. Only four of the eight possible ascus types are indicated. Note, if there is second-division segregation for *both b* and *f*, the two genes must be on different chromosome arms.

The data provided by the class 3 asci, together with the other data, tell us that the locus closest to the centromere is *b* and that *f* is on the same arm but distal to *b*. The map distance between *b* and *f* can be estimated by subtracting the *b*-to-centromere distance from the *f-to-centromere distance, using the formula*

$$\left[1/2 \cdot \frac{T}{(PD + T)}\right] \cdot 100$$

(where T = number of tetratypes, PD = number of parental ditypes), which gives the relative number of crossover strands (1/2 T) over the total number of strands.

The many complexities that one would expect to find in linkage data–such as linked loci on different arms, double crossovers, and more than two loci segregating in the cross–are also observable in *Neurospora.* In addition, crossover studies can be done in *Neurospora* with random spores. When ripe, the asci discharge their ascospores, and these can be collected, their genotypes determined, and the data handled as strand data, as they are.

The analysis of mapping data is an excellent opportunity to practice the interpretive aspect of genetic studies. Furthermore, the special insights offered by the *Neurospora* reproductive system show how certain organisms can provide us with valuable tools for genetic research. Even though the rules of genetics apply more or less equally to all organisms, their operation is much clearer in some than in others. One of the lessons of basic research is that what one learns about even the most obscure part of our world can answer questions of fundamental importance to all. For example, transmission genetic analysis in *Neurospora* has provided insights into homologous recombination mechanisms at the molecular level in all organisms.

Another example of an unusual genetic phenomenon that has been of immeasurable value in mapping is the production of polytene chromosomes in the salivary gland tissue of *Drosophila* larvae. These chromosomes are produced by successive rounds of DNA replication without associated cell division. The result is a nucleus containing giant banded chromosomes made up of about 1,000 copies of each original chromatid strand. The banding provides landmarks that can be used to trace chromosome segments that might be altered during mutation and even trace homologous regions in related species. The use of banded chromosomes for *deletion mapping* is discussed in Chapter 16.

Somatic cell hybrids offer one final example of special techniques that have been developed for mapping using unusual biological systems. The value of somatic cell hybrids for mapping was recognized when it could be demonstrated that human chromosomes were preferentially lost at random from human–mouse hybrid cell lines. This artificial segregation could then be correlated with the retention or loss of specific human characteristics, such as enzyme activity.

Production of hybrid cells: Usually, rodent cell lines derived from mice or hamsters are fused with human fibroblasts or transformed cell lines. Cell fusion can be promoted by using irradiated Sendai virus or chemical agents like polyethylene glycol. The fusion event is relatively rare, and the parental cells must be eliminated from the culture before interspecific hybrids can be isolated.

1. *HAT selection system:* The growth medium contains hypoxanthine, aminopterin, and thymidine (HAT). Aminopterin will block the de novo (i.e., "new") synthesis

of all purines and thymidine nucleotides. Normal cells can survive by *using salvage pathways* if supplemented with hypoxanthine and thymidine and if the enzymes hypoxanthine-guanine phosphoribosyltransferase (HGPRT) and thymidine kinase (TK) are functional. By using parental lines deficient in one or the other of the two enzyme activities, one can select for the rare somatic cell fusions by *complementation.*

2. *Generating mutant parental cell lines for somatic cell hybridization:* How can one obtain parental cell lines that are HGPRT or TK deficient? These lines can be generated by exposing cells to a mutagen and then selecting on a medium containing the purine analogue (and poison) 8-azaguanine (HGPRT-selection) or the thymidine analogue (and poison) 5-bromodeoxyuridine (TK-selection). Cells with a *functional enzyme* will incorporate these analogues into their DNA and be selectively killed. Cell lines that contain the mutant HGPRT or TK genes will survive.
3. *Cloning of hybrid cells:* By growing on HAT medium, the parental cells will die and hybrid cells will survive and divide. Individual colonies of surviving hybrid cells can then be isolated and clonally propagated. Since each individual cell line derives from an independent fusion event, they will probably differ in the number of human chromosomes they retain. Analysis of several clonal populations *biochemically* and *cytogenetically* is thus required to draw correspondences between specific phenotypes (e.g., enzyme activity) and which human chromosomes are present in the cell line.

Chromosome mapping using somatic cell hybrids: The correlation between a human phenotype and a human chromosome are the basis for this analysis. Cytogenetic analysis requires the ability to distinguish between chromosomes, and this is accomplished by analyzing mitotic hybrid cell populations that have been prepared so that different chromosomal regions will stain differentially. All rodent and human chromosomes show individual staining characteristics. With regard to phenotypic differences, the first genes mapped by this procedure were distinguished on the basis of differences in constitutive enzyme activity. Very often, differences between human and rodent

HYPOTHETICAL DATA FROM A SERIES OF RODENT–HUMAN HYBRID CELL LINES

Hybrid line	Human enzyme "X"	Human chromosomes present in hybrid cells 1	2	3	4	5
1	–	+	–	+	+	–
2	–	–	–	+	–	–
3	+	+	+	–	–	–
4	+	+	+	–	+	+

A plus (+) indicates the presence and a minus sign (–) indicates absence of the human enzyme or human chromosome. Note that chromosome 2 is the only one consistently present when human enzyme "X" is expressed and absent when enzyme "X" is not expressed. The gene locus for enzyme "X" can thus be assigned to human chromosome 2.

forms of enzymes can be distinguished by electrophoretic procedures. Along with the chromosome analysis, enzyme analysis is performed on parallel cultures of the hybrid cell lines. Correlations are then drawn to determine the human chromosome consistently segregating with the human enzyme activity. Often analysis of 10–15 independent hybrid clones is necessary to identify the human chromosome correlated with the presence of the enzyme (and hence the gene encoding the enzyme). Both concordance and discordance should be demonstrable between the enzyme activity and the presence or absence of a specific chromosome. (See table on p. 91.)

Regional mapping using somatic cell genetics: The somatic cell genetic technology will identify genes that are *syntenic,* that is, genes that are located on the *same* chromosome. Subchromosomal mapping can be accomplished by using human cells with known chromosomal rearrangements (deletions, translocations, etc.). For example, hybrid cell lines containing a reciprocal translocation chromosome carrying only the long arms of chromosomes 9 and 17 (t9q:17q), in conjunction with data on cell lines in which intact 17 and 9 chromosomes segregate independently, could provide information on whether a gene is located on the long or short arm of these two chromosomes.

ASSESSING CHROMOSOME LINKAGE RELATIONSHIPS

STUDY HINTS

In Chapter 10, we discussed the concept of complementation and the use of complementation tests to assess allelic relationships. Genetic fine structure illustrates how complex a genetic system can be. The existence of pseudoalleles and complex loci shows that one must think in terms of functional interactions among closely linked loci, as well as being concerned with the units of mutation. Allelism, pseudoallelism, and the cis–trans test are among the most difficult concepts in this area for beginning students. Undoubtedly part of the confusion arises from the fact that theoretical advances in the understanding of genetic systems have matured beyond the initial concepts and definitions. The term *allele* refers to the alternate forms of a gene that occupies a particular locus. The term *pseudoallele,* on the other hand, refers to mutations that are allelic in a functional sense, in that they produce a mutant phenotype as trans-heterozygotes:

$$\frac{a_1 \quad +}{+ \quad a_2}$$ Trans-linkage

Yet they can recombine with each other to produce cis-linkage on the chromosome

$$\frac{a_1 \quad a_2}{+ \quad +}$$ Cis-linkage

in which the + + chromosome now codes for a functional product and the cis-heterozygote is phenotypically normal.

Pseudoalleles are the subunits of a complex locus, and they can be explained in several ways. For example, pseudoalleles may be different parts of the same structural gene that functions properly only when it is intact as a single sequence. Thus, a useful degree of precision can be added to the idea of "allele" by distinguishing two levels of allelism: functional alleles and structural alleles. Two mutations are *functional alleles* if they are mutations in the same gene; this is equivalent to the definition of allele we have used in previous chapters. Functional alleles can, however, be of two structurally different kinds. Either the mutations can be in different parts of the gene (that is, they are not structural alleles) or they are in precisely the same nucleotide (they are *structural alleles*).

The fine-structure mapping done by Benzer with bacteriophage T_4 confirms that one can achieve amazing resolution of the genetic structure of the organism if an appropriate experimental system can be devised. The advent of DNA mapping using restriction endonucleases and other techniques of molecular biology is making fine-structure mapping even finer. In this Problem Set, we offer problems about complementation and special mapping techniques and systems. In later chapters we focus on two-point and three-point recombination mapping in diploids (Chapter 12), mapping in bacteria and viruses (Chapter 13), and mapping using structurally altered chromosomes such as deletions (Chapter 16).

IMPORTANT TERMS

Cis-configuration
Complementation test
Coupling linkage
Functional alleles
Linkage group
Pseudoallele
Repulsion linkage
Structural alleles
Trans-configuration

PROBLEM SET 11

1. Two different lozenge (an eye shape mutation) pseudoalleles in *Drosophila melanogaster* are crossed and yield wild-type recombinants for lozenge that are all *x* + for the outside marker genes *x* and *y*. What can you conclude about the sequence of lz^a and lz^b? Is lz^a to the left or to the right of lz^b?

$$\frac{x \qquad\qquad lz^a \qquad\qquad y}{+ \qquad\qquad lz^b \qquad\qquad +}$$

2. In *Drosophila,* analysis of the structure of the Bar locus and of pseudoallelic loci such as the lozenge locus has involved the use of "outside markers," which are closely linked genes on either side of the locus under study. For both types of loci, this test has led to a better understanding of the apparently high mutation rates originally reported for Bar and lozenge. How have the results of such outside marker experiments provided information on the structure and apparent mutation rate of these loci?

3. When two mutants affect the same phenotype and are in the same linkage group, the determination of whether the two mutations are allelic or represent mutations in different loci (but with the same phenotype) can become an important question. One way to test for allelism is what is called the cis–trans (complementation) test. **(a)** Describe the cis–trans test, and **(b)** indicate the evidence that demonstrates the two mutants are alleles.

4. The following are the results of a cis–trans test performed on three isolated mutants: homozygotes for mutant (*x*) have a mutant phenotype; homozygotes for mutant (*y*) also have a mutant phenotype, but somewhat different in its effect. Homozygotes for mutant (*z*) have a mutant phenotype that is different from the other two, but again appears to affect the same phenotypic character. When heterozygotes are produced by crossing homozygous (*z*) to homozygous (*y*), the offspring from this cross are normal in their phenotype (wild-type). The heterozygotes produced by crossing homozygous (*y*) with homozygous (*x*) all show a mutant phenotype (although different from either homozygote). The other alternative (*x* crossed to *z*) was performed and the offspring are normal (wild-type) in their phenotype. From these data, you can say **(a)** all are alleles of the same gene; **(b)** all are different genes; **(c)** mutants *x* and *y* are alleles, but *z* is a different gene; **(d)** mutants *x* and *z* are alleles, but *y* is a different gene; **(e)** mutants *z* and *y* are alleles, but *x* is a different gene.

5. The following chart represents the results of a mapping experiment using human–

mouse somatic cell hybridization. The human cell has active enzyme R, but the mouse does not. Four cell lines have been produced, each having a full set of mouse chromosomes but differing in which human chromosomes are still present. Each cell line is tested for enzyme R activity (minus shows it absent, plus shows enzyme R activity can be detected). From these data, what can we conclude about the chromosomal location of the enzyme R gene?

Cell line	**Enzyme R activity**	**Human chromosomes present in cell lines**
A	−	2, 4, 8, 11
B	+	3, 4, 8, 18
C	+	3, 5, 18, 20
D	−	2, 18, 12, 16

6. In the following panel of results, the presence or absence of enzyme H activity is correlated with the presence or absence of several human chromosomes in human–mouse cell hybrids. From these data, what can you conclude about the chromosomal linkage group for the enzyme H gene?

Cell hybrid	**Enzyme H**	**Human chromosomes**							
		2	**3**	**4**	**5**	**7**	**9**	**14**	**15**
A	+	+	−	+	+	+	−	+	+
B	−	−	−	−	−	−	+	−	−
C	+	−	−	+	+	+	+	−	+
D	+	+	+	−	−	+	+	+	+
E	−	−	+	−	−	−	+	−	+
F	+	−	+	−	+	+	+	+	+

7. Assume that the n locus in *Neurospora* shows 16 percent second-division segregation. What is the map distance between n and its centromere?

8. In *Neurospora,* the p locus is linked to the n locus in problem 7, and p shows 10 percent second-division segregation. How many map units separate the p and n loci?

9. In *Neurospora,* a cross of b to plus (+) strains results in the following asci. What can you conclude about map distances in this cross?

①	②	③	④	⑤	⑥
b	b	+	b	+	+
b	+	b	+	+	b
+	b	b	+	b	+
+	+	+	b	b	b
54	2	4	1	49	3

10. In *Neurospora,* a cross of *bc* and + + strains results in the following asci. Determine map distances, and discuss the results of this cross.

①	②	③	④	⑤	⑥	⑦
b+	*b*+	++	+*c*	*bc*	+*c*	*bc*
b+	*bc*	++	++	*bc*	+*c*	*b*+
+*c*	+*c*	*bc*	*bc*	++	*b*+	++
+*c*	++	*bc*	*b*+	++	*b*+	+*c*
21	3	20	5	25	22	4

ANSWERS TO PROBLEM SET 11

1. In order to obtain a nonlozenge recombinant in the lozenge region in which the flanking markers are *x* and +, lz^a must be to the right of lz^b. To visualize this, we have redrawn the chromosome and greatly exaggerated the space between the lozenge pseudoalleles.

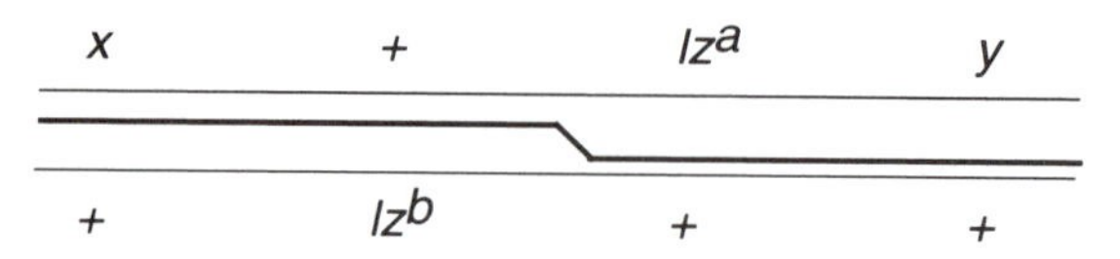

2. Outside markers are unrelated genes that are close to, and on opposite sides of, the region under test. These two loci are heterozygous, so crossovers between them are detectable. Every mutational event in the Bar and lozenge loci was accompanied by a crossover between the two outside markers, which indicates that the unusually high rates of "mutation" are not gene mutants per se but result from crossover events within the Bar and lozenge loci. The two loci differed in that the Bar reversion to wild-type (B^+) had an equal probability of being + *h* or *q* +, where *q* and *h* are flanking markers. That is, Bar was nonpolarized. The lz^+ reversions, on the other hand, were always either + *h* or *q* +, depending upon which two lozenge alleles were in the trans-position (polarized; see problem 1). The nonpolarized Bar locus is a duplication, whereas the polarized lozenge alleles represent mutations in different subloci occupying specific sites within the locus.

3. Please refer to the Overview of Genetic Mapping (Chapter 10) for a discussion of complementation.

4. **(c)** Mutants *x* and *y* are alleles, but *z* is a different gene. The heterozygotes of *x* and *y* in trans-linkage do not complement.

5. Enzyme R activity is found when chromosome 3 is present. Although both cell lines B and C also share chromosome 18, that chromosome is also found in line D, which does not have enzyme R activity. The gene must, therefore, be on chromosome 3.

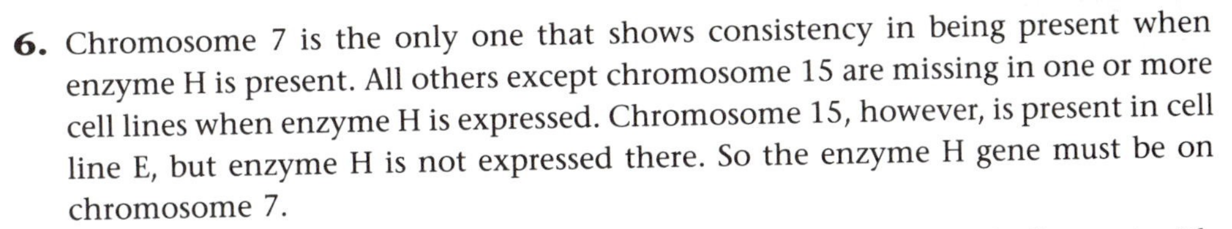

6. Chromosome 7 is the only one that shows consistency in being present when enzyme H is present. All others except chromosome 15 are missing in one or more cell lines when enzyme H is expressed. Chromosome 15, however, is present in cell line E, but enzyme H is not expressed there. So the enzyme H gene must be on chromosome 7.

7. Perhaps your immediate response is "16 map units," but we are dealing not with single strands but with an entire meiosis. Since crossing over occurs at the four-strand stage, between two of the four strands, only half of 16 (or 8 percent) of the *strands* have undergone a crossover in the centromere–to–n–locus region. In order to keep map units in *Neurospora* tetrad analysis comparable to the units in most organisms, where only strand analysis is possible, this halving of the division 2 segregation frequency is necessary. The answer is 8 map units.

8. The p locus is $1/2 \cdot 10$, or 5 map units from its centromere. From problem 7 we know that n is 8 map units from the centromere. If both loci are on the same arm, then the distance between the two loci is $8 - 5 = 3$ map units. If the two loci are on different arms, then the relationship is

n 8 • 5 p
|←—13—→|

and the p–to–n distance is $8 + 5 = 13$ map units.

9. The two (equivalent) parental ascus types showing first-division segregation are 1 and 5, with a total of 103 asci. The other types are all variants of second-division segregation, reflecting a crossover between the b locus and its centromere and different orientations of the centromeres at first and second anaphase. There are 10 of these. The number of map units separating the b locus from its centromere is $1/2 \cdot (10/113) \cdot 100 = 4.4$ map units.

10. Ascus types 3 and 5 are parental ditypes, so $20 + 25 = 45$. Types 1 and 6 are NPD, and $21 + 22 = 43$. This apparent equality indicates that the two loci are assorting independently and thus are not linked. The tetratypes seen in ascus types 2, 4, and 7 (a total of 12 asci) are all generated by second-division segregation for the c locus, so the c–to–centromere region is $1/2 \cdot (12/100) \cdot 100 = 6$ map units. Since there is no second-division segregation for the b locus in this sample of asci, you can conclude that b is very closely linked to its centromere.

12 LINKAGE AND MAPPING IN DIPLOIDS

STUDY HINTS

The relative distances between pairs of loci can be estimated from a testcross. The map distance is defined as being equal to the percentage of recombination. The following is an example of the analysis of map distances in a three-point testcross.

In *Drosophila melanogaster,* three of the loci affecting the adult cuticle are

- *h* (hairy; extra hairs on the body)
- *jv* (javelin; bristles are cylindrical, rather than being tapered to a point)
- *ve* (veinlet; wing veins shortened)

The cross between the heterozygous (*h*+ *jv*+ *ve*+) female parent and the homozygous recessive (*hh jvjv veve*) male is a testcross. The phenotypes of the progeny represent the alleles inherited from the heterozygous parent, for any recessive allele from the heterozygous parent is automatically homozygous in the offspring. The phenotypes of the progeny are as follows:

+	+	+	362
jv	+	+	4
+	+	*ve*	90
jv	+	*ve*	39
jv	*h*	*ve*	396
+	*h*	*ve*	2
+	*h*	+	34
jv	*h*	+	100
			1,027

Here are the steps you would follow to determine the map distances.

1. Determine whether the genes are linked or unlinked. They are linked in this example. If not, one would expect a 1:1:1:1:1:1:1:1 in such a testcross.

2. Determine the parental, or nonrecombinant, classes. Look for the most frequent classes: + + + and *jv h ve*. Therefore, the genotype of the heterozygous parent (without implying that this is the order of the alleles) is

$$\frac{+ \quad + \quad +}{\text{jv} \quad \text{h} \quad \text{ve}}$$

Thus the alleles are in *cis-linkages*. If the parental classes had been *jv h* + and + + *ve,* then *jv* and *h* would be in cis-linkage, but *jv* and *h* would be in *trans-linkage* with *ve*.

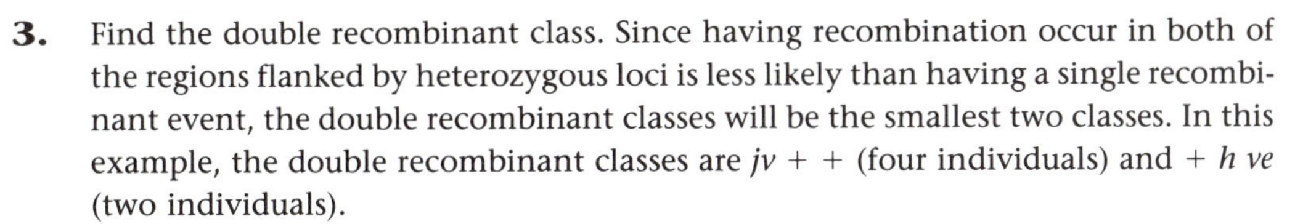

3. Find the double recombinant class. Since having recombination occur in both of the regions flanked by heterozygous loci is less likely than having a single recombinant event, the double recombinant classes will be the smallest two classes. In this example, the double recombinant classes are *jv* + + (four individuals) and + *h ve* (two individuals).

4. Determine the gene order. Gene order can be found by comparing the parental classes and the double crossover classes. The linkages in the parent and the double crossover classes will be the same, except for one locus. *The one that changes must be the middle gene in the sequence.* For example, if the chromosomes are as shown in the figure,

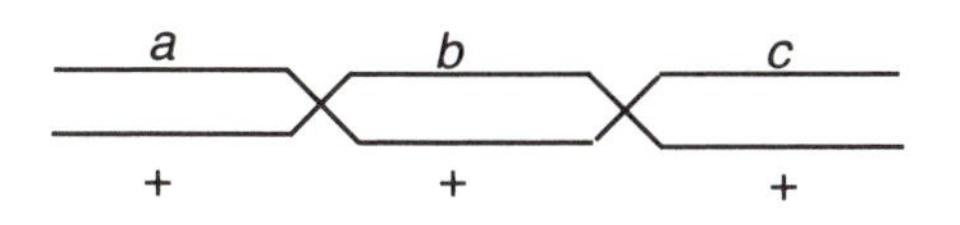

the double crossover chromosomes will be *a* + *c* and + *b* +. The middle gene (*b*) has changed linkage relationships with the other loci. In the present example the classes are

Parentals			Double crossovers		
+	+	+	*jv*	+	+
jv	*h*	*ve*	+	*h*	*ve*

The linkage of *jv* has changed with respect to the two other loci. The correct gene order is therefore *ve jv h* or *h jv ve*. These two alternatives are, in fact, equivalent, since the "left" and "right" ends of a chromosome are defined arbitrarily. The orientation is set when researchers make the first map of a species.

5. Determine the single crossover types:

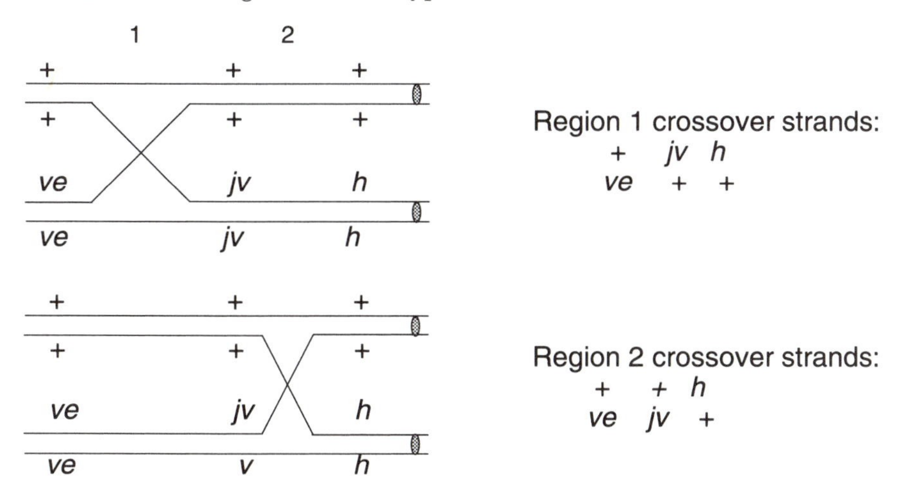

6. Determine the frequencies of crossovers in each region.

Region 1				Region 2			
ve	+	+	90	*ve*	*jv*	+	39
+	*jv*	*h*	100	+	+	*h*	34
+	*jv*	+	4	+	*jv*	+	4
ve	+	*h*	2	*ve*	+	*h*	2
			196				79

$$\frac{197}{1{,}027} \times 100 = 19.1\% \qquad \frac{79}{1{,}027} \times 100 = 7.7\%$$

The genetic map is therefore

ve ←— 19.1 —→ *jv* ←— 7.7 —→ *h*

where 19.1 and 7.7 are map units in centimorgans.

7. Determine the coefficient of coincidence. The *coefficient of coincidence* measures the difference between the observed and expected numbers of double crossovers. This reflects the degree to which one crossover event may enhance or interfere with another in an adjacent region.

$$\text{Coefficient of coincidence} = \frac{\text{observed number of double crossovers}}{\text{expected number of double crossovers}}$$

The expected double crossovers will be the product of the likelihood of recombination in each of the independent regions, that is, the probability of two independent events occurring together is the product of their individual probabilities. Since the likelihood of recombination is the same as the map distance, for this problem:

The *expected* proportion of double crossovers is .191 × .077 = .014707, which we round to .015

The *observed* proportion of double crossovers is 6 out of 1,027 = .005842, which we round to .006.

$$\text{Coefficient of coincidence} = \frac{0.006}{0.015} = 0.400$$

The coefficient of coincidence (CC) is therefore the degree to which the observed data on double crossovers fit the theoretical expectation. The degree to which crossing over has been "interfered with" by a closely adjacent crossover event is

Interference = 1 − coefficient of coincidence.

If the CC is 1, then there is no interference. If the CC is less than 1, interference is positive and fewer double crossovers have occurred than expected. If the CC is greater than 1, interference is negative and more double crossovers have occurred than expected.

In this example, the interference is 1 − .4 = .6. Therefore, 60 percent of the expected double crossovers did not occur.

IMPORTANT TERMS

Centimorgan (cM)
Chiasma (Chiasmata)
Coefficient of coincidence
Gene conversion
Interference
Linkage
Locus (Loci)
Recombination
Tetrad analysis
Three-point testcross
Two-point testcross

PROBLEM SET 12

1. Crossover map distances determined by two-point crosses are *P—C* = 7, S—M = 10, *C—M* = 8, *S—C* = 2, and *P—S* = 5. The relative positions of these four linked loci are **(a)** *P S C M*, **(b)** *S C P M*, **(c)** *S C M P*, **(d)** *P C S M*, **(e)** *C S P M*, **(f)** none of the above.

2. Assume that 6 percent of all meioses in the sugar beet have an exchange between the *A* and *B* loci. The crossover map distance separating these loci is **(a)** 3 map units, **(b)** 6 map units, **(c)** 12 map units.

3. An experimental setup that would permit you to correlate a genetic crossover with a physical exchange between the two chromosomes is indicated in which parts of the following figure? **(a)** A, C, and E, **(b)** B, C, and E, **(c)** C, D, and E, **(d)** C and E, **(e)** D and E, **(f)** E only.

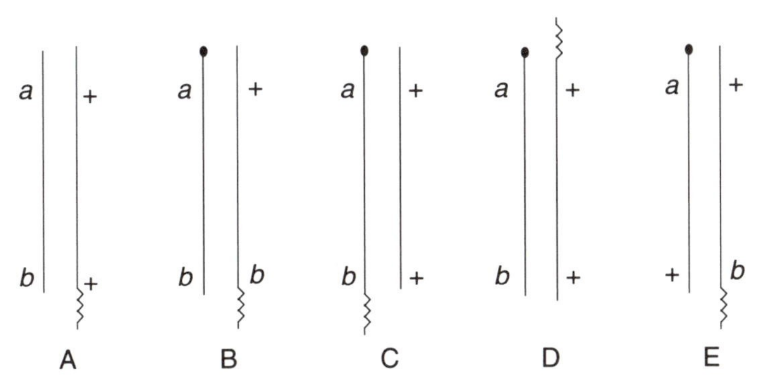

4. Diagram a three-strand double exchange in an *ABC* / *abc* trihybrid, with one exchange in the *A—B* region and the other in the *B—C* region. Give the genotypes of the four resulting chromosomes, and indicate whether the chromosome is a parental class (noncrossover), single crossover, or double-crossover strand. Is it possible to diagram another three-strand double crossover that will give products that are genetically different from those that you have listed? If your answer is yes, give the genotypes and indicate whether they are noncrossover, single-crossover, or double-crossover strands.

5. Diagram a four-strand double exchange in an *ABC* / *abc* trihybrid, with one exchange in the *A—B* region and the other in the *B—C* region. Give the genotypes of the four resulting chromosomes, and indicate whether the chromosome is a noncrossover, single-crossover, or double crossover strand. Is it possible to diagram another four-strand double crossover that will give products that are genetically different from those you have listed? If your answer is yes, give the genotypes, and indicate whether they are noncrossover, single-crossover, or double-crossover strands.

6. Trihybrid geranium plants with hairy leaves, blue flowers, and tall-growth habit are testcrossed with plants having smooth leaves, red flowers, and short-growth habit. The progeny are listed in the following table. What are the linkage relationships of these three loci?

Leaf	Flower	Height	Number
Hairy	Blue	Tall	368
Hairy	Blue	Short	20
Hairy	Red	Tall	1,797
Hairy	Red	Short	179
Smooth	Blue	Tall	191
Smooth	Blue	Short	1,752
Smooth	Red	Tall	26
Smooth	Red	Short	340
			4,673

7. Assume that in a self-fertilizing plant species an individual is a dihybrid, *TY* / *ty*. Plants that are *T* – are tall, whereas the *tt* plants are short. Plants that are *Y* – have yellow seeds, whereas *yy* plants have white seeds. We know, from testcrosses, that exchange occurs in the *T—Y* region with a frequency of 12 percent during egg formation and with a frequency of 8 percent during pollen formation. List the expected genotypes and phenotypes, together with their frequencies, for the progeny of the dihybrid.

8. A trihybrid produces the following gametes:

A T E	46	*A t e*	63	*A T e*	4	*a T e*	381
a t e	38	*a T E*	71	*a t E*	2	*A t E*	395

(a) Which are the parental (noncrossover) types?
(b) Which are the double-crossover types?
(c) Which locus is in the middle?
(d) Draw the two homologues as they would appear at synapsis.
(e) Calculate the map distances.

9. Examine the pair of homologous chromosomes in the accompanying diagram. An exchange producing chromosomes, which can be *shown* to be crossovers, will occur **(a)** in region 1 only, **(b)** in region 2 only, **(c)** in neither region 1 nor region 2, **(d)** in either region 1 or region 2, **(e)** in both regions.

A *b* *d*
a *b* *D*
1 2

10. The accompanying figure shows the genotype of a trihybrid, together with the known map distances for the two regions. Assume no double crossovers, and determine the types and frequencies of gametes produced by this trihybrid. Then determine the expected genotypes in a sample of 1,000 progeny.

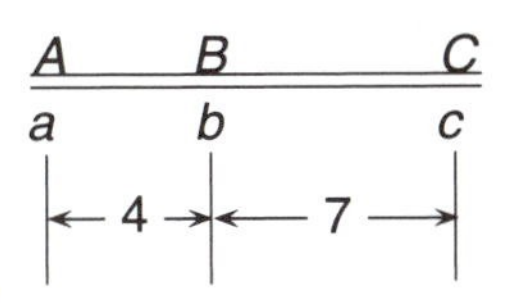

11. The figure shows the genotype of a trihybrid, together with the known map distances for the two regions.

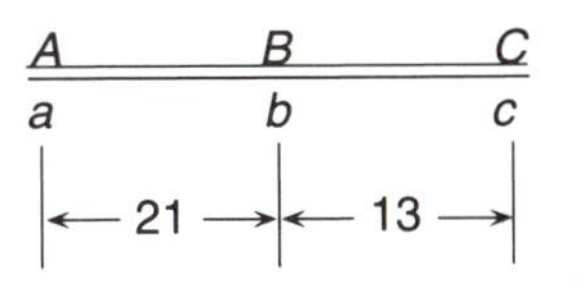

Assume that the coefficient of coincidence for the double crossover is 0.6. Determine the types and frequencies of the gametes produced by this trihybrid, and the genotypes expected in a sample of 1,000 progeny.

12. A trihybrid produces the following gametes (total = 1,000).

IBd	219	*Ibd*	212	*IBD*	32	*IbD*	31
ibD	215	*iBD*	224	*ibd*	29	*iBd*	38

Determine and thoroughly discuss the linkage relationships of these three loci.

13. The following data come from a cross made to localize a recessive lethal mutation on the X chromosome of *Drosophila*. The lethal mutation is on a wild-type chromosome, and the cross (see the accompanying diagram) is designed to localize the lethal mutation relative to the genetic markers brought in by the *y m car* chromosome (which also carries the plus (+) allele for the lethal mutation). The Y chromosome is smaller and is cytologically distinguishable from the X chromosome

y *m* *car* ♀ X *y* *m* *car* ♂
\+ + +

by its centromere's position. The lethal mutation is somewhere on the + + + chromosome.

Class			Females	Males
y	*m*	*car*	427	459
+	+	+	486	0
y	+	+	299	0
+	*m*	*car*	286	283
y	*m*	+	168	25
+	+	*car*	181	158
y	+	*car*	69	74
+	*m*	+	84	1
			2,000	1,000

(a) Determine the locations of *m* and *car* on the crossover map separately from the male and female data (assume that *y* is at 0).
(b) Explain the source of the difference in map distances.
(c) Localize the lethal locus.

14. The two chromatids attached to a single centromere in early prophase of meiosis are derived by replication of a single original chromosome and are genetically identical. When, and under what circumstances, are the two chromatids that are attached to the same centromere not genetically identical?

ANSWERS TO PROBLEM SET 12

1. Map orientation problems are simply puzzles that require that you put the distances together in a consistent way. To answer question 1, you could begin by drawing the *S*—*M* segment, which is 10 units long, as in the following figure. *C* could be either 8 units to the left of *M* or 8 units to the right, but of these two alternatives, only the order *S C M* predicts that *S* and *C* are 2 units apart. It remains only to place *P* in position.

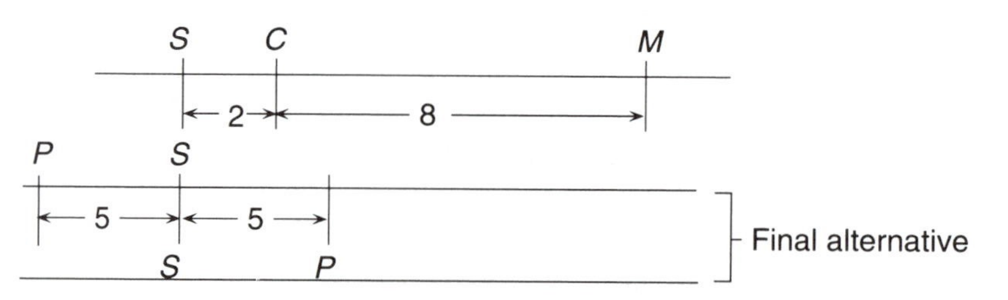

For the first alternative (*P* to the left of *S*, a total of 5 units), the distance from *P* to *C* would be 7 units. This is the observed distance. With *P* to the right of *S*, as in the second alternative, the distance between *P* and *C* would be short ($5 - 2 = 3$ units). The correct answer is, therefore **(a),** *P S C M*.

2. In each meiosis, 2 recombinant chromosomes are produced each time an exchange occurs. Two nonrecombinant chromosomes are also produced. Thus, since map distance is calculated as number of recombinants divided by total number of tested chromosomes, the map distance will be *half* as large as the percentage of meioses with an exchange (and vice versa). You can demonstrate this for yourself by drawing the meiotic products of 5 cells in which a single recombinant chromosome has occurred in only 1 of the cells. You will find that there will be 2 recombinant chromosomes and 18 nonrecombinant chromosomes. Although an exchange occurred in 20 percent of all meioses (1 in 5), the calculated map distance would be 10 map units (2 recombinant chromosomes divided by the total of 20 chromosomes, 4 chromosomes from each of the 5 cells). The answer to this question is therefore **(a),** 3 map units.

Relationships such as those described here between recombination rates and map distances may initially seem not only complex but trivial. They will seem less trivial if you realize that the relationship is important to an understanding of the process of recombination, and less complex if you learn to make a quick diagram or sketch of the products of such a problem so that you do not make a careless error by answering simply from memory.

3. **(d)** C and E. What is needed here is the ability to detect both a genetic and a cytological (physical) exchange. The genetic exchange is easy. All it requires is heterozygosity at two loci, so that parental combinations can be distinguished from the recombinants. All setups except B satisfy this requirement. The cytological exchange is more difficult to detect, since homologous chromosomes, regardless of their allelic constitutions, are normally identical under the microscope. What is needed here is heterozygosity, to provide cytological markers at each end of the genetic region under test, so that parental chromosomes can be distinguished from the crossovers by the different morphological characteristics they exhibit when examined microscopically. If these conditions are met, crossovers detected genetically should have chromosome structures different from that of the parentals. It does not matter whether the parentals are *ab* and + +, or *a* + and + *b*, or whether one parent has both the knob and the additional length. Of the five setups, A, C, D, and E satisfy the genetic requirement. The requisite cytological heterozygosity is satisfied by setups B, C, and E. Both requirements are satisfied by setups C and E.

4. There are two ways of drawing such a situation at synapsis of the first meiotic division.

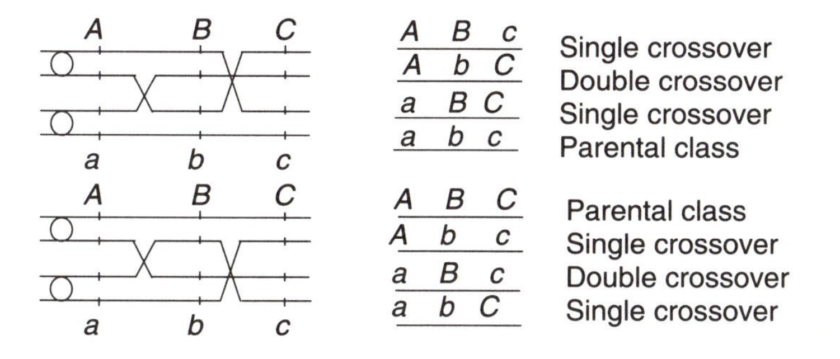

5. Only one four-strand double exchange can be drawn, given the placement of the exchanges outlined in the problem. All are single crossovers.

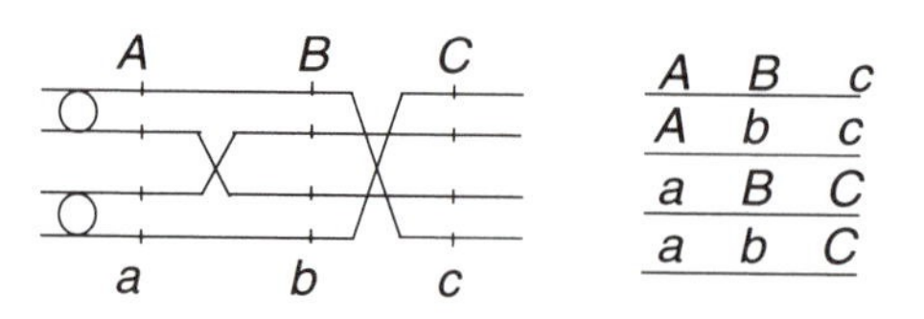

6. This problem is essentially identical to that worked out in the Study Hints section of this chapter. Following that guide, determine whether the loci are linked or unlinked. There are clear pairs of phenotypic classes, which do not fit a 1:1:1:1:1:1:1:1 ratio. The parental (nonrecombinant) classes are the largest:

hairy, red, tall and smooth, blue, short

If the testcross was made to smooth, red, short plants, these three must be the recessive markers. Let us therefore denote them as

H, hairy	*h,* smooth
R, blue	*r,* red
S, tall	*s,* short

Thus the genotype of the heterozygous parent (without implying that this is the order of the alleles) is

$$\frac{H \quad r \quad S}{h \quad R \quad s}$$

The double recombinant classes are the smallest: *HRs* and *hrS*. The middle gene is the one that differs from the parental linkage. Since *R* and *s* are linked in both the parental and double-crossover class (as is also true of *r* and *S*), the *H* locus must be the middle gene. The data can now be rewritten in the proper order, with the parental classes and single recombinants followed by the double recombinants.

1 2

$$\frac{r \quad H \quad S}{R \quad h \quad s}$$

Parental	*r*	*H*	*S*	1,797
	R	*h*	*s*	1,752
Crossover in region 1	*r*	*h*	*s*	340
	R	*H*	*S*	368
Crossover in region 2	*r*	*H*	*s*	179
	R	*h*	*S*	191
Double crossover	*r*	*h*	*S*	26
	R	*H*	*s*	20
				4,673

$$\text{Crossover in region 1} = (368 + 340 + 20 + 26)/4{,}673 = \frac{754}{4{,}673} \cdot 100 = 16.14 \text{ map units}$$

$$\text{Crossover in region 2} = (179 + 191 + 20 + 26)/4{,}673$$
$$= \frac{416}{4{,}673} \cdot 100 = 8.9 \text{ map units}$$

7. The parental chromosomes are as follows:

$$\frac{T \quad\quad Y}{t \quad\quad y}$$

The gametes produced by the males and females must be considered separately. If exchange occurs in the T—Y region at a frequency of 12 percent in eggs, then 6 percent of the eggs are *Ty* recombinants, and 6 percent are *tY* recombinants. (Note the difference between the frequencies described here and those discussed in problem 2.) In pollen, 4 percent are *Ty,* and 4 percent are *tY.*

To calculate the frequencies in the dihybrid progeny, the most direct approach is to calculate the frequencies of each combination in a Punnett square.

Gametes of female	**Frequency**	**Gametes of male**	**Frequency**
TY	.44	*TY*	.46
ty	.44	*ty*	.46
Ty	.06	*Ty*	.04
tY	.06	*tY*	.04
	1.00		1.00

♀ \ ♂	***TY* (.46)**	***ty* (.46)**	***Ty* (.04)**	***tY* (.04)**
(.44) *TY*	*TTYY* (.202)	*TtYy* (.202)	*TTYy* (.018)	*TtYY* (.018)
(.44) *ty*	*TtYy* (.202)	*ttyy* (.202)	*Ttyy* (.018)	*ttYy* (.018)
(.06) *Ty*	*TTYy* (.028)	*Ttyy* (.028)	*TTyy* (.002)	*TtYy* (.002)
(.06) *tY*	*TtYY* (.028)	*ttYy* (.028)	*TtYy* (.002)	*ttYY* (.002)

The genotypes can be tallied from this Punnett square, and the phenotypes can be calculated by adding appropriate genotype frequencies, as follows:

Genotype	**Frequency**	**Phenotype**		**Frequency**
TTYY	.202	Tall, yellow	(*T- Y-*)	.702
TTYy	.046	Tall, white	(*T- yy*)	.048
TTyy	.002	Short, yellow	(*tt Y-*)	.048
TtYY	.046	Short, white	(*tt yy*)	.202
TtYy	.408			1.000
Tt yy	.046			
tt YY	.002			
tt Yy	.046			
tt yy	.202			
	1.000			

8. This problem is set up and solved like problem 6. The answers are

(a) *a T e* and *A t E* are the parental types.
(b) *A T e* and *a t E* are the double-crossovers types.
(c) *A* is in the middle. The order in the parents is therefore *T a e* and *t A E.*
(d)

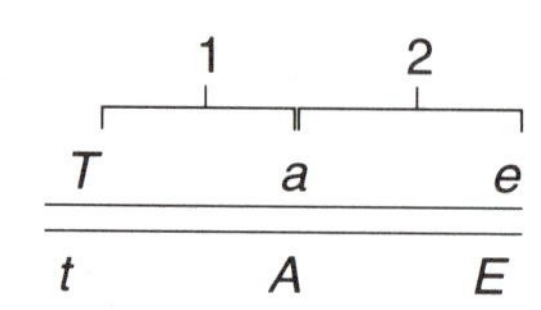

(e) Crossovers in region 1 = [(46 + 38 + 4 + 2/1000] • 100 = 9 cM
Crossovers in region 2 = [(63 + 71 + 4 + 2)/1000] • 100 = 14 cM

9. **(d)** Crossovers in region 1 and in region 2 cannot be distinguished since the *b* locus is homozygous and therefore does not change in a crossover. A crossover in region 1 produces *AbD* and *abd* strands, and a crossover in region 2 produces the same two strands. In no instance can a homozygous locus be of use in detecting linkage and mapping. Furthermore, an exchange in *both* regions (i.e., a double crossover) would produce recombinants that are indistinguishable from the parental linkages.

10. In this problem you are asked to work *backward* from the map distances to the data that generated them. If there are no double crossovers to contend with, the four map units for the *A—B* region must come directly from 4 recombinant chromosomes (or progeny) in each 100 in the sample. The frequencies of the complementary chromosome types will be approximately equal (e.g., *Ab* and *aB*). A total of 3.5 in each 100 will be *Bc,* and 3.5 will be *bC,* totaling 7 percent for the *B—C* region. In the sample in this problem, there are 1,000 total progeny. Thus we multiply the expected numbers in each 100 progeny by 10. The parental genotypes will simply be whatever is left over after the single crossovers have been calculated.

		Test cross	
Gametes	**Frequencies**	**Genotype**	**Number**
ABC	44.5 percent	*AaBbCc*	445
abc	44.5 percent	*aabbcc*	445
Abc	20.0 percent	*Aabbcc*	20
aBC	20.0 percent	*aaBbCc*	20
ABc	3.5 percent	*AaBbcc*	35
abC	3.5 percent	*aabbCc*	35
	100 percent		1,000

11. The coefficient of coincidence compares the frequency of observed (O) and expected (E) double crossovers. In this problem, the expected frequency of double crossovers is .21 • .13 = .027. The coefficient of coincidence (CC) is the frequency of observed crossovers divided by the frequency of expected crossovers: CC = O/E. The frequency of the observed double crossovers, therefore, is equal to the coefficient of coincidence multiplied by the frequency of the expected crossovers: CC • E

= O. The frequency of the double crossovers is thus .6 • .027 = .016 (.008 of each type). This yields an expectation of 16 in the sample of 1,000 progeny. The frequency of single crossovers in region 1 = (21/100) − .016 = .194, or 194 per 1,000. The frequency of single crossovers in region 2 = (13/100) − .016 = .114, or 114 per 1,000.

Note: The double crossovers need to be subtracted from the calculation of singles, because this number will be added in when the map distances are tallied from the raw data. As before, the parental class simply includes the remaining individuals.

To simplify the following, the capital letter means they have the dominant allele from the testcross; the lowercase letter means they are homozygous recessive.

Class	Genotype	Number
Parental	*ABC*	338
	abc	338
Region 1	*Abc*	97
	aBC	97
Region 2	*ABc*	57
	abC	57
Doubles	*AbC*	8
	aBc	8
		1,000

12. In these data there appears to be a total of four parental classes. The four recombinant classes are approximately equal. This indicates that one of the three loci is segregating independently of the other two (that is, two are linked, but the third is not linked). By considering the loci in pairs, it can be seen that the *B* locus is segregating at a 1:1 ratio with each of the others. Only *I* and *D* are linked.

Class	Genotype	Counts	Total
Parental	*Id*	219 + 212 =	431
	iD	215 + 224 =	439
Crossover	*ID*	32 + 31 =	63
	id	29 + 38 =	67
			1,000

Map distance = [(63 + 67)/1000] • 100 = 13 map units

13. This is a complex problem, which should definitely test how well you understand the concepts behind linkage analysis. The key is to recognize that a recessive lethal gene (*l*) will kill a hemizygous male. When such a lethal gene is located between two markers, only a proportion of the recombinants in that region will carry it, and the class will be reduced (but not eliminated) in males.

The data are in the proper order and are arranged for immediate analysis.

(a) *Female data:*

Crossover in region 1 = [(299 + 286 + 69 + 84)/2,000] • 100 = 36.9
Crossover in region 2 = [(168 + 181 + 69 + 84)/2,000] • 100 = 25.1

With *y* at 0.0, *m* is at 36.9, and *car* is at 62.0 (36.9 + 25.1) on the female crossover map.

Male data:

Crossover in region 1 = [(283 + 74 + 1)/1,000] • 100 = 35.8
Crossover in region 2 = [(25 + 158 + 74 + 1)/1,000] • 100 = 25.8

Thus *m* is at 35.8 and *car* is at 61.6 on the male crossover map.

(b) There is no significant difference between the groups.

(c) To localize the lethal mutation, one should look at the male data, since it is only there that the recessive lethal will not be masked.

In most of the classes, one or another of the complementary chromosomes is absent or reduced in frequency. They must have carried the lethal mutation. The most consistent are the parental and region 1 classes. When region 2 from the + + + chromosome is present, the fly dies. The lethal mutation must therefore be between *m* and *car*. For convenience, let us call the region to the left of the lethal 2a, and the region to the right 2b.

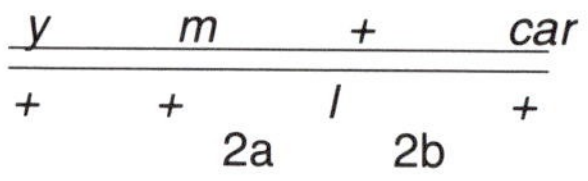

y *m* + *car*
\+ + *l* +
2a 2b

The recombinants between *m* and *car* will be of two types: those with the lethal mutation (which die) and those without the lethal mutation:

Region 2a: *y m l* + which die, and + + + *car* which survive
Region 2b: *y m*+ + which survive, and + + *l car* which die

The relative numbers of these can be determined by identifying those that have a lethal mutation (and are thus reduced) and those that do not have the lethal mutation (and are thus about the same magnitude as in the female). The recombinants that survive (and the region they represent) are

ym +	(region 2b)	25
++*car*	(region 2a)	158
y + *car*	(region 2a)	74
+*m* +	(region 2b)	1

Crossovers in region 2a = [(158 + 74)/1,000] • 100 = 23.2 map units from *m*
Crossovers in region 2b = [(25 + 1)/1,000] • 100 = 2.6 map units from *car*

The map is

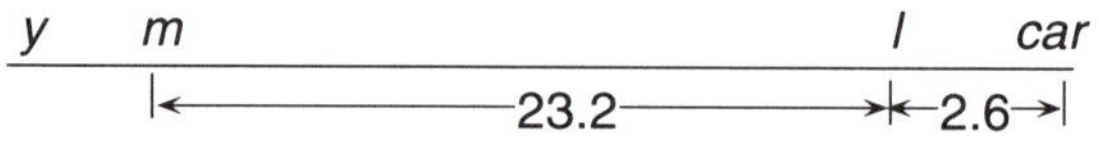

y *m* *l* *car*
|←23.2→|←2.6→|

14. One explanation is that a mutation has occurred during or shortly after the replication producing the two chromatids. This would be expected, but very infrequently. A much more likely mechanism is a nonsister strand exchange at the four-strand stage in a heterozygote (*Aa*). The exchange would have to occur between the heterozygous locus and its centromere:

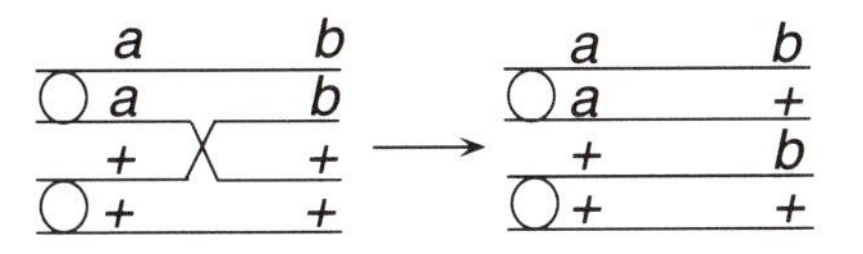

This results in two centromeres, each with its attached chromatids being partly "sister" (and homozygous for genes between the exchange and the centromere) and partly "nonsister" (and heterozygous for genes distal to the exchange) in makeup.

13 MAPPING IN BACTERIA AND VIRUSES

STUDY HINTS

Most of the material we have discussed so far in this primer has been directed at diploid animals and plants. Although the genetic code and role of DNA are essentially the same in eukaryotic and prokaryotic cells, there are important differences. A key difference is that eukaryotic cells have a nucleus (*eu-* true, *kary-* nucleus) and genetic transmission is based on the behavior of chromosomes during meiosis and mitosis. Prokaryotic cells, on the other hand, have no nucleus and are haploid. Since they do not have meiosis, they also lack the genetic transmission associated with sexual reproduction. Mendelian rules do not apply to them. In addition, there are also DNA elements, such as transposons, in eukaryotic cells that provide a special, but very important, exception to traditional genetic transfer.

There are, however, several mechanisms of genetic exchange. Although the method of DNA transfer differs in each case, it can result in a partial diploid for any genes that are carried by the DNA fragment entering the prokaryotic cell. These not only help generate genetic diversity, they also offer geneticists powerful tools to manipulate and study the prokaryotic genome.

- *Transformation* – small "naked" DNA fragments are transported into the cell directly from the cell's environment.
- *Transduction* – DNA picked up by a virus in one cell can be transferred to another cell when that virus infects it.
- *Conjugation* – exchange of DNA between a donor and a recipient cell connected by means of hairlike structures, the *pili.* The donor carries a plasmid (such as the F plasmid in *Escherichia coli*) that carries genes coding for pili formation; the recipient cell lacks the plasmid. A plasmid is a circular piece of DNA that replicates independently of the bacterial genome and does not carry genes required for normal bacterial development. There are two common situations in which this plasmid transfer can also involve bacterial genes.
  - *High-frequency recombination* (*Hfr*) – occurs when a plasmid like F becomes incorporated in the bacterial chromosome (making it an Hfr chromosome) and then carries this DNA to the recipient cell as it replicates. Since the conjugation bridge may not remain intact long enough for complete bacterial chromosome transfer, the genes most closely linked to the plasmid insertion site will have the highest probability of being transferred to the recipient cell.
  - *Sexduction* – the F plasmid incorporated in an Hfr chromosome can become excised again but carry with it a segment of bacterial DNA (called an *F′ plasmid*). When this modified plasmid replicates and transfers to a recipient cell, it also replicates and transfers the inserted bacterial DNA segment.

The proper genetic markers must be used for the recipient chromosome as well as for the transforming, transducing, and Hfr DNA. By making the partial diploid (merozygote) heterozygous at a number of loci, recombination can be studied. In these experiments, variable lengths of donor DNA enter the recipient cell, and then pairing and recombination take place between the entering linear DNA segment and the circular recipient chromosome. This circularity of the host chromosome requires an even number of multiple crossovers, with only one of the products, the intact chromosome, being recovered.

Mapping in bacteria can be accomplished, but it is usually not as straightforward as we find it in *Drosophila,* corn, *Neurospora,* or certain other organisms. On the other hand, the ability to work with very large numbers of organisms and with small pieces of chromosome makes it possible to map very closely linked genes (fine-structure mapping). Furthermore, the ingenuity of the investigators in this field is a pleasure to behold.

IMPORTANT TERMS

Auxotroph
Conditional mutation
Conjugation
Episome
F factor
Hfr
Lysis
Plasmid
Prophage
Prototroph
Sexduction
Temperate phage
Time-of-entry mapping
Transduction
Transformation
Virulent phage

PROBLEM SET 13

1. In bacterial matings 'twas found
Results that did simply astound.
Gene *B* followed *C*
But then *C* followed *B*!
The reason? The gene string is

(a) ground, **(b)** bound, **(c)** wound, **(d)** sound, **(e)** round.

2. In a transformation study, *E. coli* is incubated with transforming DNA carrying the linked genes $B^+D^+T^+$. The bacterium is $B^-D^-T^-$. Single transformations recovered include B^+, D^+, T^+; double transformants are B^+T^+ and B^+D^+ but not D^+T^+. Triple transformants ($B^+D^+T^+$) are also obtained. What are the relative positions of the three linked genes?

3. If three closely linked genes are cotransduced, two at a time (two-factor transduction), and if gene *W* can be cotransduced with gene *P,* and if gene *P* can be cotransduced with gene *L,* but if gene *W* is never cotransduced with gene *L,* what can you conclude as to the order of these three genes? Alternatively, if *A* and *B,* as well as *B* and *C,* can be cotransduced easily, but if *A* and *C* are only infrequently cotransduced, does this tell you anything about the relative position of the three genes?

4. Assume a region of the *E. coli* chromosome is paired with a length of DNA brought

1 *Lys^+* 2 *Phe^-* 3 *Trp^+* 4 *Glu^-* 5 *Val^-* 6

Lys^- *Phe^+* *Trp^-* *Glu^+* *Val^+*

in by an Hfr strain. The positions of the genes involved with the production of five different amino acids are indicated in the following diagram (chromosomal DNA above, plasmid below). If phenylalanine is the unselected marker here, then you would expect that the cells able to grow in minimal medium supplemented with phenylalanine would be **(a)** *Phe^-*; **(b)** *Phe^+*; **(c)** predominantly *Phe^-*; with perhaps a few *Phe^+*; **(d)** either *Phe^-* or *Phe^+*, with equal probability.

5. Explain, using both words and diagram(s), how the F factor can become integrated on the *E. coli* chromosome to produce different Hfr strains, whose transmission of the male chromosome shows these patterns:

Hfr 1	*T*	*Az*	*Lac*	*Gal*	*H*	*Sm*	*Mal*
Hfr 2	*Gal*	*Lac*	*Az*	*T*	*Mal*	*Sm*	*H*
Hfr 3	*Sm*	*Mal*	*T*	*Az*	*Lac*	*Gal*	*H*
Hfr 4	*Gal*	*H*	*Sm*	*Mal*	*T*	*Az*	*Lac*

Please construct the physical map of the *E. coli* chromosome, using these data.

6. At time zero, an Hfr strain with the genetic makeup b^- h^- k^- m^- (markers listed here alphabetically, not necessarily by their real gene order) is mixed with an F^- strain that is auxotrophic for all markers (that is, b^- h^- k^- m^-). Furthermore, the Hfr strain is streptomycin sensitive (str^s) and the F^- strain is streptomycin resistant (str^r). The samples were then plated onto selective medium, and the frequencies of the b^- str^r recombinants that had received the *h, k,* and/or *m* genes from the Hfr cell were estimated. The following graph shows the number of recombinants against time. What was the order of transfer from Hfr to F^-?

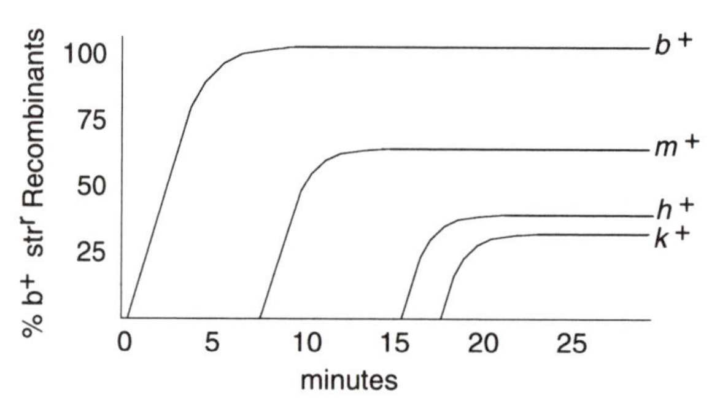

7. Referring again to the experiment in question 6, which two genes are the closest together?

8. Four different *E. coli* Hfr strains donate the genetic markers in the order shown in the following table. All of these Hfr strains are derived from the same F^+ strain. What is the order of markers on this chromosome?

Strain 1:	R	S	J	C
Strain 2:	C	A	T	D

Strain 3: L D T A

Strain 4: D L W R

ANSWERS TO PROBLEM SET 13

1. The answer is **(e),** but this "strange" idea was met with skepticism when proposed by Wollman and Jacob in 1957. See the answer to question 5 for more information on this.

2. To illustrate the principle here, first let us consider a three-factor transformation experiment in which the paired region has three heterozygous loci and four regions in which an exchange can occur. In the following figure, single transformations result from an exchange on either side of the gene.

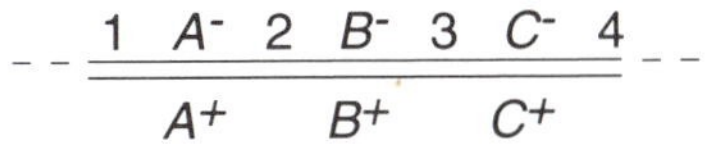

1 A^- 2 B^- 3 C^- 4

A^+ B^+ C^+

If A^+ is incorporated into the *E. coli* chromosome, it requires exchanges in regions 1 and 2. For gene incorporation of gene B^+, exchanges are needed in regions 2 and 3. For C^+, exchanges are needed in regions 3 and 4. For the triple transformation, crossovers in regions 1 and 4 are required. The double transformants are interesting here and can tell us the relative positions of the three loci. *A* and *B,* as well as *B* and *C,* can be doubly transformed by crossovers in regions 1 and 3 for *A* and *B,* and in regions 2 and 4 for *B* and *C.* For the outside loci, *A* and *C,* to be doubly transformed (without B^+) we need four exchanges, one in each region:

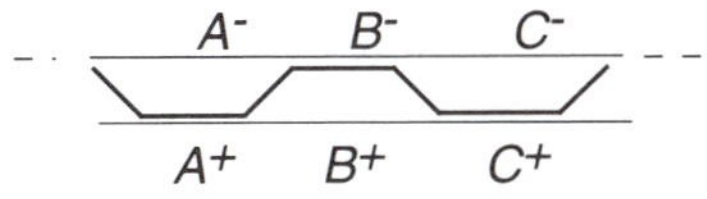

A^- B^- C^-

A^+ B^+ C^+

This is much less likely to occur and, in the study described in the problem, was not recovered, though all other possible transformants were reported. Turning now to the problem, the double transformant not seen was D^+T^+, which tells us that these are on the outside (and B^+ is in the middle).

3. In transduction, the phage vector can carry only a small piece of the donor chromosome to the cell about to be transduced. It has been estimated that only about 2 percent of the *E. coli* genome can fit into a transducing phage, so that only genes that are closely linked can be cotransduced. Genes that are separated by a length of DNA that comprises more than about 2 percent of the total chromosome will not be cotransduced. In the first part of the question, *W* and *P,* and *P* and *L,* can be cotransduced, so we conclude that they are close enough to be located on a piece of *E. coli* chromosome that is less than 2 percent of the total length; but the same cannot be said for *W* and *L,* which are not cotransduced, so the segment of chromosome that carries both *W* and *L* must exceed the 2 percent limit. This indicates that the gene sequence is *WPL.* In the second part of the question, the sequence is *A B C,* and the *A—C* distance is less than that of the *W—L* region and is close to the

2 percent limit. In order to have a transducing piece that includes both *A* and *C*, the breaks in the chromosome would have to be immediately outside the two loci, and this can only be expected to occur rarely, leading to the observed infrequent cotransduction of *A* and *C*.

4. **(c)** If we do not require the incorporation of Phe^+ onto the *E. coli* chromosome, then cells capable of growing in minimal medium supplemented with phenylalanine would need only the Lys^+, Trp^+, Glu^+, and Val^+ genes. This could be produced by a double exchange in regions 4 and 6. An infrequent quadruple crossover in regions 2 and 3, in addition to 4 and 6, would give us an occasional surviving cell with the Phe^+ allele also incorporated.

5. There are a number of sites on the *E. coli* chromosome that are recognized by the F factor and that permit the insertion of the F factor into the *E. coli* chromosome. During conjugation, a break occurs in one strand of the integrated F factor, and this strand moves across the conjugation bridge, taking with it the attached chromosome strand. When the F factor becomes integrated, it does so with one of two alternate polarities, which determine whether host material to the left of the F factor is transferred or whether material to the right of the F factor is transferred. It is obvious from the transmission pattern that Hfr 1 and Hfr 2 have opposite polarities. The determination of which gene goes first is based on both the site and the polarity of F-factor integration into the chromosome. The fact that the F factors become integrated at different positions and that we have overlapping in the transmission patterns permits us to draw the map, which makes sense of the data only if the chromosome is a circle. The map and the positions of the F factors in the different Hfr strains are indicated in the accompanying figure. The order in which the genes enter the cell is reversed.

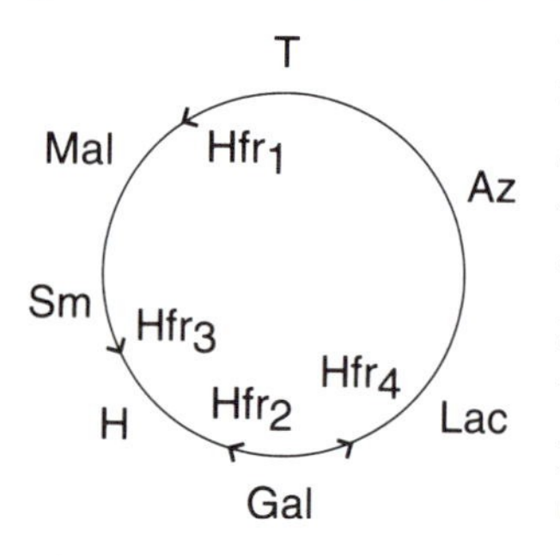

6. The genes closest to b^+ (one of the selected markers) will enter most often, whereas those that are least closely linked will run the greatest risk that transfer will be interrupted before transfer of the DNA is complete. The gene order is *b m h k*.

7. Genes *h* and *k* are transferred with the shortest time interval between them and are, therefore, the most closely linked.

8. The order of markers and the orientation of each strain are indicated by arrowheads.

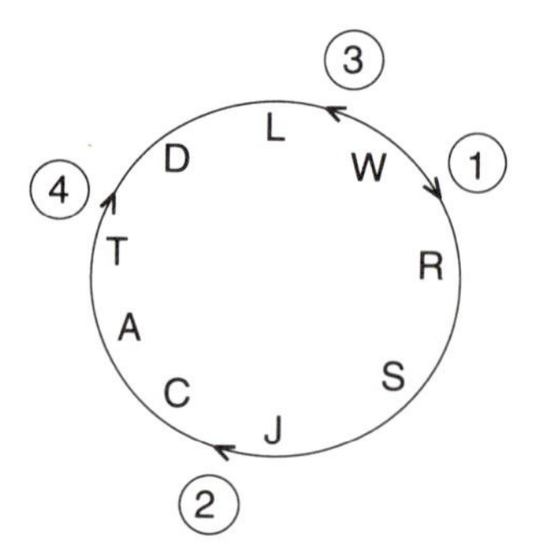

CROSSWORD PUZZLE

MAPPING IN BACTERIA AND VIRUSES

ACROSS

1. Transduction of two or more bacterial genetic markers by the same virus
5. Self-reproducing circular DNA strand that is often infectious, when present in bacteria they often confer antibiotic resistance to the bacteria
6. Transfer of bacterial genes from one bacterium to another by an F′ plasmid, also called F-duction
9. Man who did the mutation fine-structure mapping of the virus rII locus
12. Tubular structures that form on a bacterial cell when infected with the F′ plasmid, involved in conjugation
15. Fertility plasmid that produces a donor bacterium, thus allowing conjugation with a recipient
17. Movement of genetic material in microorganisms from one organism to another when connected by pili
20. Strain of *E. coli* that has a high frequency of recombination because the F plasmid becomes a part of the bacterial chromosome
21. Term for the rupturing of the membrane of a cell, e.g., as by a virulent virus after replication in a cell
23. Phage that lyses its host

DOWN

2. Use of a virus to transfer DNA from one bacterial cell to another
3. Genetic unit that is synonymous with the word *gene,* but is defined by the cis–trans test
4. Two mutants that when present together produce a wild-type
5. Bacterial virus
7. ___-acting, term that describes a genetic element that is on a separate chromosome and affects the function of another gene

(continued on p. 118)

8. Term for the alteration in a cell's makeup because of introduced DNA from another organism
10. Term used for a plasmid that is able to incorporate into the genome of an organism
11. __-acting, term for a genetic element that must be on the same chromosome as another gene in order to influence that gene
13. Mutation involving the loss of base pairs from a chromosome; also called deficiency
14. Phage genome that is integrated into a host's genome
16. Form of transduction in which a prophage becomes virulent and carries bacterial genes from one cell to another cell
18. Term for a phage that can produce a prophage
19. Form of transduction that involves the transmission of bacterial DNA by the accidental packing of a phage head with a bacterial DNA fragment
22. Hole that a virus makes on a layer of host cells, by either killing or slowing the host cell growth

14 OVERVIEW OF TYPES OF GENETIC CHANGE

Mutation can be defined as any type of heritable genetic change. But there are a number of ways in which such changes can occur, ranging from nucleotide changes within a single gene to alterations in chromosome structure or number. Here we will summarize some of the key types of mutation, and gene mutations and chromosomal mutations will be examined in more detail in the following two chapters.

I. *Gene mutation*: Gene mutations are sometimes called *point mutations* because they involve the substitution, loss, or gain of one or more nucleotides within a single gene.

   A. *Base substitution*: A mutation caused by the substitution of a single nucleotide for another.

      1. *Transition mutation*: Substitution of one purine for the other purine or substitution of one pyrimidine for the other pyrimidine.
      2. *Transversion mutation*: Substitution of one purine for a pyrimidine, or vice versa.
      3. The effect of an altered codon due to base substitution can either be a *missense* change (altering the amino acid at that point in the protein) or a *nonsense* change (generating a chain termination triplet that will stop protein synthesis).

   B. *Frameshift mutation*: A mutation caused by the loss or addition of one or more nucleotides to the DNA strand. Frameshift mutations can alter the triplet reading frame so that each codon is affected from the point of the mutation.

      1. We can illustrate this by using an analogy. Consider the following series of letters:

         S E E T H E R E D C A T A N D T H E F A T D O G . . .

         If this sequence is "read" in triplets, as occurs when the nucleotides are translated into an amino acid sequence on the ribosomes, it makes sense (although it is hardly a literary masterpiece). The loss or the gain of one letter (= one nucleotide), however, changes the sense completely. Delete the first *C*, for example, to illustrate this for yourself. The sentence becomes "SEE THE RED ATA NDT HEF ATD OG . . . " If this were a protein, the amino acid sequence and protein structure would have been altered significantly.
      2. In addition, many frameshift mutations yield chain termination triplets early in the sequence, leading to early termination of protein synthesis.

   C. *Transposable element insertion*: Mobile DNA elements can insert into a chromosome. If this occurs in a coding or regulatory gene region, it will result in a point mutation. Molecular mapping of mutant DNA sequences has shown that a large proportion of mutations are of this type.

II. *Changes in chromosome structure*: Changes in chromosome structure involve breaks in the DNA that are repaired incorrectly. If segments are lost or gained, often large numbers of genes are involved and their developmental consequences are more severe than those of most point mutations. Structural changes are defined here but are illustrated in more detail in Chapter 16.
   A. *Inversion*: A change in gene order within a linkage group.
   B. *Deletion*: A loss of a section from a chromosome.
   C. *Duplication*: The addition of a section to a chromosome so that one chromosome carries two copies of that set of genes.
   D. *Translocation*: The movement of a chromosome segment from one chromosome to a nonhomologous chromosome. This is the only structural change that involves chromosomes from different linkage groups.

III. *Changes in chromosome number*: Changes in chromosome number can involve anything from the loss or gain of a single chromosome to the addition of whole chromosome sets. The normal genetic makeup of a cell is the euploid (*eu-* true, *ploid-* multiple) chromosome number.
   A. *Aneuploid*: The loss or gain of one or more chromosomes, such as $2n + 1$, $2n + 2$, or $2n - 1$.
      1. Aneuploids result from a failure of normal chromosome separation in meiosis (nondisjunction).
      2. Most aneuploids that involve the loss of a chromosome die early in development. Most aneuploids that have gained a chromosome also die, but some that involve a small chromosome (e.g., Down syndrome in humans) can survive for at least a short time. Since many different genes are involved in such changes, the phenotypes are often characterized by a whole syndrome of effects.
      3. Since females have two X chromosomes but males have only one, dosage compensation mechanisms help equalize the genetic effects of sex-linked genes. The second X in females becomes highly coiled, forming a darkly staining spot (the Barr body) in the interphase nucleus. If there are more than two X chromosomes, all those beyond one are inactivated. For this reason, X chromosome aneuploids are generally much less severe than autosomal aneuploids.
   B. *Polyploid*: Having multiple whole sets of chromosomes: $3n$ (triploid), $4n$ (tetraploid), and so forth.
      1. Polyploids can result from processes like multiple sperm entry or failure of a polar body to separate from the egg. They can also be produced artificially by cell fusion and other techniques.
      2. Most polyploidy in animals is lethal. Plants, however, appear to tolerate polyploidy well. In fact, polyploidy is a major mechanism in plant evolution and is widely used in agricultural genetics to produce new varieties.

GENE MUTATION

STUDY HINTS

A large number of sensitive methods have been developed for measuring the rate of mutation and for isolating and characterizing the range of mutations that occur in all genomes. Novel characteristics of the genetics or the life cycles of many organisms, such as bacteria, viruses, *Neurospora,* or *Drosophila,* have been used to focus on different aspects of mutation. The giant polytene chromosomes of *Drosophila,* for example, have permitted fairly precise mapping of new mutations, using overlapping deletions. Variations in defined media have allowed specific nutritional mutations to be isolated in *Neurospora* and bacteria, and the relative simplicity of the genetic makeup of bacteria and viruses has been taken advantage of in defining the ways in which mutagenic agents may act upon the genome. The potentials offered by these and other experimental systems are described in your text.

Of particular interest is the genetic repair of DNA. The existence of repair enzymes is an important link to understanding the possible variations in the response of DNA to mutagenic agents. If a certain repair enzyme is defective, as in the human condition xeroderma pigmentosum, the genome is much more sensitive to the action of certain external agents (in this case, ultraviolet radiation). Indeed, variations in genetic repair systems may contribute in a significant way to variations in the responses of different individuals to environmental mutagens.

IMPORTANT TERMS

Ames test
Endonuclease
Exonuclease
Ligase
Mutation frequency
Mutation rate
Muton
Point mutation
Polymerase
Tautomeric shift
Thymine dimer
Transitions
Transversions

PROBLEM SET 15

1. Assume you are studying mutation rate in chickens and find that 19 chicks out of a total of 112,256 hatchlings have a rare dominant condition affecting feather shape. Only three of these chicks have a mother that had that condition. What is the mutation rate for this dominant trait?

2. In *Drosophila,* there are strains that have the two X chromosomes attached. These attached-X females also carry a Y chromosome. These strains have many experimental uses, but one of the most powerful is the measurement of the mutation rate for sex-linked visible alleles. **(a)** What kinds of offspring would you expect by crossing an attached-X female to a normal male? **(b)** How could such a cross be used to measure the mutation rate for sex-linked genes?

3. Why would it be very difficult to establish that the Hiroshima and Nagasaki atomic bombs have caused gene mutations?

4. Which would you expect to be higher, forward mutation rate or back mutation rate? Or would these be the same? Please explain your answer.

5. Consider the following data gathered during a hypothetical experiment to evaluate whether any of three chemicals causes an increase in mutation rate. Three groups of male *Drosophila* are each fed one of the chemicals; a fourth group is fed sugar water and serves as a control to measure spontaneous mutation. The males are then mated to females carrying a marked X chromosome (e.g., *Basc* carries the dominant eye shape mutant *Bar* and inversions to eliminate crossing over between the marked X and the treated X in heterozygous females). Large numbers of heterozygous F_1 females are individually mated to males carrying the marked X, and the F_2 progeny are examined. The appearance of a new lethal mutation can be detected by the absence of males in the progeny of a given cross. Since F_1 females were mated individually, each such cross is the test of a different, single X chromosome. What are the mutation frequencies in each of the following treatments?

	No. of crosses having wild-type F_2 male progeny	**No. of crosses lacking wild-type F_2 male progeny**
Control	5,379	32
Treatment 1	4,901	54
Treatment 2	5,249	213
Treatment 3	8,243	41

6. Some wavelengths of ultraviolet light are more efficiently absorbed by some molecules than by others, and different molecules have different absorption spectra for ultraviolet radiation. The accompanying figure shows the absorption spectra of DNA and protein, with different wavelengths of ultraviolet radiation. What is your expectation for the curve obtained if we change the ordinate of the graph to read "frequency of mutation"? Explain.

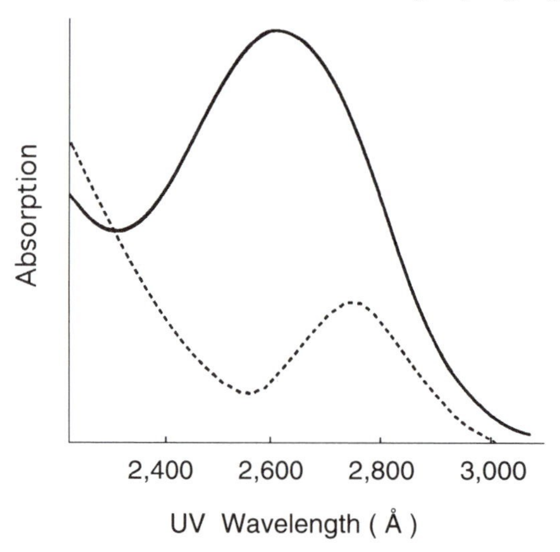

Ultraviolet radiation absorption curves for serum albumin (dotted line) and DNA (solid line). Adapted from A.H. Hollaender, *Radiation Biology* (New York: McGraw-Hill, 1947).

7. Please discuss in detail what is going on in the process delineated in the figure.

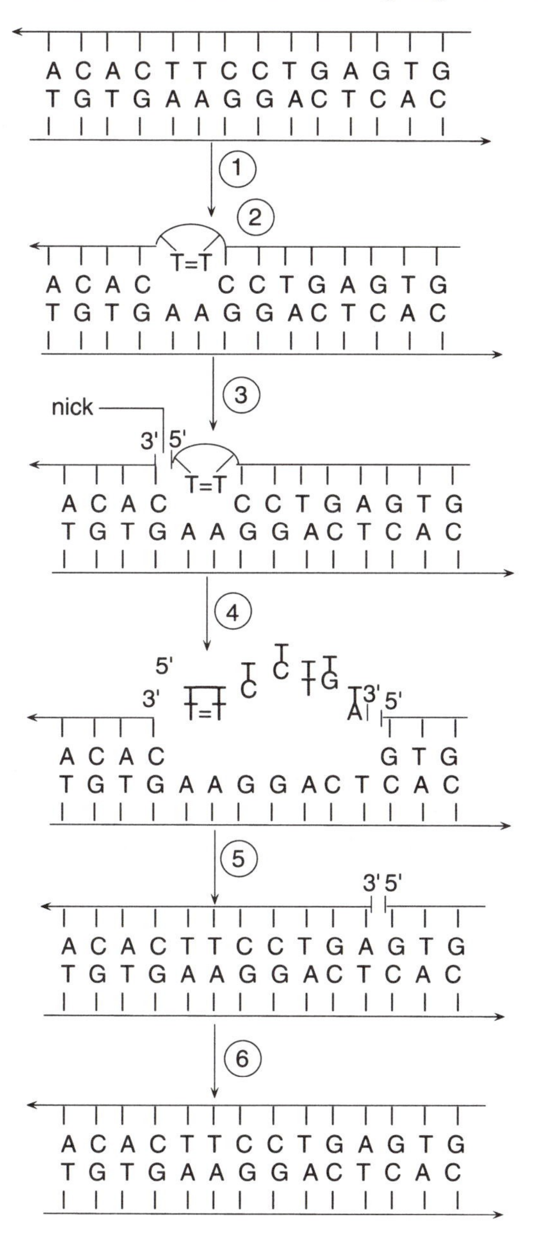

ANSWERS TO PROBLEM SET 15

1. Since three of the chicks can be assumed to have inherited the condition from their mothers, there are only 16 new mutations for the feather shape in this sample. Furthermore, since this feather shape is dominant, either of the two alleles could mutate to produce the phenotype. We must, therefore, consider all available alleles (except those in the chicks inheriting a preexisting mutant) when determining mutation rate. For 112,253 chicks (112,256 minus the 3 chicks we are discounting), there are 16 mutations occurring in 224,506 gametes, giving a mutation rate of $16/224{,}506 = 7.13 \times 10^{-5}$.

2. **(a)** The attached-X method in *Drosophila melanogaster* is designed to detect newly arisen X-linked mutations in a male. The attached-X female has a neutral Y chromosome. She produces two types of eggs, as indicated in the following diagram. **(b)** The male produces sperm containing either an X or a Y chromosome. Since we know from the male's phenotype that he carries no lethal or visible mutation on his X chromosome, any of the sons that carry such a mutant will be affected. Visible mutations are readily detected, and with appropriate care some semi-lethals might also be picked up. This type of inheritance pattern, in which the X chromosome is transmitted from father to son, is called *patroclinous* transmission.

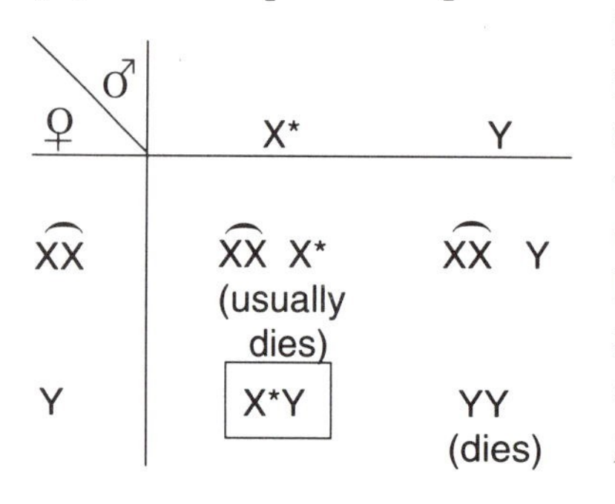

X* denotes a tested X chromosome. The patroclinous males (boxed) are scored for newly-arisen visible mutations.

3. Most gene mutations are recessive. Of those that are dominant, at least some would be lost because of lethality during development. For the more common recessive mutant to be detected within one generation of the exposure, it would have to occur in the X chromosome of the oocyte, where it could then be detected if it eventually gave rise to a son. For early detection of an autosomal recessive trait, both parents would have to have the same gene mutate independently, and a child would have to get both the (probably single) mutated ovum and the one sperm (or one of a small proportion of sperm) that carry this mutation. This requires a very large number of unlikely events. Most recessive autosomal mutants will be transmitted to a child of the exposed person and then down through subsequent generations to a number of people. Eventually, two people carrying this mutant could marry and be expected to produce an affected child, with a probability of 25 percent. It would not be possible to identify this child as a direct victim of the bombs, however, since mutations of the same gene also occur spontaneously or as a result of the influence of other environmental mutagens.

4. Forward mutation rate is higher that back mutation rate. There can be thousands of base pairs present in a single gene, and the mutations in many of these can change the gene product to yield a mutant phenotype. Yet, once the gene has mutated, there are only a limited number of changes that can occur to "repair" the gene product and return it to normal. At the extreme of considering any nucleotide change a mutation, whether it changes the protein or not, there might be literally thousands of ways to mutate a gene from "normal" to a mutant form, but only one way to return a given mutant form to normal. By analogy, "It is relatively easy to kick the TV set and break it, but not as likely that another random kick to the same set will fix the break."

5. Since males have only one X chromosome and the treated males are alive, their X chromosome must initially carry no lethal mutation. The experimental design assumes that a mutagenic chemical eaten by the male can increase the genetic errors that occur during sperm formation. The cross is diagramed, with the treated X chromosome marked with an asterisk.

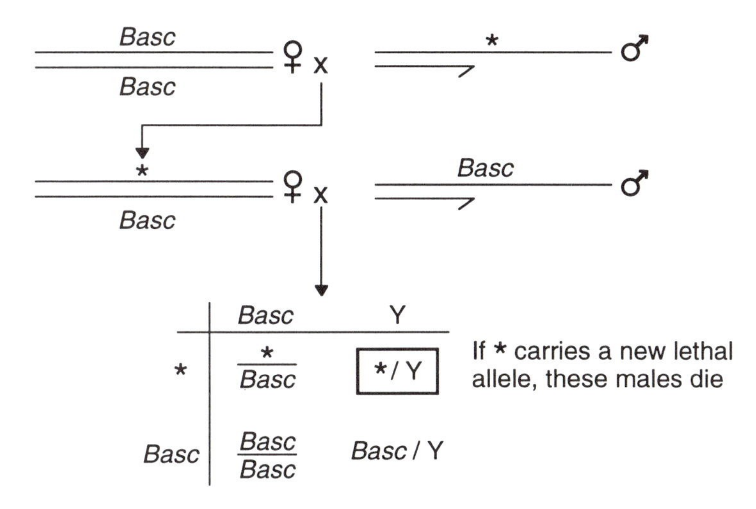

The mutation rate for each treatment is given by the number of crosses lacking wild-type males (indicated by the box in the diagram) divided by the total number of chromosomes tested: Control, 32/5,411 = 0.59 percent new lethal mutations; 1, 54/4,955 = 1.09 percent; 2, 213/5,462 = 3.90 percent; 3, 41/8,284 = 0.49 percent. Thus, chemicals 1 and 2 appear to have elevated mutation rate, but chemical 3 did not.

6. Since ultraviolet radiation is a mutagen, and since DNA is the genetic material, you might expect that the curve for the efficiency of different wavelengths of ultraviolet (UV) radiation in inducing mutation will resemble the solid-line curve (DNA) rather than the dotted-line curve (protein). This has been shown to be the case. It is also interesting to note that germicidal ultraviolet lamps emit radiation at about 2,600 angstroms, where DNA shows maximum absorption.

7. The process diagramed is the repair of thymine dimers, a form of UV-induced mutation.

(1) The DNA is hit by ultraviolet light.
(2) A thymine dimer is formed.
(3) The repair begins with a special endonuclease cutting the phosphate–sugar backbone.
(4) Exonucleases then remove six or seven nucleotides, including the dimer.
(5) The deleted portion is refilled by DNA polymerase, using the complementary strand as a template.
(6) The gaps are closed by ligase (joining enzyme).

16 CHANGES IN CHROMOSOME NUMBER AND STRUCTURE

STUDY HINTS

Errors in mitosis and meiosis can lead to *aneuploid* cells, that is, to cells that have more or fewer chromosomes than the normal diploid (the *euploid,* or literally "true multiple") set of chromosomes. Because so many genes are involved in this change in the genetic makeup of a cell, most aneuploids die or are very abnormal. Among those that do survive, however, the addition of a chromosome is apparently more readily tolerated than is the loss of a chromosome.

A related type of genomic change is *polyploidy,* in which fusion of meiotic products, multiple fertilization, or defects in spindle formation result in extra complete sets of chromosomes ($3n$, $4n$, and so forth). Although polyploidy can occur in either animals or plants, it is much more common in plants. Indeed it is an important evolutionary mechanism in plants that can yield new species. It is also a valuable tool of plant breeders who want to combine characteristics from separate but related species for agricultural purposes. Seedless polyploids are also economically useful.

It is useful to distinguish two types of polyploidy. *Autopolyploidy* (*auto-* means "self") occurs when an inhibition of normal meiosis or other processes increases the number of chromosome sets within a single species. For example, species A (where $2n$ = AA) becomes an autotriploid ($3n$ = AAA), autotetraploid ($4n$ = AAAA), or higher multiple. *Allopolyploidy* (*allo-* means "different"), on the other hand, combines chromosome sets from different species. For example, the allotetraploid between species A ($2n$ = AA) and species B ($2n$ = BB) has the chromosome makeup $4n$ = AABB. Since allopolyploids probably originate from meiotic errors in rare F_1 hybrids between two species ($2n$ = AB), it is probably less common than autopolyploidy.

The predictions of viability and phenotypic change find a special case in the X chromosomes of mammals, where X-chromosome inactivation (producing a Barr body) is a normal mechanism to balance the dosage of X-linked genes in males and females. Males have one X chromosome, whereas normal females have two. To balance the gene dose of X-linked loci, one X chromosome in females becomes condensed (is heterochromatinized) and the vast majority of X-linked genes are inactivated. In cells with an extra X chromosome, this inactivation mechanism minimizes the deleterious effects associated with such a major genomic change. Consequently, in humans, for example, many sex-chromosome aneuploids may survive with various levels of phenotypic abnormality. People with Turner's (XO) and Klinefelter's syndromes (XXY) are examples of this mechanism. In contrast, most autosomal aneuploids die.

In addition to variations in chromosome number, changes in chromosome structure also have wide-ranging effects, from altering linkage relations to yielding inviable gametes. There are four basic types of chromosome aberrations. Three of these (*inversion, deletion,* and *duplication*) involve changes within a chromosome, whereas the fourth

(*translocation*) occurs when a chromosome segment moves to a nonhomologous chromosome. The four most common types of changes in chromosome structure are illustrated in the following outline.

I. *Inversion*

A. *Paracentric inversion* (both breaks are in the same arm)

B. *Pericentric inversion* (the centromere is included within the inversion)

II. *Deletion*

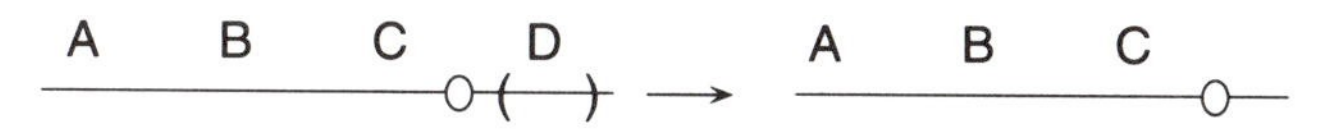

III. *Duplication*

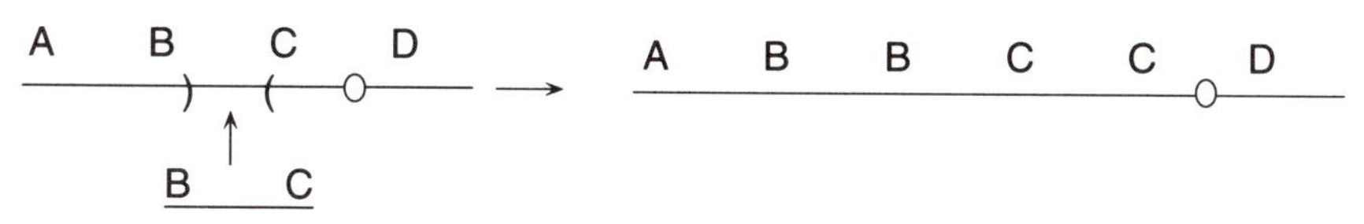

IV. *Translocation*

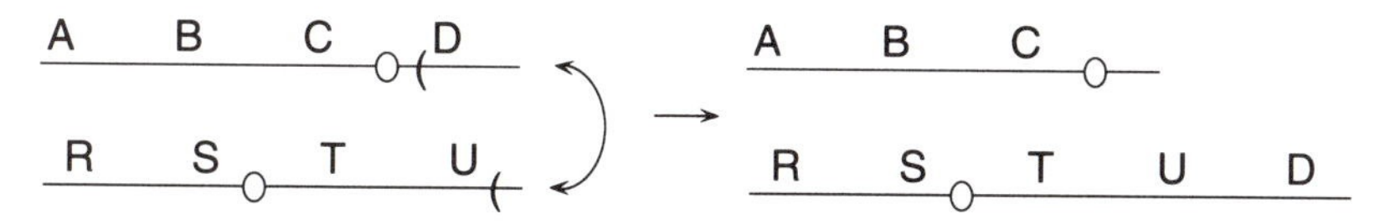

Careful study of the four major types of chromosome aberrations will let you see what changes in linkage will occur with each. Inversions change the order of genes in a linkage group, and if the centromere is included (i.e., pericentric inversion) it will also change chromosome appearance. Except for point mutations that might occur at the break points, the major consequence of an inversion is the production of duplications and deletions that can occur by crossing over in inversion heterozygotes.

Duplications and deletions (or deficiencies) add or subtract genes from a genome and thus lead to genetic imbalances. Such aberrations commonly result in serious developmental defects or early death of the carrier.

Translocation is the only one of these four structural changes that involves two different nonhomologous chromosomes. A reciprocal translocation is the exchange of fragments between two linkage groups, and as long as both translocated chromosomes segregate into the gamete, the genome will be complete. When translocated and normal chromosomes segregate together, however, gametes with duplicated or missing chromosome segments can be produced. As noted previously, these generally result in death of the zygote.

IMPORTANT TERMS

Acentric
Allopolyploidy
Amphidiploid
Aneuploidy
Autopolyploidy
Barr body
Centric fusion
Chimera
Deletion
Dicentric
Dosage compensation
Down syndrome
Duplication
Euploidy
G-branding
Gynandromorph
Inversion
Isochromosome
Karyotype
Klinefelter syndrome
Monoploid
Monosomic
Mosaic
p arm
Paracentric inversion
Pericentric inversion
Polyploid
q arm
Reciprocal translocation
Telomere
Tetraploid
Translocation
Triploid
Trisomic
Turner syndrome

PROBLEM SET 16

1. A polyploid series of related plant species has chromosome numbers of 18, 27, 36, 45, 63, and 72. What can you say about the genetic relationship of these plants based solely on chromosome number?

2. In humans, trisomic individuals are most likely to survive if they are trisomic for what kind of chromosome? **(a)** a large autosome, **(b)** a medium-sized autosome, **(c)** a sex chromosome.

3. Do you think it is possible to cross the cocoa tree with the peanut, treat the sterile F_1 plant with colchicine to double the chromosome number, and obtain an allopolyploid that will produce chocolate-covered peanuts? Assume this is a serious question from a nongeneticist and explain your answer.

4. Assume an allopolyploid with two unrelated $2n$ parent species: D with a $2n$ number of 18, and G with a $2n$ number of 14. If the allopolyploid is crossed back to species D and forms a viable plant, what would you expect to see at synapsis during the first prophase of meiosis? What is your answer if the hybrid is crossed back to the species-G parent?

5. Acentric and dicentric crossover chromosomes result from a single exchange within the loop of a **(a)** paracentric inversion heterozygote or **(b)** pericentric inversion heterozygote.

6. Changes in chromosome structure can complicate otherwise innocent-looking problems. Consider the following set of data on the results of a testcross of a hybrid for five sex-linked loci in *Drosophila melanogaster.*

Chromosomes					No. of flies
y	*spl*	*rb*	*cv*	*sn*	372
+	+	+	+	+	409
y	+	+	+	+	19
+	*spl*	*rb*	*cv*	*sn*	13
y	*spl*	+	+	+	10
+	+	*rb*	*cv*	*sn*	8
y	*spl*	+	*cv*	+	33
+	+	*rb*	+	*sn*	40
y	*spl*	*rb*	*cv*	+	41
+	+	+	+	*sn*	55
					1,000

The standard map distances for these loci, which are taken from the literature, are as follows:

y—*spl* = 3.0, *spl*—*rb* = 4.5, *y*—*cv* = 13.7, *spl*—*sn* = 18.0, *y*—*rb* = 7.5,
y—*sn* = 21.0, *c*—*sn* = 7.3.

(a) First diagram the standard linkage map and the map derived from the recombinational data.

(b) Propose an explanation for any discrepancy between the two maps.

7. Translocation heterozygotes have reduced fertility because of the unbalanced (aneuploid) products of meiosis that come from adjacent segregations. However, some of the aneuploid products of meiosis, although producing lethal or detrimental aneuploid progeny when combined with euploid gametes from a normal individual, can still produce euploid progeny when combined with aneuploid gametes from the same translocation heterozygote. How do you explain this? This phenomenon is seen more frequently in animals than in plants. Why?

8. Two diploid plant species, A and B, are closely related, whereas diploid species C is unrelated to A and B. If you were to be given allopolyploids of A × B, and of A × C, in what ways might they differ from each other **(a)** in appearance? **(b)** in karyotype? **(c)** in fertility?

9. One sometimes finds successful hybrids between two different species in which the hybrid contains only a monoploid set of chromosomes from each parent. What reasonable conclusion(s) can you come to about these hybrids?

10. An autotetraploid has 48 chromosomes. How many linkage groups does it have?

11. An allotetraploid has 48 chromosomes. How many linkage groups does it have?

12. The giant polytene chromosomes of *Drosophila* have banding landmarks that can be used to identify changes in chromosome structure. In the following sketch, we show two regions of a polytene chromosome (2D and 2E), with individual bands numbered within each region. Assume that we are studying four different chromosomes that are deficient for a small number of bands. The deficiencies are listed next to the diagram. In addition, all of the chromosomes behave as if they are

deficient for *r* locus. From the information given, what can you conclude about the bands with which the *r* locus must be associated?

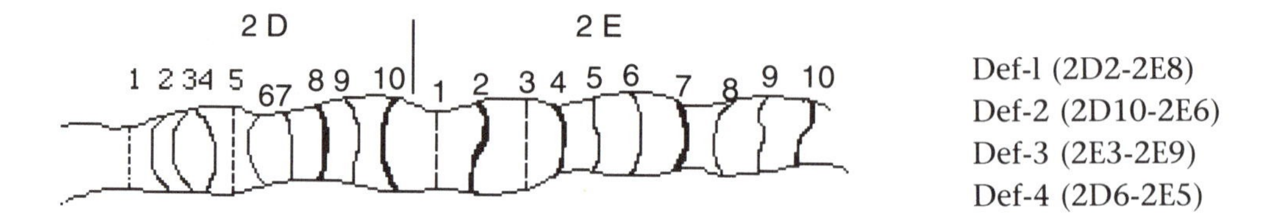

13. The following diagram shows a section of *Drosophila* salivary gland chromosome and the extent of four deletions. Def-l is genetically deficient for the genes *p, l,* and *m;* Def-2 is deficient for *p, m, t,* and *l;* Def-3 for *n, b,* and *v;* and Def-4 for *n* and *b.* What band or bands include the location of the *t* gene?

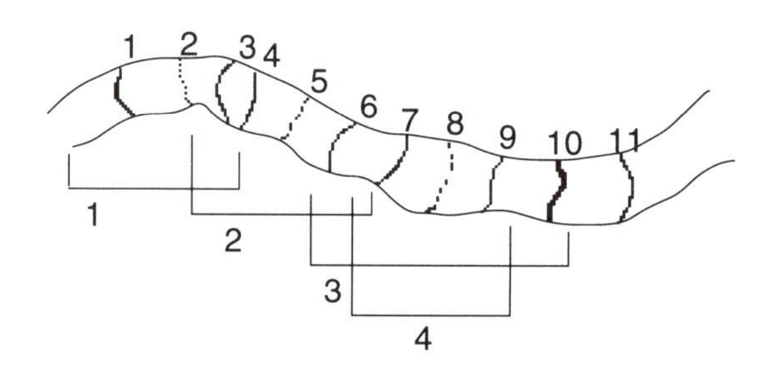

14. What are the expected chromosome constitutions of the products of mitosis and meiosis of a trisomic ($2n + 1$) cell?

15. Assume that a cross of species I (AABBCCDD) with species II (EEFFGGHH) produces a vigorous though sterile hybrid (ABCDEFGH). Is it possible to produce a hybrid between species I and II that is fertile at the time it is formed?

a b c d e f g
n o p q r s
h i j k l m
A.

a d c b e f g
s r p q o n
h i j k l m
B.

a b c d e f g
s r p q o n
h i j k l m
C.

a b f e k l m
s r p q o n
h i j d c g
D.

a b f e d c g
s r p q o n
h i j k l m
E.

a b c d e f g
n o p q r s
h i j k l m
F.

16. The chromosome structure of the ancestral species is labeled A in the preceding diagram. Indicate the evolutionary relationships for each of the species (B through F), and the mechanism(s) by which each species became chromosomally distinct.

17. The data in the following table are the hypothetical results of crosses between five known mutations (a through e) and 6 deletions (1 through 6). The results are indicated by a plus (+) for complementation and a minus (−) for lack of complementation.

(a) Determine the linear sequence of the mutations.
(b) Diagram the deletions relative to the mutant sequence.

	1	2	3	4	5	6
a	−	+	+	−	+	+
b	+	+	−	+	−	+
c	−	−	+	−	+	−
d	−	−	+	−	−	−
e	+	+	−	−	−	−

18. Please draw a bivalent from a cell in meiosis that is heterozygous for a small deletion. Please draw a bivalent that is heterozygous for a small duplication. If you were shown a bivalent and not told what kind of chromosomal mutation was involved, how could you tell whether it was a duplication heterozygote or a deletion heterozygote?

ANSWERS TO PROBLEM SET 16

1. The plant with 18 chromosomes is probably the diploid (2*n*) originator of the series. The monoploid number is then 9, and the other species are all multiples of the basic set of 9 chromosomes. The species with 27 chromosomes, for example, is a triploid ($3 \cdot 9 = 27$). Corroboration of this conclusion could be obtained by examining the chromosomes in a karyotype preparation.

2. **(c)** A sex chromosome. Lethality in trisomics appears to be associated with gene imbalance, and the larger the autosome, the more genes it probably carries and the greater the genetic imbalance. Those trisomics that survive are generally those that involve the smaller chromosomes. The phenomenon of dosage compensation, which turns off all but one X chromosome in each cell, permits survival of XXX individuals.

3. Probably not. The two genomes have to work together to produce a viable plant. It is extremely unlikely that a small tree and a peanut plant will have compatible developmental systems that would permit formation of a hybrid, but if the two genomes were compatible, the probability that the allopolyploid would exhibit the economically favorable traits of each parental species is probably very small. Our example is a bit whimsical, of course; a better one would be Karpechenko's *Raphanobrassica*. In a cross between a radish and a cabbage, the fertile allopolyploid had neither the root of the radish nor the head of the cabbage. This may well have contributed to Karpechenko's death, since he was a Russian Mendelian geneticist who did not accept T. Lysenko's Lamarckian concept of inheritance. Lysenko's ideas were supported as Marxist orthodoxy by the Soviet government, and Karpechenko's refusal to substitute dogma for science led to his arrest in 1940. He later died in prison. If the hybrid had been successful, his future might have been quite different.

4. Species D has nine pairs, and species G has seven pairs of chromosomes. Since the genetic content of D and G chromosomes is different, the allopolyploid will have nine D pairs and seven G pairs. It will form gametophytes with nine D univalents and seven G univalents, whereas the D parent will form gametophytes with nine D univalents. The hybrid will have nine pairs of D chromosomes and seven unpaired G chromosomes, so that synapsis will show nine bivalents and seven univalents. If the allopolyploid is crossed back to the G parent, at synapsis the hybrid will show seven bivalents (GG) and nine univalents (D).

5. **(a)** Paracentric. Whether the inversion is paracentric or pericentric, both are duplicated and deficient, for the regions on either side of the inversion (A and C, the complementary crossover strands) and are euploid for the inverted region (B). If the inversion is pericentric, the centromere is in the inverted portion (B), and both crossover products will have a centromere. If the inversion is paracentric, one of the crossover strands will be dicentric, and the other will be acentric (see the accompanying figure).

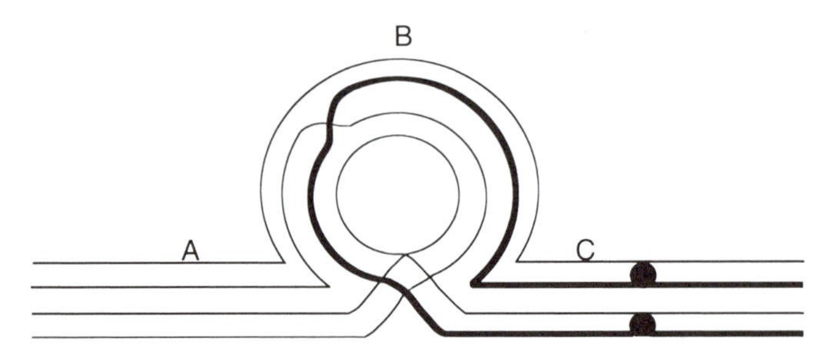

6. The greatest distance is between *y* and *sn,* so we can conclude that these are at either end of the region being tested. If we start with *y* (assume that *y* is at zero on the chromosome map), the sequence can be ascertained by adding those genes that are most closely linked:

(a) The standard map is

y		*spl*		*rb*		*cv*		*sn*
	3.0		4.5		6.2		7.3	
0		3.0		7.5		13.7		21.0

(b) Since the total number of flies is 1,000, the map distances can be calculated easily, as we described in Chapter 12. Map distances are

y—spl	= 3.2		*rb—sn*	= 9.6
spl—rb	= 9.1		*y—cv*	= 5.0
rb—cv	= 7.3		*spl—sn*	= 18.7
y—rb	= 12.3		*cv—sn*	= 16.9
spl—cv	= 1.8		*y—sn*	= 21.9

	3.2		1.8		7.3		9.6	
y		*spl*		*cv*		*rb*		*sn*
0		3.2		5.0		12.3		21.9

The crossover data must therefore have come from an individual that is homozygous for an inversion in which the breaks are between *spl* and *rb,* and between *cv* and *sn:*

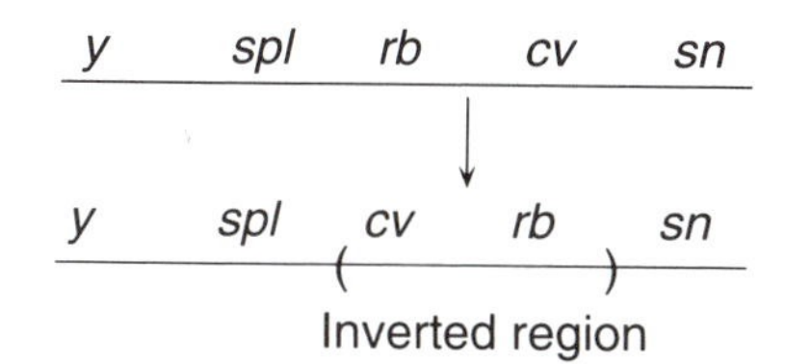

7. Each type of alternate segregation in a translocation heterozygote produces two complementary deficiency/duplication gametes (or gametophytes), with one normal and one translocated chromosome. If two such complementary aneuploid gametes (gametophytes) join, a euploid translocation heterozygote is produced. The only way an aneuploid gamete or gametophyte can give rise to a euploid zygote is for the gamete to join with the complementary deficiency/duplication gamete. An aneuploid gamete plus a euploid gamete always yields an aneuploid zygote. This phenomenon is found much more frequently in animals, since their gametes can function even if extremely aneuploid, probably because there is little or no need for gene function in these cells. In plants with alternation of generations, however, we find that the monoploid generation (gametophytes) must function. For example, genes of the male gametophyte (the pollen tube) are active. This results in the loss of many aneuploid gametophytes.

8. **(a)** The A × B allopolyploid plants would probably resemble their closely related parents, whereas the A × C hybrids would probably show a greater departure in appearance from both of the parental species.

(b) The A × B allopolyploid would have a $2n$ set of A chromosomes and a $2n$ set of probably similar B chromosomes (AA + BB), whereas the A × C allopolyploid will have a $2n$ set of A chromosomes and a $2n$ set of dissimilar C chromosomes (AA + CC). The AA + BB individuals might resemble autotetraploids, at least for some of their chromosomes. The AA + CC individuals, with very little likelihood of homology between the A and C chromosomes, would probably appear to be normal diploids when their chromosomes were examined.

(c) Assuming that flower development in the two allopolyploids is normal (that is, the A + B genomes and the A + C genomes both cooperate in the development of normal flowers), the AA + BB plants might well be less fertile than the AA + CC plants. The latter would have normal pairing of homologous A chromosomes with each other and of homologous C chromosomes with each other, leading to normal disjunction and to euploid gametophyte nuclei. The AA + BB hybrid, on the other hand, would most likely have pairing potential for four more or less homologous chromosome sets, which can interfere with the normal segregation of homologous chromosomes. Some normally monoploid gametophytes might then be disomic for some chromosomes, with the complementary products being nullisomic. This often leads to inviable or poorly viable male gametophytes and/or the next generation of sporophytes. This would result in reduced fertility of the AA + BB allopolyploids.

9. They are most likely plants that reproduce by vegetative propagation. The maintenance of this chromosome constitution is no problem in mitosis, but in meiosis the absence of chromosome pairing partners would lead to excessive aneuploidy and sterility. For this reason, the hybrids are very unlikely to be animals, since most animals reproduce sexually and undergo meiosis. Furthermore, most animal monoploids develop abnormally. Alternatively, the chromosome sets are very similar and allow extensive pairing during meiosis.

10. If the autotetraploid has 48 chromosomes, the diploid has 24, and the monoploid chromosome number is 12 (48/4). Since the number of linkage groups is equal to the number of monoploid chromosomes (i.e., the number of chromosomes in the basic set), the number of linkage groups is 12.

11. Another name for allotetraploid is *amphidiploid* (double diploid). It follows then that all chromosomes occur as pairs (diploid) and that the monoploid number for the allopolyploid is 48/2 = 24. The number of linkage groups is therefore 24. It makes no difference how many chromosome pairs are contributed by each original parent. For example, if species A has a chromosome number of $2n = 30$ and species B is $2n = 18$, species A contributes 15 linkage groups and B contributes 9. If both are $2n = 24$, each contributes 12. In all cases, the sum is 24.

12. The extent of each deficiency has been drawn on the following diagram. The bands that are shared are those that must include the *r* locus, since

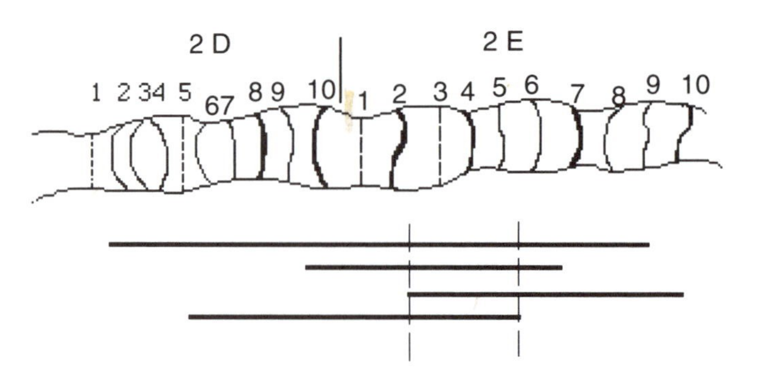

each deficiency also uncovers this mutation. The *r* locus is therefore in band 2E3, -4, or -5.

13. Since Def-1 is not deficient for gene *t*, *t* must be to the right of band 4. Def-2 is deficient for *t*, showing that *t* must be located in band 5, 6, or 7. Neither Def-3 nor Def-4 uncovers the *t* gene; that leaves only band 5.

14. In mitosis, each chromosome behaves individually, and each is expected to divide, with sister chromatids going to each pole, so that all mitotic products will be $2n + 1$, just as in the original cell. In meiosis, although the paired chromosomes segregate from each other and go to opposite poles, the unpaired extra chromosome tends to move at random to one or the other pole. The expectation is that half of the meiotic products will have $1n$ chromosomes, and the other half will have $1n + 1$ chromosomes.

15. Yes. It can be done by first making autotetraploids of species I and II by treating the diploids with a spindle poison such as colchicine. Although autotetraploids produce some sterile gametophytes, as a result of aneuploidy, some gametophytes are euploid and diploid; for example,

AAAABBBBCCCCDDDD → AABBCCDD

If a gametophyte of this constitution joins with an EEFFGGHH gametophyte, we will have instant fertile allopolyploidy.

16. In determining the evolutionary relationships among species, the first step is to look for species that are most similar to the ancestral form – that is, the one or ones that differ by the smallest number of chromosomal changes. Species C and species F differ from ancestral species A by only a single change in chromosome structure: species F by a loss of a centromere (a small deletion or centric fusion), and species C by a reciprocal translocation involving segments *rs* and *on*. Species B could, in turn, be derived from species C by a paracentric inversion:

bcd → *dcb*

Note that it could also have been derived from A directly by both a translocation and an inversion, but the simplest hypothesis would be *A* → *C* → *B* →. Species C also probably gave rise to species E; the only difference between them is a pericentric inversion of the segment *cdef*. Species D could be derived from species E by a reciprocal translocation of segments *klm* and *dcg*. Species F appears to have only the loss of the centromere to distinguish it from A, so it was apparently derived from the ancestral form separately. In summary, the hypothetical evolutionary sequence relating these species could be represented as is shown in the diagram that follows:

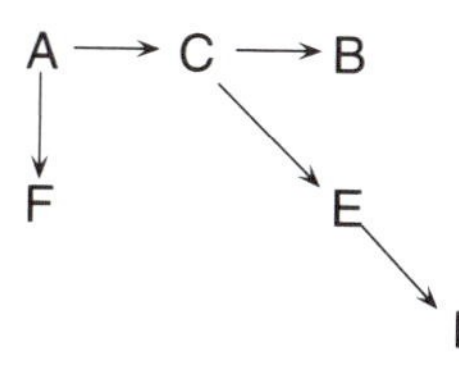

17. **(a)** The linear sequence of the mutations is a c d e b.

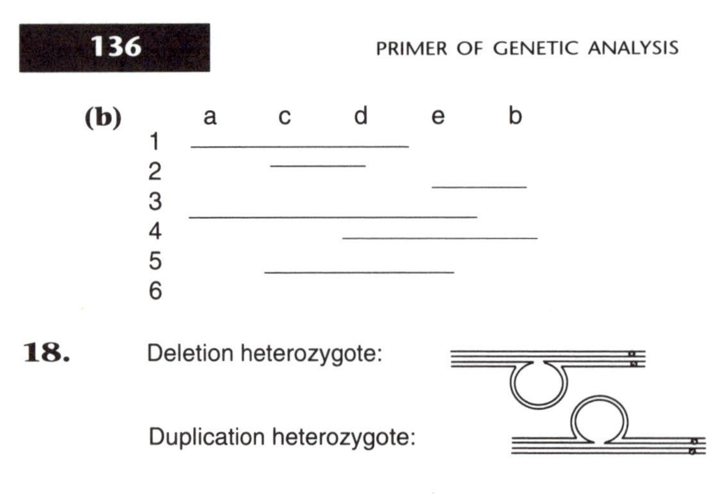

Since the two situations appear essentially identical, the only way to tell them apart in an unknown bivalent would be to use some additional information. For example, if the chromosome has landmarks, like the polytene chromosome banding pattern in *Drosophila,* it should be possible to determine what landmarks are added or missing. Alternatively, if you know the length of a normal chromosome, it might be possible to tell whether the homologue is longer or shorter than normal, indicating a duplication or deletion, respectively.

CROSSWORD PUZZLE

CHANGES IN CHROMOSOME NUMBER AND STRUCTURE

ACROSS

1. Describes the similar phenotypic expression found in individuals who have either an X chromosome or two X chromosomes
6. XO syndrome
7. A deactivated X chromosome as seen in female somatic cells
9. Condition in which patches of tissue differ in genetic makeup
13. Mosaic, for example, mice produced by the union of two different fertilized eggs
16. XXY syndrome
17. Term for having more than two sets of chromosomes
19. Term for an individual who has three copies of one type of chromosome
20. Term for having multiple copies of haploid sets of chromosomes (two or more haploid sets)
21. Polyploid having three sets of chromosomes
22. Sexual mosaic in which various tissues within an adult organism have different sexual karyotypes; often found in dipterans

DOWN

1. Describes a chromosome with two centromeres
2. Term for a change in chromosome number that involves a loss or gain of one or a few chromosomes
3. Another term for a haploid number of chromosomes
4. Describes a chromosome change in which a part of one chromosome is transferred to a nonhomologous chromosome
5. Joining of two nonhomologous chromosomes; results in the loss of one of the centromeres

(continued on p. 138)

8. Change in the order of genes on a chromosome; due to two breaks and a turning around of the broken piece
10. Type of polyploidy in which the extra chromosome sets are derived from the same species
11. Mutual translocation between two non-homologous chromosomes
12. Inversion that does not involve the centromere
14. Chromosome without a centromere
15. Describes a polyploid whose extra sets of chromosomes were derived from different species
17. Inversion that involves the centromere
18. Loss of a DNA base pair, gene, or a piece of chromosome
19. Term for a polyploid having four sets of chromosomes

17 PROTEIN SYNTHESIS AND THE GENETIC CODE

STUDY HINTS

In the Study Hints on DNA structure and replication in Chapter 3, we discussed the fact that deoxyribonucleic acid (DNA) is a double-stranded molecule in which the individual strands are antiparallel to each other and have a 5′ end and a 3′ end. The process of protein synthesis involves using one of these DNA strands (the template strand) to produce a single-stranded ribonucleic acid (RNA) molecule. This is accomplished by reading the DNA nucleotides from the 3′ toward the 5′ end, thus producing a 5′ to 3′ RNA. This process is termed *transcription,* which can be broken down into three stages: initiation, elongation, and termination.

Double-stranded DNA
5′ . . . TGCATGCATGGTTGCA . . . 3′ Coding or sense strand
3′ . . . ACGTACGTACCAACGT . . . 5′ Template or antisense strand

Transcription → reads template strand from 3′ to 5′ to produce mRNA
mRNA 5′ . . . UGCAUGCAUGGUUGCA . . . 3′

Translation → reads mRNA from 5′ to 3′ to produce polypeptides
N-terminal . . . Cys Met His Gly Cys . . . C-terminal

In the scientific literature, DNA sequences are generally reported by printing only the *coding* or *sense strand* (the one that has the same nucleotide sequence as the messenger RNA (mRNA), except having T instead of U; (B. Lewin, *Genes V,* [New York: Oxford University Press, 1994]). The use of "coding" strand for the one that is *not* used in making the mRNA molecule seems confusing (we sometimes get confused, too!). Indeed, not all texts use these terms consistently, so you need to read carefully to be certain you know what process is being discussed.

Three major types of RNA are produced. Messenger RNA (mRNA) carries the genetic code for a polypeptide sequence. The mRNA nucleotides are read in triplets (*codons*). Transfer RNA (tRNA) carries amino acids to the ribosomes during polypeptide synthesis. Each tRNA has a three-base *anticodon* specific for one of the amino acids, and a 3′ end modified to bind to that amino acid. Several different ribosomal RNA (rRNA) types bind with proteins to produce the subunits of functional ribosomes.

The ribosome is composed of two major subunits (in prokaryotes, these are the 30S and 50S subunits). Within each subunit there are two regions, each of which will hold a codon of mRNA bound with the anticodon of an appropriate tRNA molecule and its amino acid. These ribosome-binding sites are the *peptidyl* (P) and the *aminoacyl* (A) binding sites.

Protein synthesis, or translation, can also be broken down into three stages: initiation, elongation, and termination. In the following outline, we summarize *translation in prokaryotes*.

I. *Initiation*
   A. Small ribosomal subunit plus three initiation factors and a start tRNA (*N*-formyl-

methionine) are required to form a complex with the start codon near the 5′ end of an mRNA.

B. The larger ribosomal subunit binds with the initiation complex and dislodges the initiation factors. The ribosomal complex now has the anticodon on f-Met tRNA complementing the start codon (usually AUG) on the mRNA within the P-binding site.

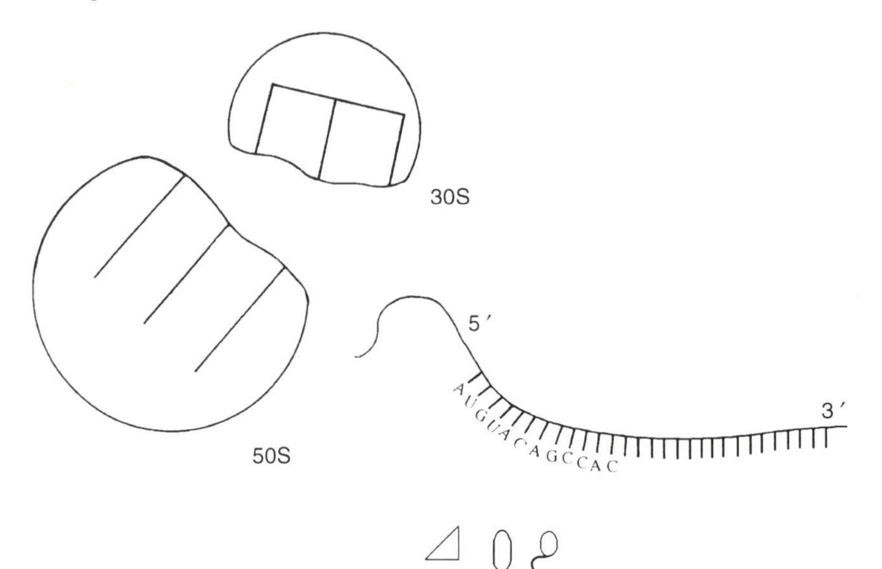

II. *Elongation:* This is a cyclical process in which new amino acids are added to a growing polypeptide chain.

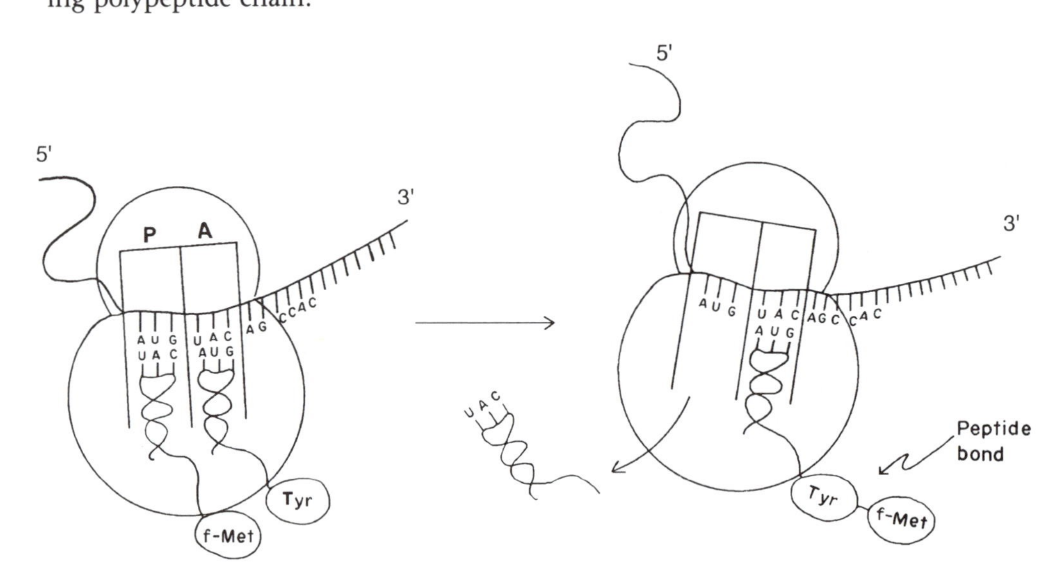

A. The P-binding site is the stronger of the two regions of the ribosome and, if it is empty, the mRNA–tRNA complex will move into that region. This is really equivalent to the ribosome's running along the mRNA molecule toward the 3′ end.

B. The A-binding site will then be occupied by the next codon on the mRNA strand. A tRNA carrying the complementary anticodon and its appropriate amino acid will enter the A-binding site and complex with the codon.

C. Peptidyl transferase, an enzyme present in the ribosome, catalyzes the formation of a peptide bond between the carboxyl group of the P-site amino acid and the amino group of the A-site amino acid, transferring the P-site amino acid to the A-site.

D. The tRNA in the P-site is then released, the remaining tRNA–mRNA complex moves to the open P-site, and the process is repeated until a termination codon is encountered.

E. As in the initiation of protein synthesis, various elongation factors and GTP hydrolysis are required.

III. *Termination*

A. Three different codons can specify termination: UAA, UAG, and UGA.

B. When one of these appears in the A-site, a termination protein causes the release of the polypeptide and breakdown of the ribosome complex.

C. The ribosomal units can then become part of another initiation complex.

The process of *eukaryotic* translation is similar to this but has the following important differences:

1. Ribosomal subunits are 40S and 60S in size.
2. Methionine in the initiation complex is not formylated, but a special initiation tRNA is used.
3. Proteins involved in initiation, elongation, and termination differ from those in prokaryotes. The process establishing ribosome assembly at the appropriate initiation codon is also different.
4. In prokaryotes, translation of the mRNA molecule can occur while its transcription is still in progress. In eukaryotes, however, mRNA is made in the nucleus and must be moved to the cytoplasm to bind with ribosomes.

Much of the early research on biochemical genetics involved the analysis of biochemical pathways that can be studied using auxotrophic (nutritionally deficient) mutants. These experiments helped establish the *one gene/one enzyme hypothesis,* later redefined to one gene/one polypeptide. After the equivalence of genes and polynucleotides had been confirmed, the stage was set for the breaking of the genetic code.

In the following problem set we have put together some simple and some more complex problems dealing with the analysis of biochemical pathways. Although some of these are rather complicated, the logic is straightforward. Do not get discouraged!

Consider the following problem. A number of different auxotrophic mutants are obtained in *Escherichia coli,* all of them requiring the addition of the same amino acid to minimal medium in order for these mutant strains to grow. When these mutant strains are grown on different minimal media, each supplemented with a different possible intermediate in the biosynthesis of the amino acid, the pattern shown in the following table is seen (the plus sign means restoration of normal growth, and no mark means no growth). Where do these intermediates fit into the pathway, and at what steps is each of the seven mutant strains defective?

	Precursor added to medium								
		Precursor						Amino	Number of precursors that
Mutant	**No supplement**	**L**	**S**	**P**	**Q**	**R**	**T**	**acid**	**support growth**
1			+	+	+	+	+	+	5
2						+		+	1
3								+	0
4			+			+		+	2
5			+	+	+	+		+	4
6		+	+	+	+	+	+	+	6
7			+	+		+		+	3

The simplest assumptions are that L, S, P, Q, R, and T are all intermediates in the synthesis of the amino acid end product and that each mutant lacks one functional enzyme that catalyzes one of the steps in the amino acid synthesis. Analysis can be approached by way of the seven different mutant strains or by way of the intermediates (precursors). Before we attempt to solve this problem, however, we shall set up a simple hypothetical enzyme-controlled biosynthesis and study its implications.

Mutant strain (gene):
Intermediate: $\rightarrow A \xrightarrow{1} B \xrightarrow{2} C \xrightarrow{3} D \xrightarrow{4} E \xrightarrow{5} F \xrightarrow{6}$ end product

Precursor A is converted to B by action of an enzyme coded for by the wild-type (normal) allele of gene 1. A mutation at this locus that results in either the absence of the enzyme or a nonfunctional enzyme will not permit B to be produced. Genes, 2, 3, 4, 5, and 6 are all normal, and the enzymes they code are present. In the absence of B, however, no product is produced, and the organism will not grow in minimal medium. Mutant strain 1 will not grow if supplied with A, since its problem is in converting A to B. Strain 1 will be able to grow if given B, which it can convert to the end product. It follows that strain 1 will also grow if placed on media supplemented with C, D, E, or F, for it can convert any of these to the end product. If we look at a different strain, such as strain 4, we can see that it can make B, C, and D, but it has a block in converting D to E. Unless it is supplied with precursor E, it cannot make F and the end product. Strain 4 will also grow if supplied with precursor F.

If we look at mutant 6, it obviously cannot be benefited by addition of any of the precursors, since it can make all of them but cannot carry out the last step in the synthesis. We can conclude, therefore, that the smaller the number of intermediates that will permit growth, the closer that mutant gene is to the end of the synthetic pathway. The greater the number of intermediates that will permit growth, the further "upstream" that gene is functioning.

Returning to the original problem, R, the intermediate used by six of the mutants, is probably just before the amino acid, and the mutant that cannot grow if given R is probably blocked between R and the amino acid. This is mutant 3. Intermediate S can support growth of five mutants and is probably the immediate precursor of R. It cannot be used by mutant 3 or by mutant 2; this tells us that mutant 2 blocks between S and R. The next most frequently used intermediate is P, which is unable to support the growth of mutants 2, 3, and 4. We can conclude, therefore, that the sequence, to this point, is

$$\rightarrow P \xrightarrow{4} S \xrightarrow{2} R \xrightarrow{3} \text{amino acid}$$

Intermediate Q permits growth of three of the mutants. Adding mutant 7 to those that cannot grow, we conclude that mutant 7 blocks between Q and P. Using the same reasoning, T comes before Q, and the mutant that blocks the transition from T to Q is mutant 5. Finally, L can permit only one mutant (6) to grow, so this gene must block the synthesis of L from some unknown precursor earlier in the sequence, and mutant 1 must block between L and T. The pathway to the amino acid is, therefore

$$? \rightarrow L \rightarrow T \rightarrow Q \rightarrow P \rightarrow S \rightarrow R \rightarrow \text{amino acid}$$

Several complications can confuse the picture. Sometimes two different genes seem to control the same reaction: For example,

$$A \xrightarrow{2 \text{ or } 3} B$$

There are several possible explanations for this. One is that it really is a single step, catalyzed by a single enzyme, but the enzyme is made up of two different polypeptides, one coded by gene 2 and the other coded by gene 3. Either mutation leads to a loss of enzyme activity. Alternatively, 2 and 3 may be alleles of the same gene. Another explanation is that there are really two steps, one controlled by gene 2, the other by gene 3:

$$A \xrightarrow{2} ? \xrightarrow{3} B$$

The intermediate molecule may be transient or unknown. Another possible explanation is that the step from A to B is a single step, controlled by a single gene (for example, mutant 3), but mutant 2 determined the synthesis of a competitive inhibitor for the enzyme coded for by gene number 3. Even in the presence of the enzyme, the product of mutant 2 will block the action of the enzyme.

Working through the logical analysis of this type of problem will give you good experience in the interpretation of data, as well as allowing you to apply your knowledge of protein synthesis and activity.

IMPORTANT TERMS

Aminoacyl binding site(A-site)
Aminoacyl–tRNA synthetase
Anticodon
Auxotroph
Codon
Elongation factors
Heterogeneous nuclear RNA
Initiation factors
Messenger RNA (mRNA)
Nonsense codons
Peptide bond
Peptidyl site (P-site)
Peptidyl transferase
Polymerase
Polysome
Protein primary structure
Protein quaternary structure
Protein secondary structure
Protein tertiary structure
Prototroph
Releasing factors
Ribosomal RNA (rRNA)
Start codon (AUG)
Transcription
Transfer RNA (tRNA)
Translation
Wobble hypothesis

PROBLEM SET 17

Special note: Although these problems are not necessarily difficult, many of them, such as problems 9 through 12, may seem complex or time-consuming. Once you feel you understand how to approach one of these problems, you might go directly to the answers to check your reasoning, rather than spending an extended period working out the details yourself. A list of genetic codes is provided in Table R.3.

1. For the following mRNA molecule, 5′ AUGUUGCGCACGUAC 3′

- **(a)** At which end will the ribosome and initiation factor bind?
- **(b)** How many codons (and which codons) will initially be within the mRNA–ribosome complex?
- **(c)** AUG codes for methionine and UAC for tyrosine. Which of these amino acids will be at the amino (N) terminal of the growing peptide chain?
- **(d)** After elongation has started, if UUG occupies the peptidyl site of the ribosome, what codon will be in the aminoacyl site?

2. Assume a DNA triplet pair:

3′ GTC 5′
5′ CAG 3′

where GTC is the template strand that transcribes mRNA.

- **(a)** What is the amino acid coded for by the triplet?
- **(b)** If a mutation occurs as a consequence of A's being in its rare tautomeric state at the time of replication, what amino acid will be coded for?

3. The amino acid sequence of a number of eukaryotic polypeptides is known. Does knowing the sequence of a polypeptide mean that all of the base sequence of its gene is also known? Please explain.

4. Suppose you have a transcribed piece of mRNA. It contains 50 percent G, 10 percent U, 28 percent A, and 12 percent C.

- **(a)** What would be the base ratios of the template strand of DNA from which this was transcribed?
- **(b)** What would be the base ratios of the sense strand of the DNA double helix?

5. Use the following DNA sequence to answer these questions.

5′ TCGTTTAAGGGCTTGTGCGCCACGGAT 3′
3′ AGCAAATTC<u>CCG</u>AACACGCGGTGCCTA 5′ template strand
1 2 3 4

- **(a)** A base is added as the result of exposure to acridine dye. At which position (2 or 4) would it have the most damaging effect on the gene product? Please explain.
- **(b)** What is the resulting change called?
- **(c)** A base substitution at the 3′ end of the codon labeled 3 **(1)** will, **(2)** will not be

more likely to produce an amino acid substitution than would a base substitution at the 5′ end of this same codon. Please choose the correct alternative and explain.

(d) The base guanine is added at position 1. What effect would it have on the gene product?

6. If a synthetic mRNA contains 60 percent adenine (A) and 40 percent guanine (G), positioned at random along the polynucleotide, what amino acids are expected to be incorporated in the polypeptide, and in what proportions?

7. A second synthetic mRNA contains 30 percent U and 70 percent G, positioned at random along the polynucleotide. What amino acids will be incorporated into the polypeptide, and in what proportions? Compare the number of amino acids coded for in this question and in question 6. What is the major reason for the difference?

8. Assume that a short polypeptide sequence contains the following amino acids:

Met Trp Asp Cys Asn His Tyr

What can you say about the relative frequency of the four nucleotides in **(a)** the mRNA molecule and **(b)** in the template strand of the DNA that codes for this amino acid sequence?

9. Assume in *Neurospora* that there are six different histidineless mutants, all nonallelic. When tested on a number of minimal media, each one supplemented with a single possible precursor to histidine, the growth pattern shown in the following table is seen, where plus (+) indicates growth. Determine the order in which the precursors occur in the synthesis of this amino acid. Indicate for each mutant the step in which it is involved.

Mutant	**Precursors added to medium**					
no.	**E**	**I**	**H**	**X**	**L**	**Histidine**
1	+	+		+	+	+
2				+		+
3						+
4		+		+		+
5	+	+	+	+	+	+
6		+		+	+	+

10. Eight independently derived, nonallelic mutants defective in vitamin B_{12} synthesis are recovered in *Neurospora*. Their growth on minimal media, each supplemented singly with one different possible precursor of vitamin B_{12}, is indicated in the following table. From the data which is given in the table, indicate the order of the precursors in this synthesis, and show which steps are blocked by each mutant strain.

Mutant no.	Precursors added to medium							Vitamin B_{12}
	H	C	P	Y	M	Q	B	
1	+	+			+	+	+	+
2						+		+
3		+			+	+		+
4	+	+		+	+	+	+	+
5		+				+		+
6								+
7		+			+	+	+	+
8		+			+	+		+

11. Assume that in *Neurospora* nine different nonallelic mutants have been isolated. Some can grow on minimal medium plus amino acid 1, some on minimal medium plus amino acid 2, and some require both amino acids in order to grow. Please describe the pathway(s) required to explain the data in the following table.

Mutant	Precursors added to medium									
	C	F	H	L	N	P	W	Amino acid 1 only	Amino acid 2 only	Amino acid 1 and 2
1[a]	+									+
2									+	+
3[a]	+				+	+				+
4								+		+
5[a]	+				+					+
6			+	+			+		+	+
7		+						+		+
8[a]							+		+	+
9			+				+		+	+

[a] These mutant strains accumulate large amounts of precursor H, when placed in media that will support growth.

12. In phage T_4, a plus (+) frameshift mutant is suppressed by a second, minus (−) frameshift mutation at a different position in the cistron. The first 10 amino acids of the wild-type protein are

Met Phe Trp Ala Leu Glu Ser Met Asp Pro

and the first 10 amino acids of the protein coded for by the double frameshift mutant are

Met Phe Cys Gly Val Gly Glu His Asp Pro

What, and where, are the base changes associated with the plus and minus mutants, and what is the mRNA sequence for the first 10 amino acids for both the original wild-type proteins and the double-mutant proteins?

13. Heating causes separation of double-stranded nucleic acid into its component strands, and cooling permits annealing (hydrogen bonding) of two strands with complementary sequences. Assume that the double-stranded DNA of a gene coding for a specific polypeptide in chickens is heated and separated.

(a) Some of this material is placed together with RNA from the nuclei of cells specialized to make a great deal of this polypeptide. The mixture is cooled, so that complementary strands can anneal. Electron microscope examination of the preparations shows single- and double-stranded nucleic acid. There are two kinds of double-stranded nucleic acid.

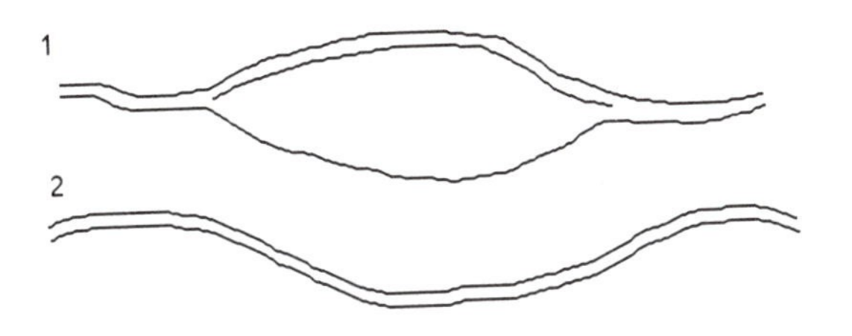

What do these two configurations represent?

(b) If the same DNA preparation has added to it RNA from the cytoplasm of these same specialized cells, and annealing of complementary strands occurs, one can see single- and double-stranded nucleic acid, and again, there are two different kinds of double-stranded nucleic acids:

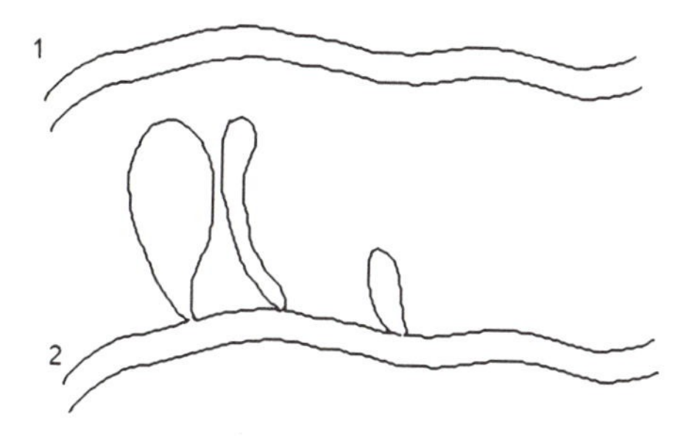

What do these two configurations represent?

14. The following diagram describes the mRNA nucleotide sequence of part of the A protein gene and the beginning of the B protein gene of phage ϕX174. The B gene is completely enclosed within the A gene. Given the following amino acid changes, describe the base changes and the consequences for the other protein:

```
     363 364 365 366 367 368 369 370 371 372 373 374 375 . . . 513
A    Ala Lys Glu Trp Asn Asn Ser Leu Lys Thr Lys Leu Ser
    GCUAAAGAAUGGAACAACUCACUAAAAACCAAGCUGUCG . . .
B                   Met Glu Gln Leu Thr Lys Asn Gln Ala Val
                     1   2   3   4   5   6   7   8   9  10 . . .120
```

(a) Asn at position 368 in the A protein is mutated to Tyr.
(b) Leu at position 370 in the A protein is mutated to Pro.
(c) Gln at position 8 in the B protein is mutated to Leu.

ANSWERS TO PROBLEM SET 17

1. **(a)** The 5′ end, AUG.
(b) Two codons: AUG and UUG.
(c) AUG methionine.
(d) CGC.

2. **(a)** The complementary RNA codon will be 5′ CAG 3′, and the table of RNA codons (Table R.3) tells us that this codes for glutamine.
(b) In the changed state, A pairs like G, so it will hydrogen bond with C when the DNA is replicated. The tautomeric shift of a base in the sense strand will result in an altered codon in the template strand. The new DNA codon is 3′ GCC 5′, and the triplet in the mRNA is 5′ CGG 3′, which codes for arginine.

3. Two characteristics prevent us from inferring the DNA nucleotide sequence from a given amino acid sequence. The first is that the genetic code is degenerate in that several different codons can code for the same amino acid. Knowing the amino acid therefore only means that you know the set of from one to six triplets at that point in the code that will produce that amino acid. Second, the genes in eukaryotes have many intervening sequences that are processed out before the mRNA leaves the nucleus.

4. **(a)** 50 percent C, 10 percent A, 28 percent T, and 12 percent G.
(b) The same as the mRNA, except that there would be 10 percent thymine rather than uracil.

5. **(a)** At position 2. The template strand is read in a 3′ to 5′ direction, so this is nearer the first part of the gene. The addition of a base here will cause a shift in the reading of the codons and result in a larger alteration in the coded sequence than a frameshift mutation farther along. We are sure you are aware, however, that since a gene is a long and complex sequence, the length we are working with on problems such as these may not, in practice, be different enough from the normal sequence to yield a major difference in protein function.
(b) Frameshift mutation.
(c) It will. DNA base substitutions at the 3′ end result in base substitutions at the 5′ end of the mRNA codon. The three possible DNA base substitutions result in A = Cys (the ACG DNA triplet yields the mRNA codon UGC); G = Arg (the mRNA codon is CGC); T = Ser (the mRNA codon is AGC). Since the DNA codon for glycine is CC (T, C, A, or G), the only tRNA molecule(s) that will form hydrogen bonds with the mRNA codons GG (A, G, U, or C) will be glycine-bearing tRNA molecules, and there will be no amino acid substitution, following a DNA base substitution at the 5′ end. This is described as the *degeneracy* of the genetic code.
(d) The frameshift mutant will have as its first three DNA codons AGC, AGA, and ATT, and the corresponding mRNA codons will be

5′ UCG UCU UAA 3′

(these code for Ser Ser terminator). The first amino acid in this sequence will remain unchanged. The second will be changed from phenylalanine to serine, and the polypeptide will be terminated here, since the third codon instructs the ribosome–mRNA complex to dissociate. This mutation is most likely to have a very damaging effect on the gene product. The closer it is to the 5′ end of the mRNA transcript, the greater the number of polypeptides lost.

6. The expected frequency of each possible triplet can be determined by listing them and multiplying the probability that each nucleotide will occur in each of the three positions. For example,

Number of A's	Triplet	Probability	Amino acid	Frequency	Codon
3	AAA	.6 • .6 • .6 = .216	Lys	.360	AA (A, G)
2	AAG	.6 • .6 • .4 = .144	Lys		
	AGA	.6 • .4 • .6 = .114	Arg	.240	AG (A, G)
	GAA	.4 • .6 • .6 = .114	Glu		
1	AGG	.6 • .4 • .4 = .096	Arg		
	GAG	.4 • .6 • .4 = .096	Glu	.240	GA (A, G)
	GGA	.4 • .4 • .6 = .096	Gly	.160	GG (A, G)
0	GGG	.4 • .4 • .4 = .064	Gly		
		1.000			

7.

Number of A's	Triplet	Probability	Amino acid	Frequency	Codon
3	UUU	.3 • .3 • .3 = .027	Phe	.027	UU (U, C)
2	UUG	.3 • .3 • .7 = .063	Leu	.063	UU (A, G)
	UGU	.3 • .7 • .3 = .063	Cys	.063	UG (U, C)
	GUU	.7 • .3 • .3 = .063	Val	.210	GU (U, C, A, G)
1	UGG	.3 • .7 • .7 = .147	Trp	.147	UGG
	GUG	.7 • .3 • .7 = .147	Val		
	GGU	.7 • .7 • .3 = .147	Gly	.490	GG (U, C, A, G)
0	GGG	.7 • .7 • .7 = .343	Gly		
		1.000			

In both this and the preceding question, there are four different combinations of the first two bases. In problem 6, with A and G, three of the four combinations (AA, AG, and GA) require either purine (A or G) in the third position to code for a particular amino acid. Since this synthetic mRNA contains only purines, only three amino acids are coded for by the six codons that start with AA, AG, and GA. The fourth combination, GG, will code for Gly, with any of the four bases in the third position. The number of amino acids coded for by A and G is four.

In problem 7, two of the four combinations (UU and UG) will each code for a different amino acid with a purine in the third position (note that this is not quite correct, since UGA codes for a nonsense stop triplet, but this has no bearing on the answer). In addition, two different amino acids are coded for if either of the two pyrimidines is in the third position. Since this synthetic mRNA is made up of a purine (G) and a pyrimidine (U), the beginning combinations UU and UG will ultimately code four different amino acids. Each of the two remaining combinations (GU and GG) at the beginning of a codon already specifies an amino acid, since the third position can be occupied by any of the four bases without changing the amino acid that is being coded. This synthetic mRNA thus codes for six amino acids.

8. From the reference table of mRNA codons (Table R.3) we can see that the seven amino acids are coded for by

Met	Trp	Asp	Cys	Asn	His	Tyr
AUG	UGG	GA(U or C)	UG(U or C)	AA(U or C)	CA(U or C)	UA(U or C)

(a) We can determine that of the $3 \cdot 7 = 21$ nucleotides, there are six A's and five G's with certainty. The number of U's and C's will vary, depending upon which of the codons for the last five amino acids are present. There are at least four U's, and there may be as many as nine. There is at least one C, and there may be as many as six. The total number of U's and C's must add up to 21 total nucleotides -11 (that is, 6As + 5Gs) $= 10$ in the mRNA.

(b) Because the template strand of the DNA is the template for the mRNA sequence, it will have the complementary array of nucleotides. It will then certainly contain six T's and five C's. There will be at least one G, and there may be as many as six. The total number of A's and G's must add up to 10.

9. There is a simple logic to working this kind of problem, and there is a pattern, when the data are presented in the tabular form that we have used. Since none of the mutants is allelic, they are probably involved in different steps of histidine synthesis. We can assume that each mutant blocks only one step in the synthesis, probably by the loss of a functional enzyme. There are two different (but related) ways of solving this problem. Select the approach that is clearer or more aesthetically pleasing to you.

I. *Mutant method:* This method concentrates on the mutants, which differ in the number of precursors they can use to make histidine.

A. The mutant (3) that cannot use any precursor blocks the last step in histidine synthesis. (This and similar remarks are valid *only* for the steps between the precursors that are given in the question.)

This gives us $\xrightarrow{3}$ His.

B. The mutant (2) that can grow on only one precursor (X) blocks the next to last step in His synthesis, and the precursor that permits growth is the one whose synthesis is blocked by a mutant.

We now have $\xrightarrow{2}$ X $\xrightarrow{3}$ His.

C. The mutant that blocks one step earlier will be able to grow if supplied with X or with a second additional precursor. The mutant (4) can grow with X and with I. The path is now

$\xrightarrow{4}$ I $\xrightarrow{2}$ X $\xrightarrow{3}$ His

D. There should be a mutant that can grow on either X, I, or a third precursor. Mutant 6 is the one, and the new precursor is L:

$\xrightarrow{6}$ L $\xrightarrow{4}$ I $\xrightarrow{2}$ X $\xrightarrow{3}$ His

E. The next precursor (working backward) should be indicated by a mutant that can grow on X, I, L, and a fourth precursor. This mutant is mutant 1, and the new precursor is E. This gives us

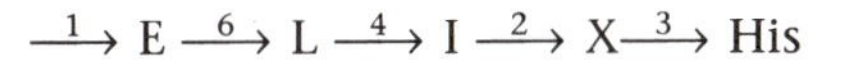

F. The remaining mutant (5) can grow on all of the four precursors just listed but can also grow when supplemented with H. This permits us to complete the synthetic pathway:

$$\xrightarrow{5} H \xrightarrow{1} E \xrightarrow{6} L \xrightarrow{4} I \xrightarrow{2} X \xrightarrow{3} \text{His}$$

(Please excuse the "subliminal" reinforcement of the structure of DNA.) Each mutant blocks only one step in the synthesis, presumably by coding for an inactive or missing enzyme. Each mutant will grow on minimal medium supplemented with any one of the precursors downstream (after the affected step), but it cannot grow on minimal medium supplemented with any of the precursors upstream (before the affected step).

II. *Precursor method:* The second method focuses on the precursors rather than on the mutants. A glance at the table in question 9 shows that each precursor is utilized by a different number of (upstream) mutants.

A. X shows the maximum number, five (mutants 1, 2, 4, 5, and 6). This should place X immediately before histidine (again, this placement is valid only for the precursors being tested) and should assign to mutant 3 the control of the $X \xrightarrow{3}$ His step.

B. The precursor that is used by $5 - 1 = 4$ mutants is I, and the two mutants that will not grow on I are mutant 3 (already assigned a step) and mutant 2. We now have $I \xrightarrow{2} X \xrightarrow{3}$ His.

C. The next (upstream) precursor will have three pluses in the table, and this is L, with mutant 4 appearing as the third mutant that will not grow unless supplemented by L (in addition to mutants 2 and 3). This gives us $L \xrightarrow{4} I \xrightarrow{2} X \xrightarrow{3}$ His.

D. The precursor that is used by two different mutants is E, and this adds mutant 6. So the mutants that are downstream of E are 6, 4, 2, and 3 in that order: $\longrightarrow E \xrightarrow{6} L \xrightarrow{4} I \xrightarrow{2} X \xrightarrow{3}$ His.

E. The precursor that is used by only one mutant is H, and we can add mutant 1 to the pathway just before E, and mutant 5 to the pathway before H: $\xrightarrow{5} H \xrightarrow{1} E \xrightarrow{6} L \xrightarrow{4} I \xrightarrow{2} X \xrightarrow{3}$ His. We now have a complete synthesis coded for by genes 5, 1, 6, 4, 2, 3, in that order.

10. Following the logic of problem 9, mutant 6 blocks just before vitamin B_{12}. Q is the last precursor, requiring the wild-type allele of mutant 2 to catalyze its synthesis from C, whose presence depends on the wild-type allele of mutant 5. Precursor M is converted to C by gene 5, whereas M appears to depend on *two* genes, 3 and 8, whose products are involved in converting B to M. The normal allele of gene 7 is involved in the H–to–B reaction, whereas the wild allele of gene 1 is responsible for the Y–to–H step. The normal allele of gene 4 acts on some unknown precursor, resulting in the production of Y. The steps are indicated as follows:

$$\xrightarrow{4} Y \xrightarrow{1} H \xrightarrow{7} B \xrightarrow{3,8} M \xrightarrow{5} C \xrightarrow{2} Q \xrightarrow{6} \text{Vitamin } B_{12}$$

With the mutants available, there is no evidence that P is even a precursor, since no mutants grow if given P. A more challenging problem is the B–to–M step, which is blocked by the mutant allele of either gene 3 or gene 8. If genes 3 and 8 were duplicate genes that had become separated, both mutant alleles would be necessary to block M synthesis. The data indicate that this is not the case. One explanation is that there are two steps, with a transient intermediate between B and M, so that we have, for example, B $\xrightarrow{8}$? $\xrightarrow{3}$ M. A second possibility is that B to M is a single step, catalyzed by an enzyme made up of two different polypeptides, one coded for by gene 3, the other by gene 8. Such enzymes are well known. The enzyme can be inactivated by mutations in either of the two genes. The third possibility is that only one gene, let us say gene 3, is responsible for the enzyme that converts B to M, but the mutant allele of gene 8 codes for a competitive inhibitor that blocks the B–to–M reaction.

11. Results like this puzzled (although not for long) the early workers in the biochemical genetics of *Neurospora*. Prior to this, all auxotrophic mutants had been found to have only one requirement, such as an amino acid, a purine or pyrimidine, or a vitamin. Here, three nonallelic mutants (1, 3, and 5) had a double requirement for both amino acids. Mutants 4 and 7 were deficient in amino acid 1 only, whereas mutants 2, 6, 8, and 9 were deficient in amino acid 2 only. The objective is to diagram the biochemical pathways to amino acids 1 and 2, indicating which mutant gene affects which step.

It is probably wise to start out with the known and familiar–that is, those mutant strains that require only one amino acid (1 or 2). Mutant 4 will grow only with amino acid 1 added, whereas mutation 7 will grow on amino acid 1 or on precursor F. So we can make a minor pathway thus:

$\xrightarrow{7}$ F $\xrightarrow{4}$ amino acid 1. Doing the same for amino acid 2, we have
$\xrightarrow{6}$ L $\xrightarrow{9}$ H $\xrightarrow{8}$ W $\xrightarrow{2}$ amino acid 2

The remaining mutants, 1, 3, and 5, require both amino acids, and one explanation is that they are involved in a sequence that is common to both amino acids. Examination of the data for these mutants gives us the pathway $\xrightarrow{3}$ P $\xrightarrow{5}$ N $\xrightarrow{1}$ C, and we can try to devise schemes in which C is common to both amino acids. One such scheme is

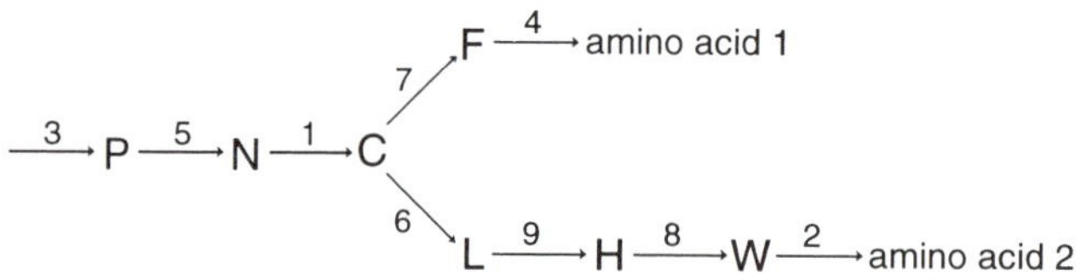

$$\xrightarrow{3} P \xrightarrow{5} N \xrightarrow{1} C \begin{cases} \xrightarrow{7} F \xrightarrow{4} \text{amino acid 1} \\ \xrightarrow{6} L \xrightarrow{9} H \xrightarrow{8} W \xrightarrow{2} \text{amino acid 2} \end{cases}$$

This pathway is consistent with the growth results in the table in question 11, but it does not help to explain the unusually large accumulation of precursor H in mutant strains 1, 3, 5, and 8. Sometimes, in this type of research, a blocked step can cause the accumulation of the intermediate on the "good" side of the block; for example, in

$$\longrightarrow A \longrightarrow B \not\longrightarrow C$$

where the B-to-C step is blocked, B may continue to be synthesized and may accumulate in large amounts. In some cases the high level of B may also cause higher levels of earlier precursors such as A or of the other molecules for which B may be a precursor. There are a number of inherited blocks in human biosynthetic pathways in which intermediates whose normal pathway is blocked pile up in the tissues and/or urine. This may result in excessive production of related molecules, causing conditions such as phenylketonuria and alkaptonuria. In this case the H molecule's continuing modification appears to be blocked. A pathway that is compatible with the data is

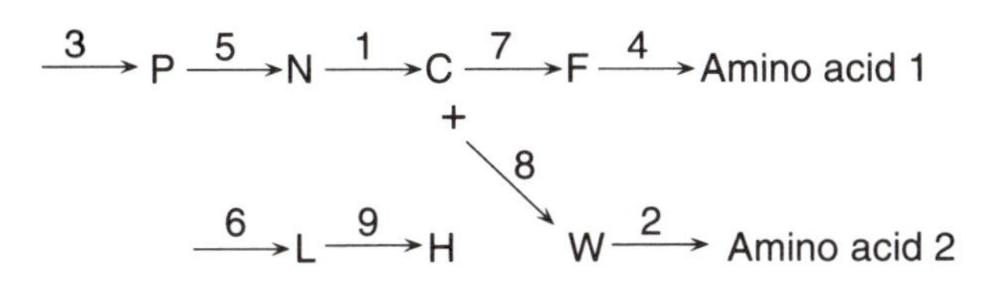

With mutants 1, 3, and 5, there is no C, and since H normally combines with some of the C molecules (under control of the gene affected in strain 8) to form W, H piles up in the organism. If the condensation of C and H is blocked by mutant gene 8, H also piles up, since, again, it is not being incorporated into W.

Can you come up with a different scheme that will be compatible with these data?

12. This is really not a very difficult problem, but it can be messy, and good organization will facilitate its solution. Here is one way to go about it:

(a) List the amino acids in order, together with all possible RNA codons, for both proteins. Arrange them so that the frameshift consequences are easily seen, for example:

	1	2	3	4	5	6	7	8	9	10
Wild	Met	Phe	Trp	Ala	Leu	Glu	Ser	Met	Asp	Pro
Possible codons	AUG	UU U/C	UGG (↑ U or C (+))	GC U/C/A/G	CU U/C/A/G; UU A/G	GA A/G	UC U/C/A/G; AG U/C	AUG (↓ or ↓ (—))	GA U/C	CC U/C/A/G
Possible codons	AUG	UU U/C	UG U/C	GG U/C/A/G	GU U/C/A/G	GG U/C/A/G	GA A/G	CA U/C	GA U/C	CC U/C/A/G
Mutant	Met	Phe	Cys	Gly	Val	Gly	Glu	His	Asp	Pro
	1	2	3	4	5	6	7	8	9	10

(b) Clearly, one frameshift mutation changes the third through the eighth codons; the first, second, ninth, and tenth codons are assumed to be the same in both mRNAs (and DNAs).

(c) Amino acid 3, Trp to Cys. For this change, the addition of U or C between the second and third bases of the Trp codon is needed. There is no deletion of a base that can result in the Trp-to-Cys mutation.

(d) Amino acid 4, Ala to Gly. The plus (+) mutation now causes the third base (G) of the wild-type Trp codon to act as the first base of the codon for the fourth amino acid in the mutant polypeptide (as shown by the lines we have drawn in the figure). Since the first two bases of the wild-type Ala codon must be GC, this makes the fourth amino acid of the mutated protein coded for by GGC (Gly).

(e) Amino acid 5, Leu to Val. The third base in the Leu codon can be any one of the four, but since Val is coded for by GU (UCAG) and the first base of the Val triplet is the last base of the Ala triplet, the original Ala in the wild-type protein was coded for by GCG. Furthermore, since the second base in the Val codon must be U, this permits us to rule out CU (UCAG) as the codon for Leu in the wild-type.

We might pause here to make an observation that will facilitate answering this question. Until we arrive at the end of the mutated sequence and are once more on the original reading frame, the codon for any mutated amino acid will be derived from the last base of the codon for the wild-type amino acid just before it, combined with the first two bases of the wild-type amino acid in the same position. For example, the mutated Val (5) is coded for by G (the last base of wild-type amino acid 4) plus UU (the first and second bases of the wild-type amino acid 5).

(f) Amino acid 6, Glu to Gly. Since Gly is coded for by GG (UCAG), it follows that the last base of the original Leu (5) was G, and since Glu is coded for by GA (AG), the new Gly must be coded for by GGA.

(g) Amino acid 7, Ser to Glu. Since Glu is coded for by GA (AG), we concluded that the last base of the wild-type Glu in position 6 was G. It is also obvious that the wild-type Ser is not coded for by UC (UCAG) but by AG (UC), and since the mutated Glu derives its last two bases from the first two bases of Ser, the codon for Glu is, unambiguously, GAG.

(h) Amino acid 8, Met to His. The last base of the Ser codon can be U or C, but since His, which gets its first base from the last base of the wild-type Ser, is coded for by CA (UC), the original Ser was coded for by AGC. The His codon must be CAU, deriving A and U from the AU of Met.

(i) Amino acid 9 and amino acid 10, Asp Pro. . . on both proteins. We are now obviously back to the original reading frame, which means that there is a minus (−) mutant lurking in the vicinity, compensating for the plus (+) mutation in the third codon. In the absence of the minus (−) mutation, the next four bases of the plus (+) mutant are GGA$^{U}_{C}$.

For Asp, we need GA$^{U}_{C}$. We can obtain this triplet by losing either of the two G's as a minus (−) mutation. This gets us back on the original reading frame, and amino acids Asp Pro and so on.

It may now be obvious to you that the analysis of amino acid changes in frameshift mutations permits us to remove the veil of degenerate ambiguity that clouds the codons. In the example given earlier, the possible codons for amino acid 4 through amino acid 7 of the wild-type included as few as two and as many as six possible triplets for these four different amino acids. It also gives us the exact codons for amino acid 4 through amino acid 8 of the mutant protein. Studies of this type were one of the several different experimental approaches to the determination and verification of the genetic code.

To summarize, the first 10 amino acid and codon sequences for the wild and mutated polypeptides are:

	1	2	3	4	5	6	7	8	9	10
Wild	Met	Phe	Trp	Ala	Leu	Glu	Ser	Met	Asp	Pro
	AUG	UU(U/C)	UGG	GCG	UUG	GAG	AGC	AUG	GA(U/C)	CC(U/C/A/G)
Mutated	Met	Phe	Cys	Gly	Val	Gly	Glu	His	Asp	Pro
	AUG	UU(U/C)	UG(U/C)	GGC	GUU	GGA	GAG	CAU	GA(U/C)	CC(U/C/A/G)

In answer to the question, the base changes associated with the plus and minus mutations occurred in the DNA, of course. The entire DNA (template-strand) sequence (which, remember, is complementary to the mRNA sequence) is as follows, with the plus (+) and minus (−) mutants indicated:

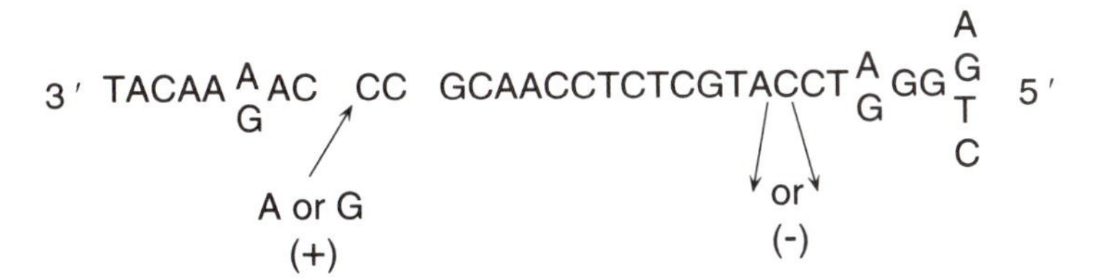

13. (a) Molecule 1 results from annealing of the template strand of DNA and its RNA transcript. The third strand, the coding DNA strand, can pair with the template strand in regions where there is no transcription, and hence no pairing of DNA and RNA strands. Molecule 2 results from reannealing of the two DNA strands, the complementary template and coding strands of the gene.

(b) Molecule 1 results from the hydrogen bonding of the template and coding strands of the gene, whereas molecule 2 results from hybridization of the template strand with the RNA transcript from the cytoplasm. The primary RNA transcript has been tailored in the nucleus, so that three internal segments have been excised and the four separate pieces rejoined to form the end product, mRNA. The loops occur where the template strand of DNA has no complementary RNA to pair with. These sections of the DNA are called *introns*. The coding strand of the gene is annealed with the template strand outside the limits of the gene and also at the introns.

14. (a) Asn (AAC) can code for Tyr (UAC) by a transversion of the first base. This will change the middle base in the third amino acid (Gln) of the B protein to U, which will make that codon CUA, which codes for Leu. So this mutation results in an amino acid change in both proteins.

(b) Leu (CUA) can code for Pro (CCA) by a transition of the middle base. This will change the last base in the fifth amino acid (Thr) of the B protein to C but will not change the amino acid, since ACU and ACC both code for Thr.

(c) Gln (CAA) can code for Leu (CUA) by a transversion of the middle base. This will change the first base in the codon for Lys (amino acid 373) of the A protein to U, making that codon UAG, which is a terminator or nonsense codon. As a result, only the first 372 amino acids of the A protein will be translated, and the last 141 amino acids will be missing. This probably will have a serious effect on the protein's functioning and on the ability of the virus to reproduce, since the A protein is involved in the replication of the replicative (double-stranded) form of the viral DNA.

CROSSWORD PUZZLE

PROTEIN SYNTHESIS AND THE GENETIC CODE

ACROSS

1. Sequence of three nucleic acid bases found on tRNA
3. ___ methionine that is the start amino acid for prokaryotes
5. ___ factor, a protein that brings about the cleavage of the protein from the ribosome
10. Small RNA sequences with associated proteins that are involved in intron removal
13. Complex enzyme that can read DNA sequences and build complementary RNA sequences
14. RNA polymerase that is required in eukaryotes for tRNA transcription
15. Carries the anticodon
17. Protein needed in some prokaryotes for transcription termination
18. Three bases in RNA that code for an amino acid
19. Bond between two amino acids
23. RNA that carries the DNA code for an amino acid sequence
25. Aminoacyl-___ enzymes that are specific for activation or charging tRNA with an amino acid
27. Term for the movement of the ribosome along the mRNA

DOWN

2. Carries the anticodon
4. Start amino acid in translation of all known proteins
6. 5′–5′ 7-methyl guanosine attached to eukaryote mRNA
7. Several ribosomes attached to one mRNA
8. 3′ Sequence placed on mRNA of eukaryotes before they leave the nucleus

(continued on p. 158)

9. Term for making an RNA sequence from a DNA
11. Three-base codon that does not have a complementary anticodon
12. Refers to the code in which more than one codon can code for one amino acid
16. Removal of introns from RNAs
19. Sequence on DNA that allows initiation of transcription
20. tRNA binding site on the ribosome, where peptide bonds are formed
21. Sequence of amino acids
22. Idea that the 5′ end of the anticodon can form hydrogen bonds with more that one 3′ codon base
24. Structure of polypeptides such as the alpha helix and the beta pleated sheet
26. Start codon

GENE REGULATION AND DEVELOPMENT

STUDY HINTS

The problem of how genes are turned on and off at the proper time is a fascinating one. Geneticists still have a lot to learn in this area. Particularly in eukaryotic cells, the interaction between the nucleus and cytoplasm and the coordinated activation of functionally related genes at different times or in different tissues make development a complicated process, even in the simplest organisms. The operon model in prokaryotes is probably not directly analogous to the control systems of higher organisms, but it is an excellent place to begin getting the feel of the logic of regulatory systems.

There are two main types of operons: inducible and repressible. *Inducible operons* are normally turned off, since a repressor protein is bound at the operator site (*O*) and thus blocks RNA polymerase, which binds at the promoter site (*P*). Active transcription is blocked. Inducible operons can be activated by some substrate (the inducer) that binds with, and deforms, the repressor protein. Typically the inducer is some substance that is acted upon by the enzymes coded in the operon, so that the operon is turned on only when its products are needed by the cell. The repressor gene (*i*) may be quite distant from the operon, but the promoter and operator must be adjacent to the structural genes (SG_1, SG_2, etc.) that they control. The key is that the repressor gene's diffusible product is synthesized in an active form.

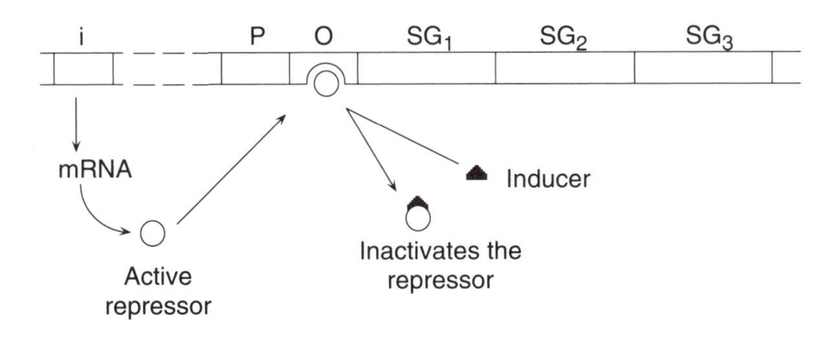

We can contrast this arrangement with that for a repressible operon. As with the inducible operon, function is controlled by an operator site that interacts with a repressor. However, unlike the inducible operon, the repressor protein is initially synthesized in an inactive form. It is activated by a sufficient buildup of the end product made by the enzymes that are coded for by the operon. The end product binds with the inactive repressor to produce an active repressor that turns off the operon. Thus a repressible operon is generally turned on until the product it codes for is in sufficient quantity. A key difference between the two types of operons is therefore the initial activity or inactivity of the repressor protein.

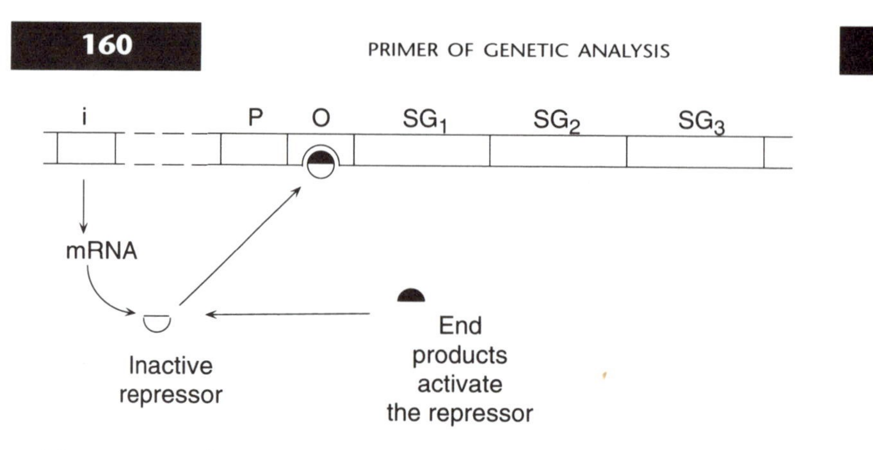

Mutations in the operator or in the repressor (sometimes called the *regulator gene*) can lead to constitutive activity – that is, to continuous transcription. For example, if the repressor protein is structurally abnormal, it will be unable to bind with the operator of an inducible operon, and the operon will be active even in the absence of the inducer. Alternatively, if the operator gene mutates to a form that will not allow binding of a repressor, as shown in the following figure, the operon will show constitutive activity.

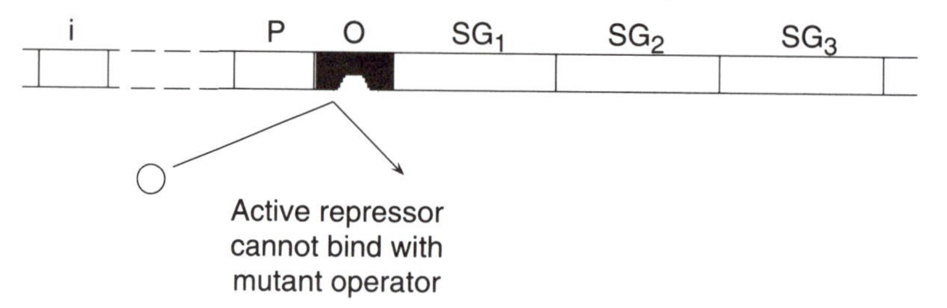

Patterns of operon action can therefore be described largely in terms of the nature of the repressor protein and of the operator DNA sequence to which it binds. For example, mutations at the inducer binding site or mutations that interfere with the transition from an active to an inactive repressor protein shape will lead to an uninducible state whether the inducer is present or not. Mutations in the promoter gene or the structural genes can also lead to the absence of transcription and subsequent expression of one or more genes in the operon, even if other components are normal. If you understand the function of each of the parts in this interrelated system of genes, though, these variations should not be too confusing.

IMPORTANT TERMS

Allosteric proteins
Attenuation
Canalized characters
Catabolite repression
Constitutive mutation
Differentiation
Epigenesis
Histones
Imaginal discs
Informosomes
Operator
Operon
Polar mutation
Preformation
Promoter
Trans-acting factors

PROBLEM SET 18

1. What is the major difference between an inducible operon and repressible operon in *Escherichia coli*?

2. Various theories have been advanced to account for the extreme specialization exhibited by animal tissues. For example, it was suggested that all genes were lost from a cell during development except those genes expressed in each kind of cell. Another theory was that all of the original genes were present but that specialization involves imposing controls whereby particular genes are activated or repressed in different tissues. R. Briggs and T. J. King (1952) demonstrated the first strong evidence for one of these theories. Which theory was supported, and what was the evidence?

3. The turnover rates of mRNA differ considerably when prokaryotes are compared with eukaryotes. In prokaryotes, many mRNAs function for only about 2–5 minutes, whereas in eukaryotes many function for several hours or more. It has been suggested that the turnover rate of specific mRNAs may be genetically determined by the gene that codes for the mRNA. What conclusions can you draw about the relationship between mRNA turnover rates and the adaptive responses that prokaryotes and eukaryotes show to their environments?

4. In the African clawed toad, *Xenopus*, individuals homozygous for the anucleolate mutant die. These embryos produce what appear to be normal blastulas and gastrulas, and they form neural tissue, but they cannot go on to form normal tadpoles. They die early. How can you explain the delayed "time of action" or expression of the anucleolated gene?

5. Assume that it is possible to produce a chromosomal aberration in *E. coli* in which the structural genes for one operon (histidine synthesis) replace the structural genes of the lactose operon, which normally codes inducibly for β-galactosidase permease, and gal acetylase. Normally, the histidine operon is repressible by an end product. The new operon would result in histidine synthesis that is **(a)** induced by lactose, **(b)** inhibited by excess histidine, **(c)** inhibited by excess lactose, **(d)** induced by excess histidine.

6. Consider the following inducible operons:

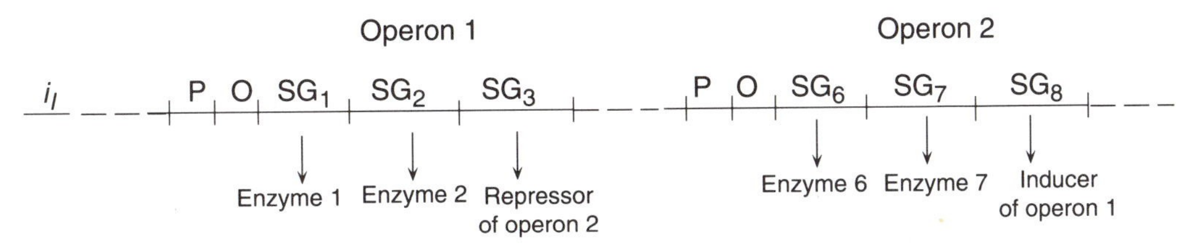

Suppose you supply the inducer of operon 1 for only a short time. Also assume the inducer and repressor have a short half-life. Starting with the first operon turned on and the second operon turned off, how would you describe the continuing action of these two operons after the original (exogenous) inducer has been used up?

7. Referring to the arrangement in problem 6, if the first operon is supplied with its inducer by some artificial means, what is the continuing action of these two operons?

8. Again referring to the system in problem 6, if the inducer for operon 2 is supplied exogenously, what is the continuing action of the two operons?

ANSWERS TO PROBLEM SET 18

1. In inducible operons, transcription occurs only in the presence of a substrate. In repressible operons, on the other hand, transcription is initiated in the absence of the end product, and repression occurs when the end product has been built up to a sufficient level.

2. Briggs and King demonstrated the constancy of the genome (our second theory). Their experiment involved the successful transplantation of a somatic-cell nucleus from a blastula cell into an enucleated egg of the leopard frog *Rana pipiens*. They found that the recipient egg could develop into a complete, normal individual. Thus the somatic cells must have had a full complement of genes.

3. The rapid adaptation that prokaryotes, such as bacteria, show to changing environments depends on their ability to turn genes on and off in a rapid response. There would be little adaptive advantage to a rapid genetic response if the mRNA that it produced persisted for hours. In eukaryotes, however, much of the homeostatic response to changes in the environment is accomplished by cells and tissues showing specialized arrays of proteins. Under these conditions of epigenetic regulation, many mRNA species could be more efficient (or at least not maladaptive) if they functioned longer. Other factors might include protein half-life and rate and duration of mRNA synthesis.

4. The anucleolated genotype is deficient for some of the ribosomal RNA genes, and the loss of these rRNA molecules prevents the formation of functional ribosomes. During early stages of development, the embryos are using the large number of maternal ribosomes that are produced during oogenesis. New rRNA and new ribosomes normally appear at the gastrula stage, and this is where the lethality is expressed. Since the embryo cannot make new ribosomes, it loses its ability to make new proteins, and it dies.

5. The aberration described in this problem could be represented as shown in the figure on p. 163. The lactose operon (II) is inducible; that is, that it will be turned on by the presence of lactose in the cell. Even though the histidine operon is normally repressible by an end product, it is the activity of the operator that determines the action of the operon. The new operon has the inducible O_{II}, and histidine synthesis would therefore be induced by lactose. The correct answer is **(a).**

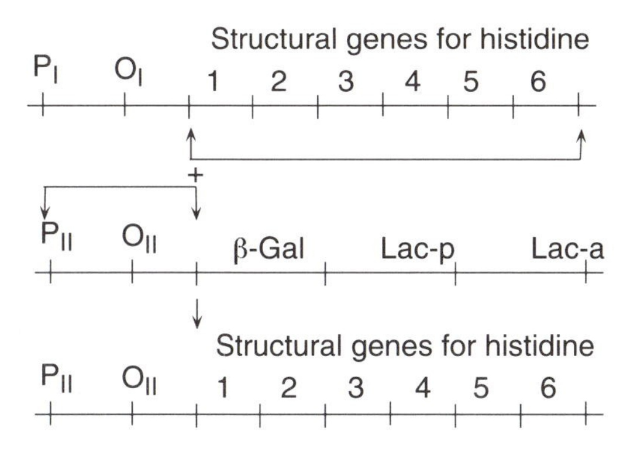

6. If the first operon is turned on at the beginning of the problem, it will synthesize the first two enzymes and a repressor of operon 2. Since the inducer of operon 1 is gradually used up, operon 1 will eventually turn off. The repressor of operon 2 will no longer be produced, and operon 2 will be turned on. This will yield a new inducer for operon 1, and the two operons will cycle on and off.

7. If the first operon is supplied with its inducer constantly, the lack of inducer synthesis by the second operon is unimportant. Thus operon 1 will stay turned on. Operon 2 will not be turned on.

8. If the inducer for operon 2 were supplied exogenously, synthesis of enzymes 6 and 7 would begin, and inducer of operon 1 would be produced. This would turn on operon 1, which would then begin transcribing mRNA, leading to the synthesis of enzymes 1 and 2. In addition, repressor of operon 2 would be produced. If inducer for operon 2 were supplied exogenously, it should block the action of the repressor of 2. Both operons should then be on continuously.

19 OVERVIEW OF MOLECULAR BIOLOGY TECHNIQUES

By the early 1970s, basic research in microbial genetics and biochemistry had provided a fundamental understanding of how genetic information was stored and transmitted, as well as how genes are expressed in response to specific physiological conditions. This knowledge was vital in harnessing microorganisms to replicate foreign DNA, generating the flood of technical applications that are often grouped under the banner of "recombinant DNA technology." This overview represents a primer in the basic tools used by biologists to introduce foreign DNA into bacterial cells. These protocols make extensive use of enzymes produced by microorganisms and used in replicating, repairing, or protecting their own genetic material. It also describes some of the applications of these methodologies in basic research, medicine, and industry.

I. *Restriction endonucleases.* Restriction enzymes are site-specific nucleases that produce internal breaks at distinct nucleotide sequences in double-stranded DNA.

A. In bacteria, they are part of a protection system against viral infection that allows the host to recognize invading DNA and degrade it. Each restriction enzyme will bind to a particular base sequence and cleave DNA at that base sequence. The DNA of the host is modified by methylases, enzymes that add methyl groups to certain bases within the endonuclease recognition pattern. This protects the host genetic material from digestion. Hundreds of different restriction endonuclease/methylase modification systems have been identified from a large variety of different microorganisms. For example:

*Eco*RI (from certain *E. coli* strains)

Recognition sequence *GAATTC*
CTTAAG

*Hin*dIII (from certain *Hemophilus* strains)

Recognition sequence *AAGCTT*
TTCGAA

B. Note that the recognition sites for these enzymes are often *palindromes,* reading the same way in the $5' \rightarrow 3'$ direction on both strands (some call these the Watson and Crick strands). Note also that some restriction enzymes cleave each individual DNA strand off-center, producing a "staggered" cut that leaves a single-stranded tail:

$5' \rightarrow 3'$
G↓AATTC
*Eco*RI *CTTAA↑G*
$3' \leftarrow 5'$

G↓   *AATTC*
CTTAA   *↑G*

As discussed later, these overlapping, or "sticky," ends are useful in joining DNA molecules that have been cut with the same enzyme.

II. Use of *vectors* in the propagation of foreign DNA in bacteria

A. *Plasmid vectors.* By the use of specific nucleases and DNA ligase, it is possible to introduce foreign DNA molecules (i.e., genes) into plasmids, transform this recombinant plasmid DNA into bacteria, and use the bacterial cell to produce large amounts of the foreign DNA (DNA cloning, described later). Plasmids have several features that make them highly suitable for the replication of foreign DNA in bacterial cells (i.e., as *cloning vectors*).

1. They are capable of autonomously replicating, since they possess their own origin of replication. Most plasmids used for DNA cloning replicate many times per bacterial chromosome replication cycle and are present in multiple copies per cell.
2. They replicate as small, supercoiled, covalently closed circles. This topological property allows them to be easily purified away from the large bacterial chromosomal DNA.
3. They often carry genes that can serve as selectable markers, such as antibiotic resistance. Thus, the bacterial cells that carry them can be isolated by growing on the appropriate culture media.

Bacteria thus become factories for the production of foreign DNA, or in specialized cases, for proteins that can be expressed from the introduced genetic material.

B. *Bacteriophage vectors:* Bacteriophages (bacterial viruses) are capable of infecting a host bacterial cell, entering into a lytic cycle, and producing large numbers of progeny phage from a single infection. They can also serve as cloning vectors for propagating foreign DNA. The foreign DNA can be biochemically inserted to replace regions of the phage genome normally used in the lysogenic pathway. After infection, the recombinant bacteriophage will replicate in the bacterial cell, produce numerous progeny phage (containing the foreign DNA of interest), and go on to infect other cells. This will produce lysed regions upon a bacterial lawn that contain large amounts of the same recombinant phage viral particles.

III. Construction of a *recombinant DNA library* in a bacterial vector and introduction into bacteria

A. Vector DNA and the foreign DNA of interest (e.g. genomic DNA from a eukaryotic organism) are purified biochemically and then cut with a restriction endonuclease that produces complementary staggered ends in both molecules. An enzyme is chosen that will produce only one distinct cut within the bacterial vector, in a region of the molecule that does not contain any gene essential for its replication or selection. The eukaryotic DNA, which has a much greater sequence complexity, will be cut hundreds or thousands of times, that is, whenever the recognition site for the enzyme appears in the genome. The two DNA populations are mixed under conditions in which the staggered ends of vector and genomic DNA can base pair, and DNA ligase is added to the reaction. The ratio of the insert and vector DNA is also adjusted, so that most of the time, one vector molecule will base pair to one molecule of foreign DNA. If bacteriophage vectors are used, the ligation products may be combined with a bacterial cell extract containing viral coat proteins, and the recombinant DNA is "packaged" – allowed to self-assemble into infective phage particles.

One trick used to assure that only recombinant phage are packaged is to engineer the *phage vector DNA* to be too small to be efficiently packaged by itself; only phage containing a "stuffer" fragment of foreign DNA will be packaged into an infective phage particle. If plasmids are used, some of the *plasmid* molecules will also recircularize during the ligase reaction, either by themselves or after a piece of eukaryotic DNA has been joined to them. There are also a number of tricks to limit the number of plasmid molecules that do not pick up a foreign DNA insert; one is to treat the vector DNA with the enzyme alkaline phosphatase, which removes 5′ phosphates and prevents the vector molecule from recircularizing without an insert.

B. The recombinant molecules produced in vitro can now be used to infect or transform *E. coli* cells. For *viral vectors,* individual phage infections upon a bacterial lawn in a petri dish will give rise to clones of viruses clustered within zones of lysed cells (*plaques*). For *plasmids,* a recipient strain of *E. coli,* sensitive to a particular antibiotic, is chemically treated to increase the probability of DNA uptake and then exposed to recombinant plasmids. As discussed previously, plasmid vectors contain genes that can confer selective advantage, such as a gene that encodes an enzyme that degrades antibiotics. Successful tranformants are selected by plating on the appropriate medium (e.g., agar plates containing an antibiotic) such that only cells containing the plasmid will survive.

   In library construction, each bacterial colony growing on the selective media will have resulted from a single plasmid transformation. Each individual colony will be replicating the same foreign DNA. Each initial transformation, however, will be unique to a particular DNA molecule, and the vast majority of the transformed cell population will therefore be replicating *different* pieces of foreign DNA. The same situation holds for *viral vectors.* A single recombinant phage will infect a cell, be replicated, lyse that cell, and go on to infect neighboring cells in a chain reaction that propagates a single phage's recombinant DNA insert within a single plaque. The initial infecting bacteriophage, however, will be derived from an independent packaging event and each plaque will contain phage harboring different pieces of foreign DNA. It is thus possible to construct a "library" of individual *clones,* each of which contains a different part of the genome of the eukaryotic organism. Virtually all of a particular eukaryotic genome may therefore be represented in a recombinant DNA library consisting of hundreds of thousands of individual recombinant DNA clones.

C. *cDNA libraries:* In addition to producing recombinant DNA libraries that are representative of an organism's entire genome, it is possible to construct libraries that are representative of only those genes that are expressed in a particular tissue or during a specific stage in an organism's developmental history. This is accomplished by first isolating messenger RNA from a tissue of interest. If, for example, we were interested in isolating a DNA sequence that encoded the hormone insulin, pancreatic tissue would be the logical source of mRNA. Once biochemically purified, the mRNA can serve as a template for producing a "copy DNA" or *cDNA.* The process of efficiently converting mRNA into a double-stranded DNA molecule that can be inserted into a bacterial vector is accomplished in vitro, using many of the bacterial enzymes described. The critical first step in cDNA synthesis, however, requires *reverse transcriptase,* an enzyme purified from the RNA viruses of vertebrates. This enzyme is an RNA-dependent DNA polymerase; that is, it can copy an RNA sequence into DNA, an activity critical in the viral reproductive cycle.

IV. Identifying a specific DNA clone within a recombinant DNA library

A. Given that a particular DNA sequence may be represented once among thousands of individual clones, how does one go about isolating from a recombinant DNA library a particular clone carrying a specific gene sequence? A number of methods are used, but the most common approach is to produce a single-stranded nucleic acid "probe" that is capable of base-pairing with complementary sequences present in the cloned DNA. The "probe" nucleic acid sequence is tagged with a covalently attached radioactive or fluorescent molecule, a molecular label that allows small amounts of the probe to be detected. There are several ways to produce labeled single-stranded nucleic acids complementary to a particular gene sequence. For example, if a small portion of the protein sequence of a particular gene product has been determined, it is possible to synthesize a small, labeled *oligonucleotide* (18–30 oligomers) based on the determined peptide sequence

$$NH_2 - \text{Met} - \text{Glu} - \text{Lys} - \text{Gly} - \text{Trp} - \text{Phe} - \text{COOH}$$

$$5' \quad - \text{ATG} - \text{GA}^{\text{A}}_{\text{G}} - \text{AA}^{\text{G}}_{\text{A}} - \text{GG}^{\text{U,C}}_{\text{A,G}} - \text{UUG} - \text{UU}^{\text{U}}_{\text{C}} - 3'$$

Note that since the code is degenerate, several codon possibilities may exist for any given peptide sequence. This is not normally a problem since a number of different oligonucleotides can be synthesized and used in combination as a probe.

B. The bacterial colonies or plaques representing the library can be replica-plated onto a piece of filter paper, such that their orientation in relation to the agar plate can be identified. The bacterial cells or phage transferred to such filters can now be lysed and the DNA in them denatured to form single-stranded molecules. By incubating the filters with the probe under controlled salt and temperature conditions, the probe will only base pair with complementary sequences in the genomic DNA library. This technique is termed *nucleic acid hybridization.* If, for example, the probe molecule were radioactively labeled, the plaques or colonies that hybridized to the probe could be detected by covering the filters with photographic film and allowing the radioactivity to expose the film (*autoradiography*). Developing the film will reveal the positions of the colonies or phage containing the gene of interest. Returning to the original plate, the individual colonies or phage containing this foreign gene can now be grown, the recombinant DNA isolated biochemically, and the foreign DNA molecule studied.

V. Polymerase chain reaction

A. In 1985, a technique called the *polymerase chain reaction* (PCR) was introduced. It significantly reduced the time and effort necessary to generate large amounts of a specific DNA sequence. A short pair of oligonucleotides (*primers*) complementary to opposite strands of a given DNA sequence could be synthesized, and the region of a source DNA flanked by these oligonucleotides amplified a million-fold in an in vitro reaction. The source DNA sample could be extremely heterogeneous (e.g., genomic DNA) and the concentration of the sequence of interest within the source material exceedingly small (theoretically, a single molecule).

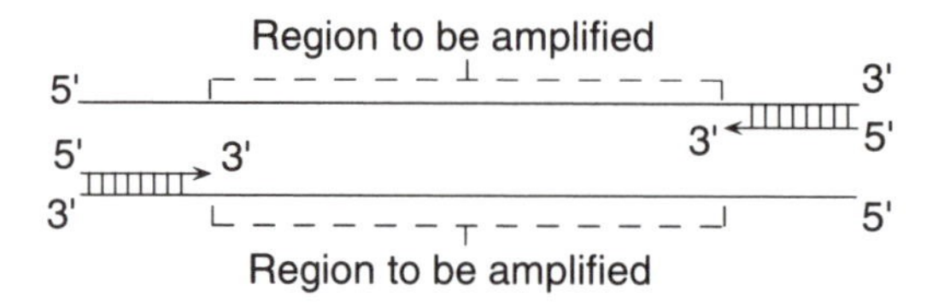

B. The protocol requires two oligonucleotide *primers* that are capable of base-pairing with the nucleotide sequences at the two 3′ ends of the DNA sequence of interest. These primers, along with deoxynucleotide triphosphate (dNTP) precursors, are introduced into the reaction in great molar excess relative to the source DNA molecule. The reaction is heated to denature the source DNA, then cooled to permit the single-strand primers to anneal to complementary regions of the sample. DNA polymerase is now added, and the oligonucleotides, which serve as primers, are elongated in the 5′ to 3′ direction. This round of denaturation, annealing and synthesis leads to a doubling of the DNA in the region flanked by the primers. Indeed, successive cycles of synthesis will lead to an exponential increase of product: 1, 2, 4, 8, 16, and so on. As a result of asynchrony in synthesis, molecules that are longer than the region flanked by primers will be synthesized in each cycle, but they will accumulate linearly and be a minor component of the final product.

C. When first introduced, this protocol was performed with *E. coli* DNA polymerase I. Since this enzyme is heat labile, it was destroyed after every denaturation, requiring that fresh enzyme be added during each cycle of synthesis. This inconvenience was eliminated with the isolation (and gene cloning) of DNA polymerases isolated from thermophilic bacteria, such as those inhabiting hot springs. These enzymes can be added just once to the reaction tube and the entire process automated on thermal cyclers, water baths that can be programmed to switch progressively between the different temperature conditions required for denaturation, annealing, and DNA synthesis.

VI. Applications

A. *Restriction enzyme mapping:* If large amounts of a particular DNA sequence are available, one obvious benefit is the ability to subject that DNA to detailed biochemical analysis. Often the first step in characterizing cloned DNA is the construction of a *restriction map.* The DNA in question is usually digested with a variety of different enzymes, both singly and in various enzyme combinations. The fragments produced from the digestions are then separated by gel electrophoresis. In this technique, the negatively charged DNA molecules move to the positive pole and are separated on the basis of size, with the smaller molecules migrating fastest through the gel matrix. By comparing the sizes of the digestion products from different combinations of enzymes, the restriction sites in the DNA can be positioned relative to one another. The order of specific enzyme sites constitutes a "map," which is a characteristic of any given DNA molecule.

B. *DNA sequencing:* If the linear array of nucleotides in a DNA molecule could be determined, then one could potentially examine what proteins it encodes or what structural features may be important for gene expression. In 1977, two techniques were introduced that permitted researchers to determine the linear order of nucleotides present in a DNA molecule. The technique that is now most widely applied, termed *the chain termination method,* is again based on an enzymatic reaction using DNA polymerase. As in PCR, the source DNA is denatured by heating and a synthetic oligonucleotide primer complementary to a small region of the DNA is allowed to base pair to one of the denatured strands. A DNA polymerase reaction can be initiated off the primer oligonucleotide; the enzyme will then synthesize a complementary copy of the DNA of interest. In addition to the four dNTPs needed for synthesis (which can be radioactively labeled to detect the DNA being synthesized) a nucleotide analogue, missing a 3′ hydroxyl group (a *dideoxynucleotide,* or ddNTP) is also included in the enzymatic reaction. Four separate enzymatic reactions are therefore set up, all of which include the four dNTPs, a primer, and the source DNA of interest, but which differ in the type of dideoxy-analogue added: ddATP added in reaction 1, ddGTP added in reaction 2, ddCTP added in reaction 3, and ddTTP added in reaction 4. The purpose of this is to produce a controlled interruption of enzymatic replication at particular sites; in each of the four reactions, the synthesis will be randomly terminated at A, G, C, and T, respectively. If a dideoxynucleotide is inserted instead of the normal nucleotide (e.g., ddCTP instead of dCTP) the chain cannot be continued, since the analogue lacks the 3′ hydroxyl terminus necessary for the next phosphodiester bond. Remember that, in each reaction, termination will occur randomly at only one type of nucleotide, producing a discrete fragment. Each of the fragments in the four reaction tubes will differ in length from the next largest fragment by a single nucleotide. By sizing the length of the fragments produced in each reaction on an electrophoretic gel, the sequence of the DNA can be read 5′ to 3′, starting from the bottom of the gel (the closest to the priming site) and moving upward:

5′ CGGGAGCGCTCGCTCGAGCTTTCAGA-3′

Remember that the chain termination technique requires an oligonucleotide primer of known sequence to initiate synthesis (although this is not required for the other method of DNA sequencing mentioned). For many vectors, the DNA sequences flanking the restriction enzyme sites used for DNA cloning have been determined. Short oligonucleotides complementary to these sequences are available commercially and are essentially *universal primers* that can be used to sequence any DNA introduced into the restriction enzyme insertion site.

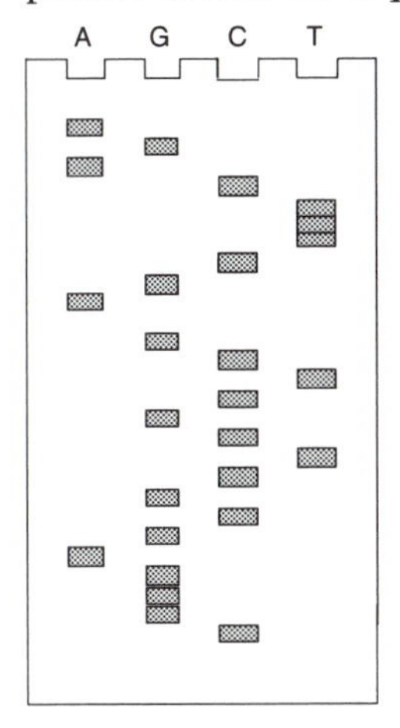

C. *Probes for gene expression:* Cloned DNA sequences are powerful tools that can be used to monitor gene expression in a particular tissue. It is possible to isolate mRNA biochemically from any given tissue type, then use nucleic acid hybridization techniques to quantify the amount of that RNA that is complementary to a cloned gene probe. A variant of this technique, *termed in situ hybridization,* can also be applied directly to tissue sections, which can then be examined under a microscope, allowing researchers to identify individual cells within the tissue sample expressing the gene of interest. The use of in situ hybridization in the localization of specific gene sequences also will be discussed in the next chapter.

D. *Production of recombinant proteins in microorganisms:* It is possible to link the coding sequences for eukaryotic proteins to transcriptional control sequences and produce mRNAs that will be translated into proteins in microorganisms, such as bacteria or yeast. These fusion genes can be placed under the control of inducible "strong" promoters (such as the control sequences for the lactose operon) so that large amounts of protein can be synthesized and purified from recombinant cell extracts. This process has had enormous impact in the production of pharmaceuticals. Compounds of therapeutic value that are made in minuscule amounts in their host organism (insulin, growth hormones, cytokines, clotting factors) can now be synthesized in large amounts for use in the treatment of disease. In some respects, using proteins produced by recombinant DNA techniques may be safer than using the same molecules isolated from the host organism; the spread of human immunodeficiency virus (HIV) infection to hemophilia sufferers treated with pooled human plasma is one example.

E. *Transgenic organisms:* Standard genetic crosses generate recombinant DNA via meiosis, and this process has long been used by humans to select for desirable phenotypic characteristics. Techniques are currently being developed, however, that permit the introduction of cloned genes into the genome of a target eukaryotic organism, with the goal of influencing host phenotype. As in bacteria, introduction of the DNA is often vector mediated, although the vectors usually are capable of integrating into the host genome. The transformation events may involve introducing genes normally associated with the host genome (e.g., genes involved in fruit ripening) or genes that were never part of the genetic repertoire of the host species (e.g., genes involved in pesticide resistance, or nitrogen fixation). Creation of transgenic plants and animals has broad agricultural and industrial applications, and some examples are beginning to reach the marketplace. Like many technological breakthroughs, the ability to generate transgenic organisms has also stimulated public controversy over perceived misapplications of the technology or potential unforeseen environmental consequences.

F. *Diagnosis of genetic disease and somatic gene therapy:* Over the past decade, nucleic acid hybridization techniques using recombinant DNA probes have been increasingly used to identify individuals at risk for particular genetic diseases. If a mutation at a specific gene locus can be defined, differences between normal and abnormal alleles in their nucleotide sequence can be used to identify genotypes. In many cases, however, the transmission of an abnormal allele may be traced from parents to offspring by detecting differences not in the defective gene itself, but in DNA sequences closely linked to the disease gene. Variations in DNA sequence are often detected by population differences in restriction enzyme sites within a specific chromosomal region. This methodology, *restriction fragment length polymorphism* (RFLP) analysis, is a general tool for assessing variation in populations and will be discussed in more detail in the next chapter. If individuals at risk for genetic disease can be identified, however, can transgenic techniques be applied to rectify the defect via introduction of a functional gene? Treatment regimens and clinical trials attempting such intervention are still in their infancy but have been initiated for a number of single-gene disorders (e.g., cystic fibrosis).

20 DNA MAPPING AND HUMAN GENOME ANALYSIS

STUDY HINTS

The techniques described in the previous chapter can be used to map DNA at several different levels of resolution over several different degrees of scale. At the nucleotide level, a linear array of bases can be ordered by DNA sequencing. Using the techniques currently available in most laboratories, each sequencing experiment resolves contiguous regions on the order of hundreds to thousands of nucleotides. Using cloned DNA and batteries of restriction enzyme digests, a linear array of restriction sites can also be mapped relative to one another. Mapping via restriction enzymes can be used to order DNA regions thousands to tens of thousands of nucleotides in length. Finally, techniques of nucleic acid hybridization can be applied to locate genes to specific chromosomal regions, where any chromosomal band identifiable cytologically contains millions of base pairs.

What advantages are there to localizing genes? One advantage that has already gained clinical importance is the ability to follow the transmission of certain genetic diseases using tightly linked genetic markers. The recent application of recombinant DNA technology to gene mapping is having tremendous significance in identifying DNA sequences that are highly variable in populations, and whose inheritance pattern can be used to follow the transmission of closely linked disease genes. Moreover, once a search has been narrowed to a specific region of the genome by establishing linkage to a particular marker, molecular techniques can be brought to bear on finding and cloning the disease gene itself. Once it is cloned, the organization and expression of the gene in both normal and affected individuals can be examined, providing a means of investigating the molecular basis (cause, or etiology) of the disease process.

A great deal of progress has been made in the assignment of specific genes to chromosomes. These advances have been primarily due to the joint application of two techniques: somatic cell genetics and nucleic acid hybridization using cloned DNA. Combined with more classical forms of kindred analysis, localization of human genes to specific chromosomes has been increasing geometrically since 1970.

I. *Kindred analysis*

   A. Examination of pedigrees represents the oldest method of establishing linkage. As mentioned earlier, it is hard to use for humans because of the relatively small number of progeny produced from any one family and the inability to structure test crosses (selective matings) as in animal and plant systems. This increases the difficulty of determining patterns of phase relationships between disease and marker loci – that is, whether linked genes in heterozygotes are in a cis- or a trans-orientation. Nevertheless, examples of linkage can be obtained from human pedigree data through the use of mathematical modeling techniques (e.g., lod score method), even in cases where the linkage phase is unknown. Information from several different families may also be combined for analysis.

B. *Lod scores*: Linkage data are often expressed statistically in terms of a *likelihood ratio*. This ratio represents the probability that two alleles are linked at some given recombination frequency (θ), divided by the probability that the alleles assort independently (i.e., a recombination frequency that equals 50 percent). The data are usually presented as the log of the likelihood ratio (*lod score*):

$$Z(\theta) = \log_{10} \frac{\text{(Probability of family for } \theta = 0.01\text{, etc.)}}{\text{(Probability of family for } \theta = 0.50\text{)}}$$

At a specified frequency of recombination (θ), a likelihood ratio of 1,000:1 (lod score = 3) is usually considered "proof" of linkage.

What relation does a genetic map based on linkage data have to physical features in a chromosome? Map units are based on recombination frequencies, which are clearly inconstant. As a rule of thumb, however, 1 percent recombination represents a DNA segment on the order of 500–1,000 kilobase pairs.

II. *Applications of recombinant DNA technology to gene localization and characterization*

A. *Southern blot hybridization:* The value of somatic cell hybrids for mapping has been discussed in Chapter 10, where the presence or absence of a particular human gene product (e.g., an enzyme) was correlated with the presence or absence of a human chromosome in the hybrid cell. Clearly, any demonstrable physical difference between rodent and human cells may be useful for mapping. Indeed, the nucleotide sequence of the gene itself can be used to identify its chromosomal position. A large number of human genes have been recovered from recombinant DNA libraries. An insert of cloned DNA in standard vectors (which is usually no more than 10–50-kb in size) inherently contains no information on where this DNA is located in the human genome. Cloned DNA, however, can be localized to specific chromosomes *by nucleic acid hybridization*. The nucleotide sequences for many evolutionarily related genes common to both rodents and humans have usually diverged sufficiently so that they can be differentiated from one another. If the DNA from a hybrid cell line is isolated, denatured, and annealed to a single stranded probe derived from cloned human DNA, its presence can be detected and correlated with the human chromosome complement of that cell. The rodent and human genes in question can be quite similar in sequences and still be discriminated from one another. There is usually sufficient divergence between the rodent and human genome that a restriction enzyme can be identified that will differentiate between the genes on the basis of restriction fragment length polymorphisms. The technique used to detect these differences is termed a *Southern blot*, named for its inventor, E. M. Southern. DNA from the hybrid cell line is isolated, purified, and digested with the appropriate enzyme. The resulting fragments are then separated by gel electrophoresis, and the DNA in the gel is denatured and transferred (as in the library screenings discussed previously) to a piece of paper. The DNA is permanently attached to the paper by baking, and the paper is then incubated in the presence of labeled probe DNA, where the probe can hydrogen bond to complementary sequences. Unpaired probe molecules are then washed away, and the probe molecules that are specifically annealed to the DNA on the filter is detected (e.g., by autoradiography, in the case of radiolabeled probes). Correlations are then drawn between the human chromosomes (or parts thereof) that are present in a particular cell line and the presence or absence of restriction fragments characteristic for the human gene.

B. *In situ hybridization:* It is now possible to map a gene directly using probe nucleic acids annealed directly to mitotic chromosomes. This technique was first developed using radioactive probes. Briefly, mitotic chromosomes are mounted on a microscope slide. After denaturation of the DNA in the chromosome and partial disruption of chromatin structure, usually by alkali treatment, a drop of radiolabeled nucleic acid probe is added to the slide and incubated under precisely defined temperature and buffer conditions. Any probe that does not specifically anneal to the chromosomes is washed away and the slide is coated with a photographic emulsion. After an appropriate exposure period, the slide is developed. Silver grains in the exposed photographic emulsion (*autoradiography*) will mark the area of a chromosome where the radioactive probe has specifically bound, and differential staining techniques will identify the chromosome. Refinements in this technology (use of fluorescent probes, new hybridization conditions to increase the rate of hybridization, use of single-stranded RNA probes) now allow for the routine detection of one gene copy per chromatid.

C. *DNA typing: DNA polymorphisms* are now being used extensively for analysis of population structure, determining evolutionary relationships and establishing paternity, and (in humans) in forensic applications. Many of the polymorphisms detected involve mutations that arise in *tandem arrays of repeated DNA.* As a result of either slipped pairing during DNA replication or unequal crossing over, variations in the number of tandemly repeated sequences present at a particular chromosomal site will accumulate in populations over time. Restriction digests using enzymes whose recognition sites bracket the repeated regions will produce fragments of different lengths, depending on the number of repeats present in the array (2, 3, 4, etc.). These fragments can be detected after electrophoresis and Southern hybridization analysis, using probes complementary to the tandem repeat loci. Alternatively, if primers flanking the repeats can be identified, the variable regions can be amplified by using the polymerase chain reaction. These techniques allow one to analyze DNA from extremely small quantities of starting material, which is crucial when dealing with rare or precious samples (e.g., museum artifacts) or physical evidence obtained from a criminal investigation.

IMPORTANT TERMS

Autoradiography
Bacteriophage
cDNA
Cosmid
Dideoxyribonucleotides
DNA sequencing
Haplotype
In situ hybridization
Lod score
Minisatellite
Northern blot
Nucleic acid hybridization
Plaque
Plasmid
Polymerase chain reaction (PCR)
Restriction endonuclease
Restriction fragment length polymorphism (RFLP)
Reverse transcriptase
Somatic cell genetics
Southern blot
Variable number of tandem repeats (VNTR)
Vector
Yeast artificial chromosome (YAC)

PROBLEM SET 20

1. DNA is isolated from a bacteriophage that has a genome consisting of a single double-stranded linear DNA molecule. The phage DNA is then digested with three different restriction enzymes in various combinations. Single enzyme digests with the enzymes A, B, and C gave the following fragment sizes when the digested DNA was examined by gel electrophoresis (kb = kilobase pairs = 1,000 base pairs):

Enzyme A:	Enzyme B:	Enzyme C:
10 kb	3 kb	15 kb
40 kb	47 kb	35 kb

Restriction enzyme digests using two enzymes in combination produced the following fragments:

Enzymes A + B:	Enzymes A + C:	Enzymes B + C:
3 kb	10 kb	3 kb
7 kb	15 kb	15 kb
40 kb	25 kb	32 kb

Use the preceding data to order the restriction sites for these three enzymes relative to one another in the bacteriophage genome.

2. Suppose the preceding viral DNA was individually digested with two additional enzymes. After single digests, the following fragments were observed through gel electrophoresis:

Enzyme E:	Enzyme F:
a broad band of 25 kb	14 kb
	a broad band at 18 kb

Are these results in conflict with the digests described in 1? Explain.

3. The preceding viral DNA was subjected to double enzyme digests, where enzymes A, C, and E were used in combination with enzyme F. The following fragments were produced:

Enzymes A + F:	Enzymes C + F:	Enzymes E + F:
4 kb	3 kb	7 kb
10 kb	14 kb	11 kb
a broad band at 18 kb	15 kb	14 kb
	18 kb	18 kb

On the basis of the information provided in problems 1, 2, and 3, position the restriction sites for enzymes E and F relative to the sites for enzymes A, B, and C.

4. Plasmid DED3 is a recombinant DNA vector used to propagate foreign DNA in bacteria. When plasmid DNA is subjected to digestion with the following restriction enzymes, fragments with the following sizes are produced:

*Eco*RI:	*Bam*HI:	*Hin*dIII:
10.5 kb	3.2 kb	10.5 kb
	7.3 kb	

*Eco*RI + *Hin*dIII:	*Eco*RI + *Bam*HI:
1.9 kb	0.6 kb
8.6 kb	3.2 kb
	6.7 kb

*Bam*HI + *Hin*dIII: 1.3 kb
1.9 kb
7.3 kb

How many sites exist for each individual enzyme? Position the *Eco*RI, *Bam*HI, and *Hin*dIII cleavage sites on a restriction map.

5. Plasmid DED3 carries two genes that confer resistance to both ampicillin (amp^r gene) and tetracycline (tet^r gene) in host bacteria. The plasmid DNA has a single site for the enzyme *Eco*RI, located in the middle of the tet^r gene. Suppose we were interested in constructing a recombinant DNA library of frog genomic DNA, where the frog DNA had been cut with *Eco*RI then purified, mixed with *Eco*RI-digested DED3 plasmid DNA, treated with DNA ligase, and finally used to transform bacteria. How do we recover transformed bacterial colonies? How might we assess what percentage of transformed bacterial cells also contained cloned frog DNA?

6. One of the first vertebrate genes examined in detail was the gene encoding the protein insulin, which is synthesized as a protein precursor (preproinsulin) and enzymatically processed to the active enzyme. In 1980, Efstradiatis, Gilbert, and collaborators isolated the chicken preproinsulin gene sequences from a chicken genomic DNA library, to compare it to both cDNA clones and genomic DNA clones that had previously been cloned from rats. What is the rationale for screening both cDNA and genomic libraries for preproinsulin clones? What would be a logical tissue source (in both rats and chickens) for the isolation of the nucleic acids used to construct preproinsulin cDNA and genomic libraries?

7. When the chicken preproinsulin genomic clone is cut with the restriction enzymes *Bam*HI and *Taq*I, a 162-bp fragment that contains a piece of DNA encoding amino acids 5–57 of the preproinsulin sequence is produced. When this fragment is radiolabeled and used as a probe in a Southern blot hybridization experiment, it detects a single 0.8-kb band in *Bam*HI digested chicken genomic DNA. When the same region from the rat cDNA clone is used as a probe of *Bam*HI-digested rat genomic DNA, two bands are observed, 5.3 kb and 1.25 kb. What hypotheses might you make to account for this difference between the two organisms?

8.

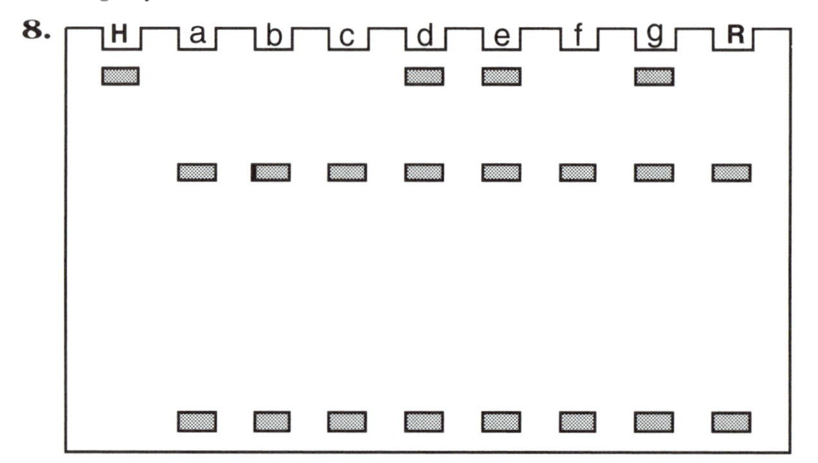

When human genomic DNA is digested with *Eco*RI and probed with an insulin gene sequence specific for the 5′ end of the insulin coding sequence, a 14.8-kb

*Eco*RI fragment is detected in Southern blot experiments. *Eco*RI-digested rat DNA, however, produces two fragments of 7.0 kb and 0.8 kb, which hybridize to this insulin probe. A somatic cell mapping experiment is performed to localize the human insulin gene to a specific chromosome. *Eco*RI-digested DNA isolated from seven human-rodent cell lines (a, b, c, d, e, f, and g) are run on a gel, in conjunction with *Eco*RI-digested human (H) and rat (R) genomic DNA. After Southern blotting of the gel and hybridization with an insulin probe, the following patterns are observed on the resulting autoradiograph of the blot on p. 176.
The human chromosomes contained in the seven hybrid cell lines are as follows:

Cell line a: 1, 4, 6, 9, 10, 14, 22
Cell line b: 1, 6, 9, 12, 13, 18, 21
Cell line c: 5, 8, 17, 19, 20, 22
Cell line d: 2, 3, 4, 6, 11, 15
Cell line e: 1, 6, 7, 11, 14, 16
Cell line f: 3, 5, 7, 10, 13, 20
Cell line g: 3, 5, 6, 11, 13, 14, 22

On which chromosome can the human insulin gene be found? Can you suggest a method for localizing the human gene other than somatic cell mapping?

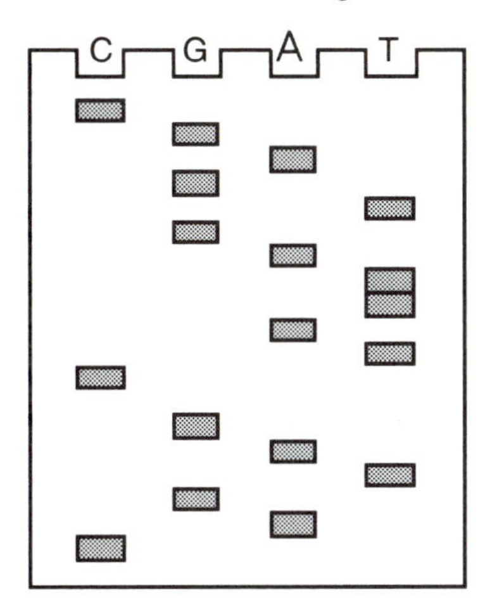

9. For the preceding portion of a DNA sequencing gel produced by using the Sanger dideoxynucleotide technique, answer the following:

(a) True or false. The preceding gel is capable of separating single-stranded DNA that differs in size by a single nucleotide.
(b) What is the sequence of the strand being synthesized?

10. In the enzymatic reaction used to produce the DNA fragments shown in the "A" lane of the gel, which of the following were present in the reaction mixture? **(a)** ddATP; **(b)** ddTTP; **(c)** dCTP; **(d)** dATP; **(e)** a primer molecule.

11. In human populations, abnormal hemoglobin (Hb) variants are usually named for the geographic area in which they were discovered. Defects in the molecule may be found in either the alpha- or beta-globin proteins, which are encoded by separate genes. An abnormal hemoglobin called Hb Constant Spring causes a severe form of anemia. Although the beta-globin protein is normal, the alpha chains of Hb Constant Spring are 31 amino acids longer than the normal alpha chain. A DNA sequencing gel of the wild-type alpha-globin gene, showing the DNA encoding the last 5 amino acids on the coding strand plus some 3′ flanking DNA sequences, is shown on the left side of the figure. The sequence of the corresponding region of the Hb Constant Spring gene is shown for comparison on the right.

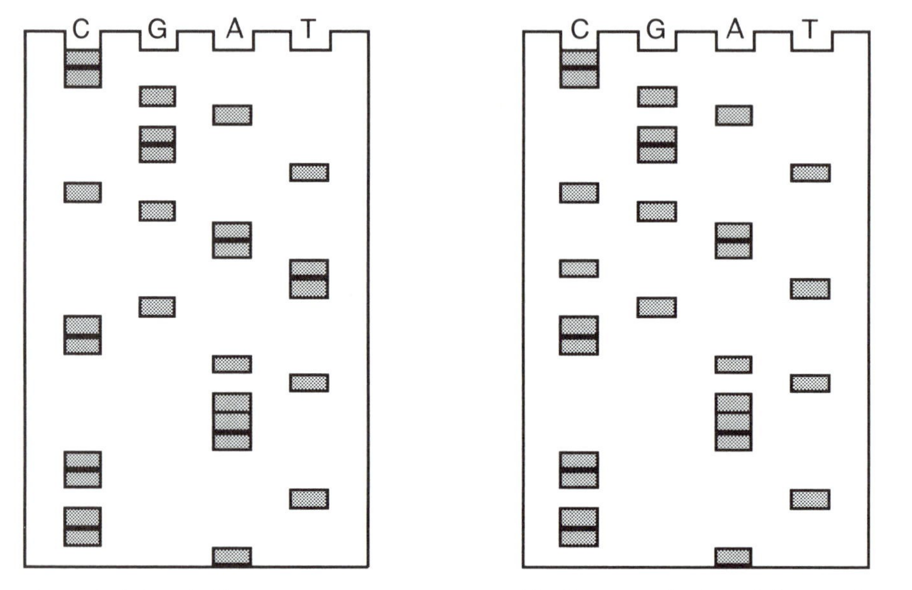

Normal alpha globin gene:

Hb Constant Spring alpha globin gene:

What is the nature of the mutation? Can you explain why the mutant alpha-globin protein is 31 amino acids longer?

12. The alpha-globin proteins from the abnormal hemoglobin variants Hb Seal Rock and Hb Icaria are identical in size to the Hb Constant Spring alpha-globin. Protein electrophoresis, however, indicates that the Icaria protein is more positively charged than the Hb Constant Spring protein, and the Seal Rock protein is more negatively charged than the Hb Constant Spring protein. What is the nature of the mutation in the Hb Seal Rock and Hb Icaria alpha-globin genes? Table R.5 in the Appendix gives information on amino acid characteristics.

13. A restriction fragment length polymorphism for the enzyme *Bam*HI is identified in human populations by using a specific recombinant DNA probe, termed R2. For this region of the genome, the two external *Bam*HI sites are always observed, whereas two other internal sites, indicated by asterisks on the figure, may or may not be present, depending on the individual. The region of genomic DNA that hybridizes with the R2 probe is indicated by the shaded box.

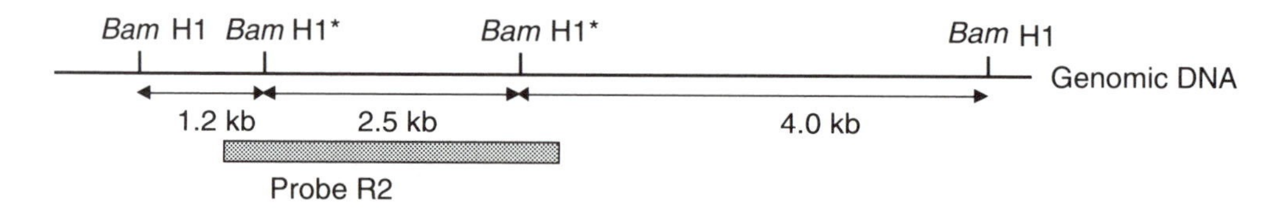

(a) How many different RFLP variant types (haplotypes) would be expected in human populations? Give the fragment pattern for each haplotype that would be detected by Southern blot analysis.

(b) How many different haplotype combinations would one expect to see in human populations, given that each haplotype can be considered an "allelic" variant?

14. In the following pedigree, an autosomal dominant trait showing complete penetrance is depicted. Mapping experiments have established that the disease locus segregating in this pedigree is linked (within approximately 1 map unit) to the polymorphic locus detected by the R2 probe discussed previously. A Southern blot hybridization experiment using the R2 probe is performed on *Bam*HI-digested genomic DNA from individual family members. The results of the autoradiograph and the source of the genomic DNA in each lane are shown directly below the pedigree (e.g., lane 1 on the autoradiograph represents III-1, lane 5 represents II-1, lane 8 represents I-1).

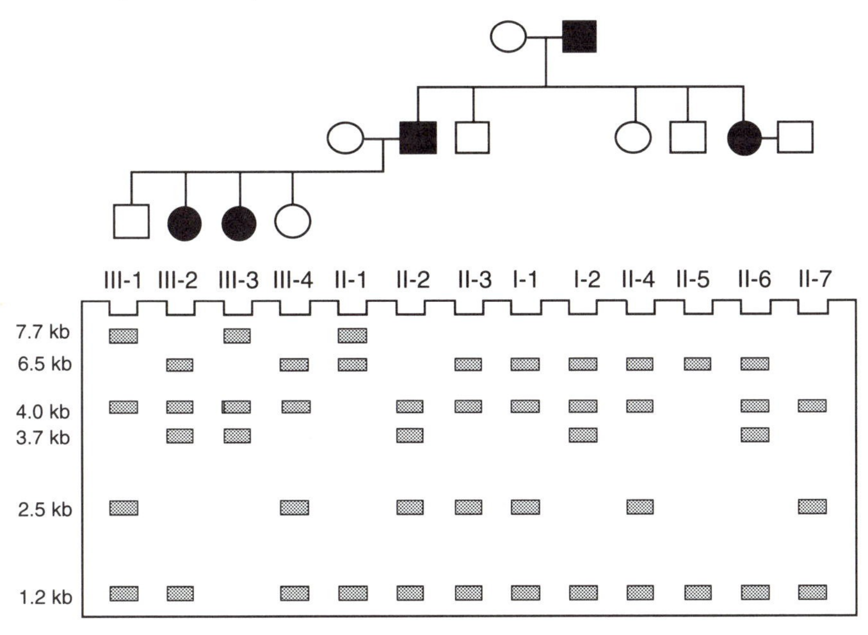

(a) What combinations of haplotypes are present in each of the individuals in the pedigree?

(b) In this family, which haplotype is segregating with the deleterious gene? Would we necessarily expect the same haplotype to be segregating with the mutant allele in an unrelated family with this disorder? Explain.

15. A tigress is acquired by a local zoo and is a candidate for their captive breeding program. Although the identity of the tigress's mother is believed known, there is ambiguity as to her paternity, and some potential male mates could be closely related. DNA typing is performed on the tigress, her mother, and three potential mates to prevent excessive inbreeding. The results of a screen using a minisatellite probe are as follows (M = mother; T = tigress; M1–3 = potential mates):
Which of the three potential males would be most suitable? Justify your reasoning.

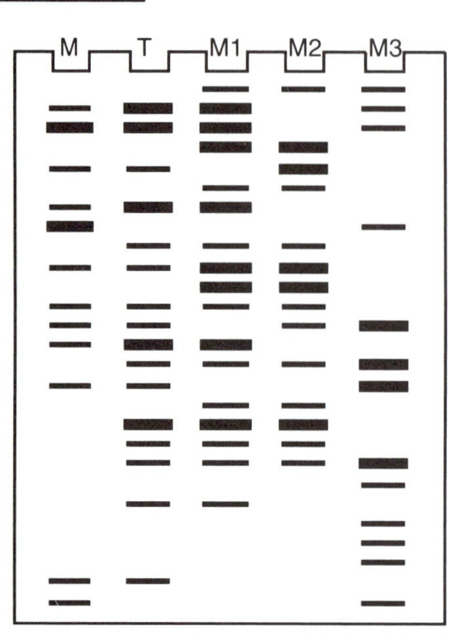

ANSWERS TO PROBLEM SET 20

1. The first consideration is to note that the three single digests all produce fragments that total 50 kb, representing the size of the entire phage genome. All of the digests generated two fragments. Since a linear molecule is being cut, this indicates that there is a single site for each of the three enzymes. The restriction site for enzyme A, therefore, must lie 10 kb from one end; similarly the restriction sites for enzymes B and C must also lie 3 kb and 15 kb from an end. Since the molecule has two ends, however, the next goal is to order these sites relative to one another.
The double digest using enzymes A + B helps resolve the position of these two sites relative to one another. The 10-kb fragment produced by enzyme A is broken into two smaller fragments by enzyme B, and their sizes total the size of the intact 10-kb fragment. This indicates that both the B and A sites lie near the same end of the molecule. Digests with enzyme A + C, however, leave the 10-kb fragment intact, but cut the 40-kb A fragment instead. This indicates the C site lies 15 kb from the other end of the molecule. The B + C digest supports this placement. A map can therefore be drawn:

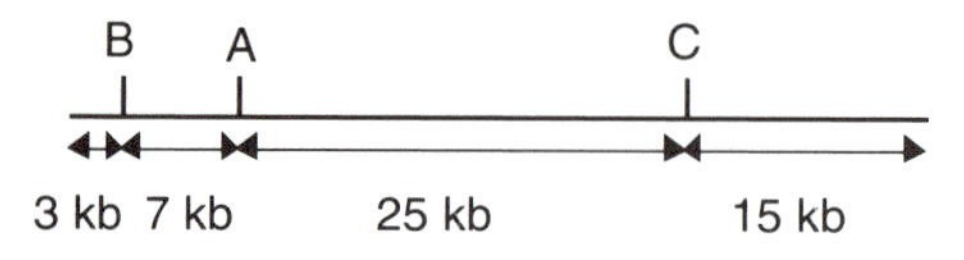

2. The results in question 1 indicate that the total length of the phage genome is 50 kb. Since the fragments resolved in these digests do not total 50 kb, a logical assumption is that these digests produced certain fragments that overlapped in size. The broad bands at 25 kb and 18 kb could represent doublets: two different fragments that have approximately the same size. If this is the case, the total fragment lengths for each digest still total 50 kb.

3. Since enzyme E produces two fragments of identical size, the single site for E must lie in the center of the 50 kb genome. Since enzyme F produces three fragments, there must be two sites for this enzyme, which need to be positioned relative to the other known sites. The single site for enzyme A cuts the 14-kb F fragment into two fragments 4 kb and 10 kb in size, so one of the F sites must lie 4 kb internal to the A enzyme site. Similarly, the single C site cuts one of the 18-kb F fragments into a 3-kb fragment and a 15-kb fragment, suggesting that the other F site lies 3 kb internal to the C site. This would produce the following map:

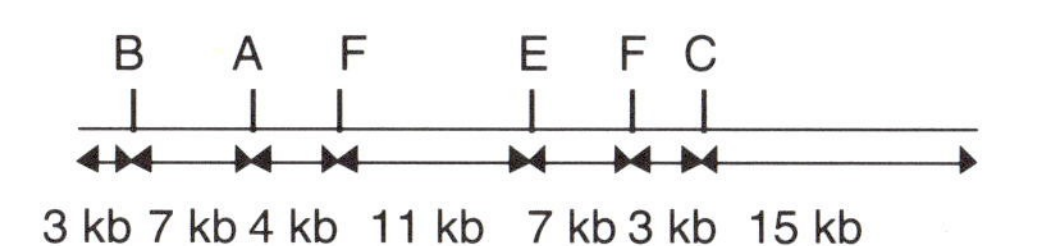

The A + F digest is consistent with these assignments.

4. When a circular plasmid is cut once, a linear molecule is produced; when it is cut twice, two pieces are produced. There are, therefore, single restriction sites for the enzymes *Eco*RI (R) and *Hin*dIII (H), and two sites for the enzyme *Bam*HI (B). The linear molecule is 10.5 kb in length. Furthermore, the *Eco*RI and *Hin*dIII sites must lie 1.9 kb apart on the circle, to yield the two fragments observed in the *Eco*RI + *Hin*d III double digest:

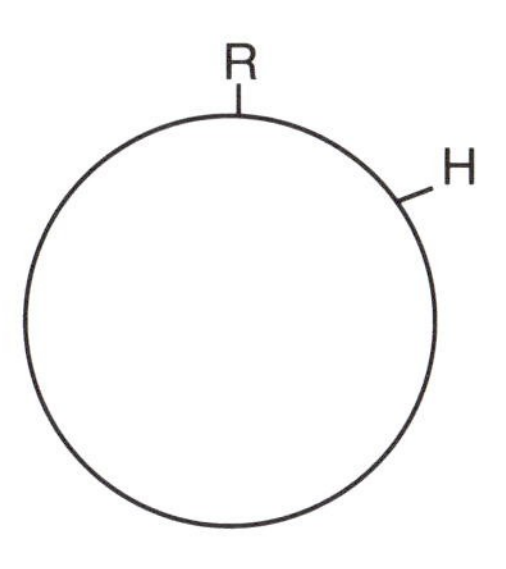

A comparison between the *Bam*HI single digest and *Eco*RI + *Bam*HI double digest indicates that the *Bam*HI fragment that is cleaved by *Eco*RI is the 7.3-kb fragment. This indicates that one *Bam*HI site is positioned 0.6 kb from the *Eco*RI site. If this site were located *proximal* to the *Hin*dIII site, we would expect to recover a 1.3-kb fragment on a *Bam*HI + *Hin*dIII double digest; if it were *distal* to the *Hin*dIII site, we would expect to recover a 2.1-kb fragment. The data indicate that a 1.3-kb fragment is produced, so one *Bam*HI site must lie between the *Eco*RI and *Hin*dIII sites. Using the same reasoning, the second *Bam*HI site must lie 6.7 kb to the opposite side of the *Eco*RI site.

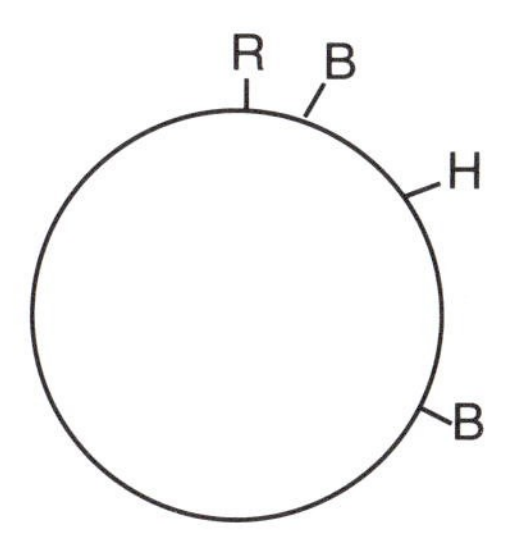

If the map is linearized at the *Eco*RI site (to yield two "half sites," R′) and portrayed as a line, the distances between the various sites are as follows:

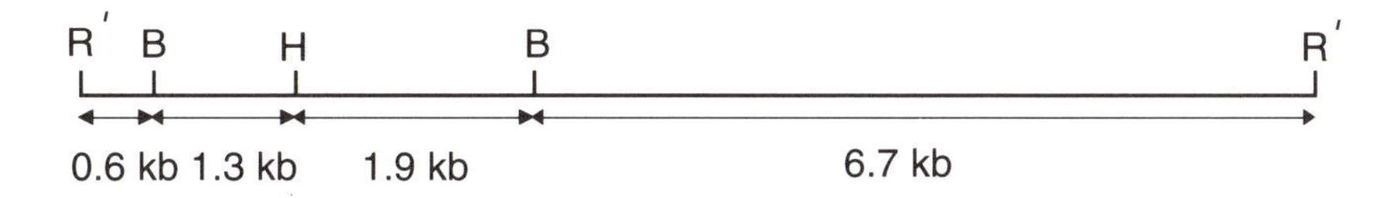

5. The ligation reaction is mixed with competent bacteria, and the transformation mixture is plated on agar plates containing ampicillin. All cells that had been transformed with the plasmid should now be capable of surviving on ampicillin plates. Not all of these plasmids, however, would be expected to contain an insert of frog DNA, particularly if the plasmid were free to self-ligate. This is the reason that plasmid DNA is often treated with alkaline phosphatase prior to ligation. To identify what fraction of the transformed cells contain a plasmid carrying an insert, an aliquot of the library could be replica-plated on tetracycline plates. If a piece of frog DNA were ligated into the *Eco*RI site, this would produce an insertional mutation in the tet^r gene. Colonies that grow on media containing ampicillin but not tetracycline would likely be carrying recombinant DNA inserts.

6. cDNA libraries are generated from mRNA and full-length clones will contain the entire coding region in a contiguous piece along with flanking 5′ and 3′ noncoding sequences. The location of coding regions in the genome, however, may not be contiguous but separated by introns. Accordingly, a transcription unit in genomic DNA could span tens, or even hundreds of thousands of base pairs. If one is interested in expressing the gene of interest in a microorganism, it would make sense to use a cDNA sequence. There is important information, however, that can be obtained from comparisons of genomic and cDNA sequences. For example, cDNA clones do not contain promoter sequences, since mRNA transcription begins 3′ (relative to the coding strand) of the RNA polymerase binding site. Comparisons of cDNA and genomic clones can therefore provide information on promoter or other sequences important in gene regulation, as well as information on intron location.

Since most tissues in vertebrates contain a complete genome, virtually any tissue that contains nuclei could be used to isolate DNA for genomic libraries. One experimental concern might simply be to use a tissue with low endogenous nuclease activity where the cells could be easily disrupted. To build a cDNA library, however, it would be necessary to use a tissue source where the gene in question was actually being transcribed. For insulin, it would be necessary to use pancreatic cells, since insulin is produced only in the specialized beta cells in this tissue. In the actual cloning of insulin, researchers used an insulinoma, cancer cells derived from the pancreas that could be grown in vitro and that produce large amounts of insulin.

7. One hypothesis is that a *Bam*HI site exists in rat cDNA that is not present in the same region of the chicken gene. Given the degree of evolutionary divergence between the two vertebrates, a difference in restriction pattern would not be surprising. This sequence divergence could be the result of changes in the coding sequence of the protein, or an intron in the rat gene that is absent from the chicken gene.

A second alternative is that there may be two genes for insulin in rats, and only a single gene in chickens. One way to test this alternative would be to perform a Southern blot analysis with a number of different enzymes, to see whether any enzyme produces only a single band. By using probes derived from a different region of the coding sequence one could also test to see whether the same multiple bands contain both 5′ and 3′ coding sequences. In fact, the second alternative holds for rats; a recent gene duplication has led to two insulin sequences in this rodent.

8. In this panel of somatic cell hybrids, the fragment bearing the human gene is present in cell lines d, e, and g. Relative to the chromosomes contained in these seven lines, concordance and discordance for the human chromosome complement indicate the following:

Hybrid cell line	Insulin probe	Human chromosomes in hybrid cell lines																					
		1	2	3	4	5	6	7	8	9	10	11	12	13	14	15	16	17	18	19	20	21	22
a	−	+	−	−	+	−	+	−	−	+	+	−	−	−	+	−	−	−	−	−	−	−	+
b	−	+	−	−	−	−	+	−	−	+	−	−	+	+	−	−	−	−	+	−	−	+	−
c	−	−	−	−	−	+	−	−	+	−	−	−	−	−	−	−	−	+	−	+	+	−	+
d	+	−	+	+	+	−	+	−	−	−	−	+	−	−	−	+	−	−	−	−	−	−	−
e	+	+	−	−	−	−	+	+	−	−	−	+	−	−	+	−	+	−	−	−	−	−	−
f	−	−	−	+	−	+	−	+	−	−	+	−	−	+	−	−	−	−	−	−	+	−	−
g	+	−	−	+	−	+	−	+	−	−	−	+	−	+	+	−	−	−	−	−	−	−	+

The pattern of concordance and discordance matches only chromosome 11; whenever it is present, the human gene is observed in Southern blots; whenever it is absent it is not observed.

For somatic cell mapping, tissue culture cells that contained chromosome rearrangements or deletions could also be used to refine the map to a particular region of the chromosome. For example, once it was determined that the gene was present on chromosome 11, hybrid cells carrying deletions of chromosome 11 could be used to refine the map. Of course, once a nucleic acid probe is available, it would also be possible to map the gene directly to a chromosome using in situ hybridization.

9. **(a) true.** The DNA sequencing technique requires that nucleotides that differ in chain length by a single base be discriminated from one another. From the bottom of the gel upward, each DNA fragment is larger by one nucleotide.

(b) The gel is read from the bottom upward, since the labeled fragments are the result of an enzymatic reaction using DNA polymerase, which synthesizes DNA in the 5′ to 3′ direction. The sequence reads

5′-CAGTAGCTATTAGTGAGC-3′

10. **(a), (c), (d), (e).** In the DNA polymerase reaction used to produce the "A" lane on the gel, all four deoxyribonucleotide triphosphates must be present to synthesize a new molecule. Similarly, DNA polymerase requires a primer to initiate synthesis; it cannot start a chain de novo. To terminate specifically at adenine residues, the nucleotide analogue ddATP must also be present in the reaction. The only component not required would be ddTTP; this would be present in the "T" lane reaction, where it would serve to terminate synthesis whenever it was incorporated into growing chain.

11. The DNA sequence for the wild-type protein and the protein it encodes is:

. . . ACC TCC AAA TAC CGT TAA GCT GGA GCC . . .
. . . Thr Ser Lys Tyr Arg TER

TER denotes a chain termination (stop) triplet. The DNA sequence for the mutant protein and the protein it encodes is

ACC TCC AAA TAC CGT CAA GCT GGA GCC . . .
Thr Ser Lys Tyr Arg Gln Ala Glu Ala . . .

In Hb Constant Spring, a stop codon has mutated into the sense codon for the amino acid glutamine. The former 3′ noncoding region of the mRNA is now translated. Although this region of the gel is not depicted, we predict that another stop codon is present in the mRNA 93 nucleotides downstream of the mutated stop codon, producing a protein 31 amino acids larger than wild-type alpha globin.

12. Since the Seal Rock and Icaria proteins are also 31 amino acids longer than those of the wild-type, we predict that they also contain a stop codon that has mutated to a sense codon. For Seal Rock to mutate into a more negatively charged molecule than Constant Spring, it would have to be acidic. Similarly, for Icaria to be more positively charged, it would have to be more basic. Consulting the genetic code, we find that TAA → GAA would produce a glutamic acid residue in place of the stop codon (Seal Rock) and TAA → AAA would produce a lysine (basic) residue in place of the stop codon (Icaria).

13. **(a)** For the two polymorphic sites, either the site exists in the DNA or it doesn't. This will produce four distinct haplotypes, with the following fragments hybridizing to R2:

+/+ (Both present)	+/− (Left present, right absent)	−/+ (Left absent, right present)	−/− (Both absent)
1.2 kb	1.2 kb	3.7 kb	7.7 kb
2.5 kb	6.5 kb	4.0 kb	
4.0 kb			

The characteristic fragments observed on Southern blot analysis allow us to make haplotype assignments. For example, let haplotype A represent the +/+ pattern, haplotype B represent the +/− pattern, haplotype C represent the −/+ pattern, and haplotype D represent the −/− pattern.

(b) We can consider this problem as essentially an allelic series with four variants. Since every individual will have two chromosomes, there are four types of homozygotes and six types of heterozygotes for these four haplotypes, or ten combinations: $[n(n-1)]/2$, where n = number of haplotypes.

14. **(a)** On the basis of the haplotype analysis given in problem 13, certain bands can be seen to be diagnostic of a given pattern. Haplotype A, for example, is characterized by the 2.5-kb fragment, haplotype B by the 6.5-kb fragment, haplotype C by the 3.7-kb fragment, and haplotype D by the 7.7-kb fragment. We can now make the following assignments:

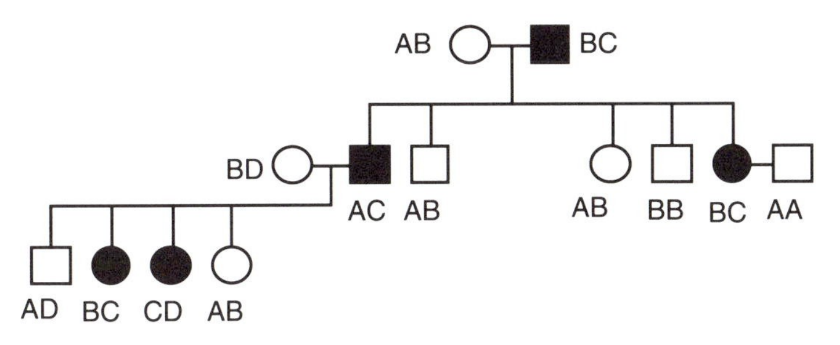

(b) In the preceding pedigree, the deleterious locus is segregating with the C haplotype. Since these loci are located within one map unit of each other, we would expect them to be separated by recombination in about 1 percent of the meiotic products. This would be useful for tracing the deleterious allele within this family, but in the population at large any of the haplotypes could be associated with the mutant disease locus. Because of recombination within the population, we would need to establish which haplotype was associated with this disease locus for each individual family.

15. An examination of the DNA "fingerprint" given in this problem indicates that every band in the tigress's DNA can be matched either to her mother (who is a known first degree relative) or to male 1. This would eliminate male 1 as a good candidate for mating. Similarly, a comparison between male 1 and male 2 suggests a high degree of relatedness between them, since they share 12 out of 19 bands. A comparison of the pattern in male 3 shows the highest degree of polymorphism relative to the other two males, and he also does not appear to share many bands with the tigress's mother. Of the three candidates available, male 3 would be the best choice for the captive breeding program designed to minimize inbreeding.

BASIC POPULATION GENETICS

STUDY HINTS

The rules of Mendelian inheritance are largely based upon the probabilities of inheriting a given allele or combination of alleles from a specified genotype. With the exception of sex-linked traits, each individual carries at most two alleles of each gene, and the number of possible combinations is limited. Expectations can be calculated easily, since the probability of getting a certain allele from a known heterozygous diploid is $p = q = .5$.

In a population the focus changes from the *genome* to the *gene pool* and to estimates of allele frequency, yet the rules of probability still apply. Allele frequencies p and q are determined by pooling all of the genes carried by all of the individuals in the population. Working problems that require estimating allele and genotype frequencies from the Hardy–Weinberg relationship are fairly straightforward if you keep in mind that you are dealing with probabilities and if you are consistent with your use of symbols.

For example, if we let the frequency of the *A* allele be p and the frequency of the *a* allele be q in a population in Hardy–Weinberg equilibrium, $p + q = 1$, as long as *A* and *a* are the only alleles at that locus segregating in the population. In other words, the probability of picking an *A* allele at random plus the probability of picking an *a* allele at random account for all possible events. The genotypes that can be produced and their probabilities are shown in the following table. Remember the rule that the probability of two independent events occurring together is the product of their individual probabilities.

	Sperm	
	(p)	(q)
	A	*a*
Eggs		
(p) *A*	*AA* (p^2)	*Aa* (pq)
(q) *a*	*Aa* (pq)	*aa* (q^2)

(p) (q) = allele frequencies
A *a* = alleles in the population
Table body = genotypes and genotype frequencies

The sum of the probabilities of each genotype also accounts for all possible genotypic events and therefore equals 1.

$$\underset{AA}{p^2} + \underset{Aa}{2pq} + \underset{aa}{q^2} = 1$$

Thus, if you can determine the frequency of homozygous recessive (*aa*) individuals in a population, for example, you have q^2. Its square root (q) can then be used to determine the frequency of the *A* allele, since $p = 1 - q$. Given p and q, you can then estimate the frequencies of each genotype.

The Hardy–Weinberg law describes the fact that allele frequencies and genotype frequencies will not change if certain assumptions are made about the population. These assumptions include an infinite population size, random mating, and the absence of mutation, migration, and selection. In the following outline we have summarized some of the effects on allele frequencies and genotypes that one might expect if any of these assumptions were to break down.

Genetic consequences when Hardy–Weinberg assumptions break down:

I. *Small population size:* If the population size is small, random sampling errors (chance) can cause unpredictable changes in allele frequencies that are independent of any other factors acting on the population. Over time, this sampling error (or genetic drift) can lead to fixation (homozygosity) of one allele or the other.

II. *Deviations from random mating:* Deviations from random mating mean that certain matings occur at frequencies different from those predicted purely on the basis of genotype frequencies in a population. There are several types of deviation from random mating, and these can have different effects upon the population.

   A. Assortative mating: Mating between phenotypically or genotypically similar individuals increases homozygosity for the loci involved in mate choice (and those loci in linkage disequilibrium with them) without altering allele frequencies.

   B. Disassortative mating: Mating between dissimilar individuals increases heterozygosity without altering allele frequencies.

   C. Rare-male mating advantage: Density-dependent mating generally tends to increase the frequency of the rare allele and thus increase heterozygosity.

   D. Inbreeding: Mating between close relatives increases homozygosity for the whole genome without affecting allele frequencies.

III. *Mutation:* Mutation changes allele frequencies at a rate that depends upon many factors, such as repair ability, mutator factors, chemical and physical mutagens, and the frequency of the alleles themselves. The role of the allele frequency can be illustrated by considering the mutational equilibrium:

$$a \underset{v}{\overset{u}{\rightleftharpoons}} a'$$

where a and a' are two allelic forms, and u and v are the forward- and back-mutation rates. Changes from one allele to the other in a population will be a function of the frequency of each type of allele and the magnitude of u and v. In general, the rate of forward mutation is significantly greater than the rate of back-mutation, since there are many ways to create an error in coding, but fewer ways to correct it. The rate of mutation from a to a' is u times p, where p is the frequency of the a allele. The reverse mutation rate is v times q. Thus the change in gene frequency (Δq) is $up - vq$. If mutation rates remain constant, a stable equilibrium will be reached when

$$up = vq$$
$$u(1 - q) = vq$$

and the equilibrium frequency of q (symbolized $\hat{q}$) is

$$\hat{q} = \frac{u}{u + v}$$

IV. *Migration:* Migration, or gene flow, alters allele frequencies in a way that depends upon the difference between the allele frequency in the migrant (q_m) and the recipient (q_r) populations and upon the proportion of migrants (m). The new allele frequency in the recipient population is

$$q'_r = (1 - m)q_r + mq_m$$

The change in allele frequency, Δq, is therefore

$$\Delta q = \Delta q'_r - q_r = (1 - m)q_r + mq_m - q_r$$
$$= m(q_m - q_r)$$

Allele frequency is stable ($\Delta q = 0$) when $q_m = q_r$.

V. *Selection:* Selection changes allele frequencies and thus genotype frequencies, as a function of the relative fitness of a given genotype. Types of selection are discussed in the Study Hints section in Chapter 22.

IMPORTANT TERMS

Assortative mating
Balanced polymorphism
Deme
Disassortative mating
Founder principle
Gene frequency
Gene pool
Genetic drift
Hardy–Weinberg equilibrium
Heterosis
Inbreeding
Inbreeding depression
Linkage disequilibrium
Mutation equilibrium
Panmixia
Population
Selection
Unstable equilibrium

PROBLEM SET 21

1. Suppose that an autosomal trait, hairy earlobes, is found in 3.24 percent of a certain population. Pedigree analyses of this population confirm that all matings between individuals with hairy earlobes produce only hairy-lobed offspring. If the population is in Hardy–Weinberg equilibrium, what percentage of the population will be heterozygous carriers of this trait?

2. If albinism occurs in only 1 individual in each 10,000 in the population, how many matings at random in this population will be between heterozygous carriers? (Note that "mating at random" refers only to the fact that mates are not chosen on the basis of the trait in question.)

3. A class of 75 students was tested for their ability to taste the chemical phenylthiocarbamide (PTC). A total of 48 were found to be tasters, expressing the dominant trait. What is the frequency of the recessive (t) allele in this sample? How frequently would one expect a homozygous nontaster (tt) to date a homozygous taster (TT), if dating were at random with respect to this phenotype?

4. Phenylketonuria is a metabolic disorder that can cause irreversible brain damage in early life if not treated by controlling the dietary intake of phenylalanine. In a certain hypothetical population, phenylketonuria occurs in 1 in each 40,000 individuals. First calculate the frequency of homozygotes and heterozygotes. Then determine what proportion of all possible matings will have the potential of giving rise to at least one affected individual in the family. Note that the question asks about matings, not the frequency of the resulting affected offspring.

5. In a certain strain of *Drosophila melanogaster,* the dominant mutation "Lobe eye" (L) has 90 percent penetrance in both homozygotes (*LL*) and heterozygotes (*Ll*). In a population containing both Lobe eye and wild-type flies, 5,760 of 10,000 are found to express the Lobe phenotype. What proportion of the population is expected to be heterozygous for this dominant condition?

6. Suppose that a population of diploid cross-fertilizing (that is, outbreeding) individuals in Hardy–Weinberg equilibrium is suddenly forced into an exclusively self-fertilizing mode of reproduction. Considering a single locus with two alleles in this population, what will happen to the distribution of the **(a)** genotype, **(b)** phenotype, **(c)** gene frequencies?

7. If the gene frequency of the *M* blood allele is .6 and the gene frequency of the *N* allele is .4, what is the probability that a man accused of being the father of a child with blood type N will be cleared by MN blood-type testing? Assume that the man is picked at random from a population and is not the actual father of the child.

8. If the frequency of female *Drosophila* exhibiting the recessive sex-linked trait crossveinless (*cv*) in an experimental population is .36, what is the expected frequency of male *Drosophila* exhibiting this trait?

9. Dyslexia (congenital word blindness) is an autosomal dominant trait in humans. If the frequency of this trait is about 19 affected people in each 100 in a certain population, how frequently would one expect matings to occur between a heterozygote and a homozygous affected individual?

10. In a certain population of self-fertilizing *Hydra,* 84 percent are heterozygous for a tentacle-number mutation. After three generations of total inbreeding, what will be the percentage of heterozygosity in this population?

11. A hypothetical population of grasshoppers is in Hardy–Weinberg equilibrium for two unlinked mutations. One is a recessive mutation causing shortened legs. The common long-legged phenotype is found in 99 percent of the individuals. The second mutation affects electrophoretic mobility of the lactate dehydrogenase (LDH) enzyme. The slow-mobility allele is homozygous in only 16 percent of the grasshoppers. What proportion of the population is expected both to have the short-legged phenotype and to be heterozygous for mobility alleles of *Ldh*?

12. Please consider the following pedigree for a relatively common inherited condition in which the frequency of the dominant allele is .23 in the population at large. First determine the mode of inheritance. Then calculate the probability that the child IV-2 inherits the condition. If IV-2 is unaffected and marries at random with respect to this condition, what is the probability that the couple's first child will show the trait?

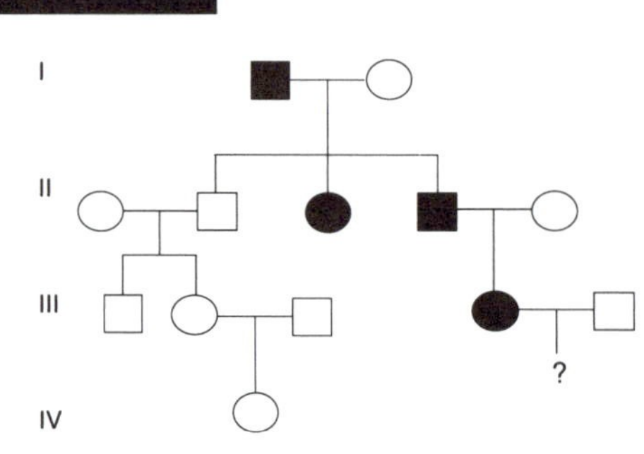

13. Assume that there are two populations of house mice living in adjacent farms. They differ in the allelic frequency of the recessive albino mutation. In population 1 the frequency is .3 and in population 2 it is .6. Assume further that the harvest is better in farm 1, so that mice from population 2 migrate in. The proportion of immigrants is .2. What is the change in allelic frequency for the albino gene in the new population of farm 1?

ANSWERS TO PROBLEM SET 21

1. Since all matings between affected individuals produce only affected offspring, the trait must be a recessive. Since the population is in Hardy–Weinberg equilibrium, one simply uses the observed frequency of affected individuals (q^2) to calculate q, p, and $2pq$. Since $q^2 = .0324$, $q = .18$. Since $p = 1 - q$, p must be .82. Then, $2pq = 2(.18)(.82) = .2952$.

2. Since $q^2 = .0001$, $q = .01$. Thus $p = .99$, and the frequency of heterozygotes ($2pq$) = .0198. A mating between two heterozygotes is composed of two independent events (the likelihood that the father is a heterozygote and the likelihood that the mother is a heterozygote). Such a mating is then a function of the product of the two independent probabilities (.0198 • .0198). The expected frequency of heterozygote mating is therefore .00039, or .039 percent.

3. First one needs to calculate the frequency of homozygous recessive nontasters. There are 75 − 48 = 27 nontasters, who make up 27/75, or 36 percent, of the population. If nontasting (*tt*) is q^2, then $q^2 = .36$, so $q = .6$ and $p = .4$. The frequency of homozygous tasters (*TT*) = $p^2 = .16$. A dating pair could be *tt* ♀ × *TT* ♂ or *TT* ♀ × *tt* ♂. Thus the probability of such an occurrence is 2(.36)(.16) = .1152.

4.

	Fathers		
Mothers[a]	***AA***	***Aa***	***aa***
AA	X	X	X
Aa	X	$(2pq)(2pq)$	$q^2(2pq)$
aa	X	$q^2(2pq)$	$q^2(q^2)$

[a] Where *A* = normal, and *a* is phenylketonuria.

Since $q^2 = 1/40{,}000$, $q = 1/200$, or .005. Thus $p = 1 - q = .995$. The frequency of homozygous normal individuals is $p^2 = (.995)^2 = .990025$. The frequency of heterozygous individuals is $2pq$. This is $2(.995)(.005) = .00995$. We already have $q^2 = .000025$. Only the four matings that involve individuals carrying at least one recessive (a) allele could produce a homozygous aa offspring. The total proportion of potentially affected families is simply the sum of these four probabilities.

$$\begin{aligned} (2pq)(2pq) &= .000099003 \\ q^2(2pq) \cdot 2 &= .000000498 \\ q^2 \cdot q^2 &= \underline{.000000001} \\ & \quad .000099502 \end{aligned}$$

5. Let us define p and q so that p is the frequency of the dominant Lobe allele and q is the frequency of the wild-type allele. The individuals carrying the Lobe mutation will therefore be represented by $p^2 + 2pq$, since they will include both homozygous dominant and heterozygous individuals. One must determine the proportion of genetically homozygous wild-type individuals (q^2) in order to calculate p and q. To calculate $p^2 + 2pq$ from the data in the problem, one must first take into account that some flies in this class will not have expressed this trait, since penetrance in our strain is only 90 percent. That is, $p^2 + 2pq$ times the penetrance is the number of carriers that will show the trait. Thus, to determine the actual size of this class,

$$\text{Mutant carriers } (p^2 + 2pq) = \text{number showing the condition / penetrance}$$
$$= \frac{(5{,}760/10{,}000)}{.9} = .64$$

Thus $q^2 = 1 - (p^2 + 2pq) = .36$, and $q = .6$. Since $p + q = 1$, $p = .4$.

The proportion expected to be heterozygous for the condition is $2pq = 2(.6)(.4) = .48$. Remember, however, that this does not mean that all individuals within this 48 percent of the population will express the trait. The value $2pq$ only gives the genotypic frequency. Phenotypic frequency must again take penetrance into account. Since this problem asked for genotypic frequency, the correct answer is 48 percent. If we had asked for phenotypic frequency, however, the answer would have been (.48)(.9).

6. **(a)** Inbreeding associated with self-fertilization increases the frequency of homozygotes and decreases heterozygosity. Half of the progeny of each heterozygote in each generation of such extreme inbreeding are homozygous, and there is no way of regenerating new heterozygotes without some form of outcrossing (in the absence of some other factor such as mutation).

(b) The two homozygous phenotypes would eventually be the only phenotypes.

(c) Inbreeding changes only phenotype and genotype frequencies. It does not alter allele frequencies.

7. Let p = the frequency of the M allele = .6, and let q = the frequency of the N allele = .4. The man would be unambiguously cleared only if his MN blood type is M (that is, if he is an MM homozygote), because only then could he not contribute an N allele to the genotype of the child. Thus the frequency of MM genotypes, $p^2 = (.6)^2 = .36$, gives the probability that this test alone will exonerate him.

8. If the frequency of *Drosophila* females exhibiting the sex-linked trait is .36, the allele frequency (q) must be .6 (i.e., $\sqrt{q^2}$), since females have two X chromosomes.

The males, on the other hand, have only one X chromosome. The probability that they carry the recessive mutation is therefore the same as the frequency of mutant X chromosomes in the population (.6).

9. The frequency of those showing the dominant trait (.19) includes both homozygous dominant individuals and those heterozygous for the dominant. In order to determine allele frequencies, therefore, we must determine the proportion of the population that is homozygous for the recessive allele.

Since $p^2 + 2pq = .19$, $\quad q^2 = 1 - (p^2 + 2pq) = .81$
Thus, $q = \sqrt{.81} = .9$, and $p = 1 - .9 = .1$

The frequency of heterozygotes is $2pq = 2(.9)(.1) = .18$. The frequency of homozygous affected individuals is

$p^2 = (.1)^2 = .01$

(Note that the sum of these two values accounts for the observed frequency of dominant-phenotype individuals in the population.) Since a mating between a heterozygote and a homozygous affected individual could occur in either of two ways (homozygous man with heterozygous woman, and vice versa), the overall probability of the mating is

$2(.18)(.01) = .0036$

10. The proportion of heterozygotes is halved each generation, since a heterozygote that self-fertilizes produces 1/4 *AA* and 1/4 *aa* individuals, leaving only half as many heterozygotes as the population began with. In the *Hydra* population, there will be 42 percent heterozygotes after one generation of self-fertilization, 21 percent after two generations, and 10.5 percent after three.

11. The frequency of the recessive allele for short legs is the square root of the frequency of the homozygotes (1 in each 100 in the population). Thus $q^2 = .01$, and $q = .1$, so $p = 1 - q = .9$. For the *Ldh* slow-migrating allele, since 16 percent are homozygous the frequency of $Ldh^S = .4$ and the frequency of the fast allele, Ldh^F, is .6.

The probability of obtaining a short-legged individual that is heterozygous for the *Ldh* locus is the product of the separate probabilities for each event. The probability for short legs is .01 and for being $Ldh^S\ Ldh^H = 2(.4)(.6) = .48$. The overall proportion of the population fulfilling both genetic requirements is $(.01)(.48) = .0048$.

12. The trait is a dominant, because each affected individual has at least one affected parent, and normal offspring do not transmit the trait to their children. The trait is also autosomal, because there is an example of father-to-son transmission (male I-l to son II-4). The probability that female III-4 will transmit the dominant allele to the child is therefore 1/2.

If child IV-2 is unaffected, the transmission of the trait to the next generation would depend upon the genotype of the other parent, with probabilities differing if the person is homozygous or is heterozygous. Since the frequency of the mutant allele is .23, the probability that the parent is homozygous is $(.23)^2 = .0529$. The probability of being heterozygous is $2(.23)(.77) = .3542$. The likelihood of transmission from a homozygote to a child is 1; the likelihood of transmission from a heterozygote to a child is 1/2. Thus the total probability that a child of IV-2 will inherit the trait is (.0529)(1) + (.3542)(.5) = .23.

13. The change in allele frequency $\Delta q = m(q_m - q_r)$, where m is the proportion of migrants (.2 in this case) and q_m and q_r are the allele frequencies in the migrant and recipient populations, respectively. Thus,

$$\begin{aligned}\Delta q_1 &= .2(.6 - .3)\\ &= .2(.3) = .06\end{aligned}$$

In other words, since the frequency of the albino allele is higher in population 2, the migrants will carry a higher proportion of the albino alleles and will increase the frequency in the recipient population by .06 (that is, from .30 to .36).

CROSSWORD PUZZLE BASIC POPULATION GENETICS

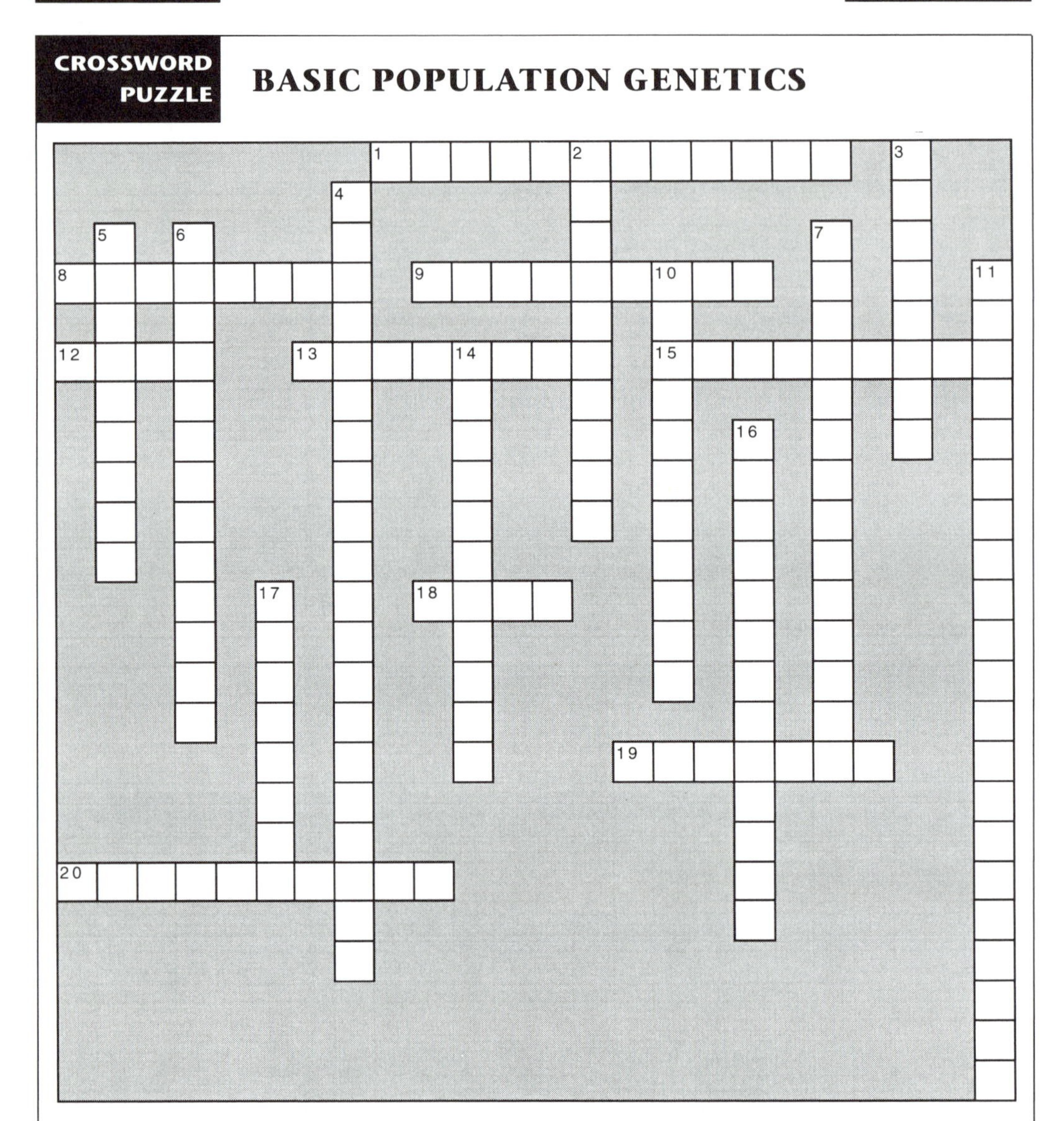

ACROSS

1. Changes in gene frequency caused by limited population size
8. Total genetic makeup of a population
9. Term for the heterozygous genotype's being more fit than either homozygous genotype
12. Local inbreeding population that has an equal chance of mating with other members of the opposite sex in the same local population
13. Equilibrium that, if disturbed and then allowed to return to equilibrium, will not return to the original equilibrium point
15. Only method, other than selection, in which directional change in gene frequency occurs
18. ___ mutation: condition whereby a mutant genotype changes to wild-type
19. Principle that refers to a form of genetic drift in which only a small sample of a population's gene pool is used in establishing a new population
20. Group of individuals of the same species in a defined area

DOWN

2. Matings between relatives in a population
3. Detectable and heritable change in genetic material, unrelated to recombination or segregation
4. Term for selection maintaining large amounts of genetic variation within populations
5. Elimination of genotypes due to competition or preference for another genotype

6. Proportion of one allele relative to other alleles of a specific gene in a population
7. __ mating: selective mating of unlike phenotypes
10. Refers to chance deviation from expected ratios due to small sample size
11. Occurrence of alleles from different genes together more often than predicted by their individual frequencies
14. __ mating: nonrandom mating based on phenotypic similarity
16. Principle that states that gene frequencies in a population will not change over time if specific conditions are met by that population
17. Term meaning random mating

SELECTION AND EVOLUTION

A biologist might define evolution as a change in allele frequencies over time. These changes in allele frequencies could be the result of selection, drift, mutation, migration, or combinations of these population variables. These Study Hints are primarily concerned with selection and the effects of relative differences in reproductive success.

The problem-related questions in this section might best be introduced by first considering a general selection model:

Genotype	A^1A^1	A^1A^2	A^2A^2	**Total**
Frequencies before selection	p^2	$2pq$	q^2	1
Fitness	W_1	W_2	W_3	
Relative contribution to next generation	p^2W_1	$2pqW_2$	q^2W_3	$\overline{W}$
Frequencies after selection	$\frac{p^2W_1}{\overline{W}}$	$\frac{2pqW_2}{\overline{W}}$	$\frac{q^2W_3}{\overline{W}}$	1

In this model, p and q are the allele frequencies of A^1 and A^2, respectively, and the W's refer to the relative fitness of each genotype. The relative fitness is a measure of the relative contribution that a genotype makes to the next generation, and it can be measured in terms of the intensity of selection (s), where $W = 1 - s$ and s can differ for each genotype ($0 \leq s \leq 1$).

In the case of selection against the homozygous recessive, for example, $W_1 = W_2 = 1$ and W_3 would be less than 1 (e.g., .2, where the intensity of selection is .8). Substitution into the table for the case where the allele frequencies are identical before selection ($p = q = .5$) yields the following results:

Genotype	A^1A^1	A^1A^2	A^2A^2	**Total**
Frequencies before selection	.25	.50	.25	1.00
Fitness	1.00	1.00	.20	
Relative contribution to next generation	.25	.50	.05	.80
Frequencies after selection	.25/80 =.31	.50/80 =.62	.05/80 =.07	

Allele frequencies calculated from these genotype frequencies are $p = .62$ and $q = .38$. Selection against the A^2A^2 homozygote has reduced the frequency of the A^2 allele in the sample.

For selection against the dominant, the relative fitnesses of the genotypes are $W_1 = 1 - s$ for A^1A^1 individuals, $W_2 = 1 - s$ for heterozygotes, and $W_3 = 1$ for A^2A^2 homozygotes. The proportion of zygotes after selection is, therefore, $p^2W_1 = p^2(1 - s)$, $2pqW_2 = 2pq(1 - s)$, and $q^2W_3 = q^2$. The gene frequency of p after selection is

$$p_1 = \frac{p(1 - s)}{1 - s + sq^2}$$

The change in p is

$$\Delta p = \frac{-psq^2}{1 - s + sq^2}$$

This formula tells us that the rate of change in allele frequency is faster than in selection against the recessive, because the dominant gene is not hidden in heterozygotes.

For selection against both homozygotes (sometimes called *overdominance,* or *heterosis*), the relative fitnesses of the three genotypes are $W_1 = 1 - s$, $W_2 = 1$, and $W_3 = 1 - t$, where s and t can be different intensities of selection. The proportion of zygotes after selection is therefore $p^2(1 - s)$, $2pq$, and $q^2(1 - t)$. To determine the change in allele frequency of the recessive, q, the gene frequency after selection (q_1) is

$$q_1 = \frac{q - q^2t}{1 - p^2s - q^2t}$$

The change in q is

$$\Delta q = \frac{p^2sq - q^2tp}{1 - p^2s - q^2t}$$

Finally, for selection against the heterozygote, the relative fitnesses of the three genotypes are $W_1 = 1$, $W_2 = 1 - s$, and $W_3 = 1$. The proportion of zygotes after selection is p^2, $2pq(1 - s)$, and q^2. The new gene frequency of q is

$$q_1 = \frac{q - spq}{1 - 2pqs} \quad \text{and} \quad \Delta q = \frac{spq(2q - 1)}{1 - 2pqs}$$

If $q < 1/2$, then Δq is negative, and q decreases under selection. If $q > 1/2$, then Δq is positive, and q increases. There is consequently an unstable equilibrium at $q = 1/2$. Note also that only in the case of selection against both homozygotes is a stable polymorphism possible.

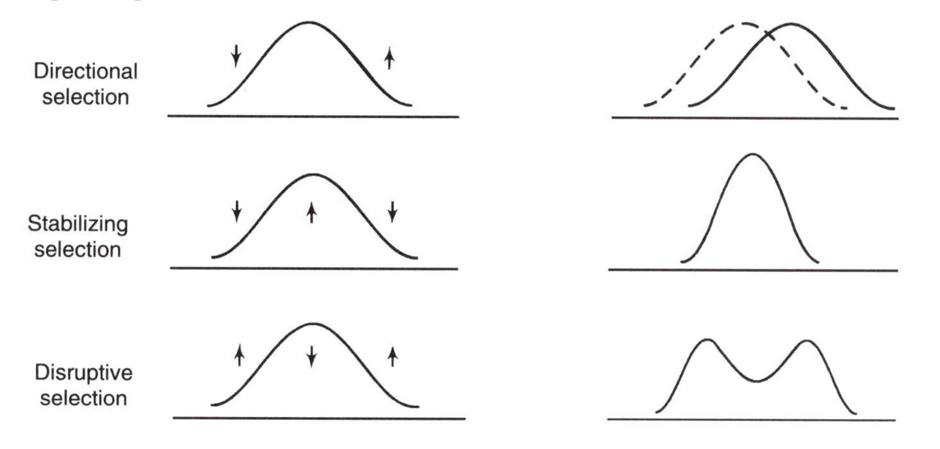

When we turn our attention from selection on a single locus to selection on the polygenic determinants of a phenotype, we find that it is convenient to divide selection into three basic categories. These and their consequences for the phenotypic mean and variance are diagramed on the preceding page.

Directional selection shifts the mean toward one extreme. Although it may temporarily increase phenotypic variance, the primary effect is on the phenotypic mean. Stabilizing selection, on the other hand, reduces phenotypic variance without changing the mean, whereas disruptive selection favoring two or more phenotypes in the same population increases phenotypic variance.

Much of the concern of population genetics centers on attempts to measure parameters such as selection intensity, relative fitness, and the interactions among genotypes and environment that influence these measures. Although this is a complex application of genetic principles, it is clear that there are many types of selection, that heritability is important in predicting the rate of response to selection, and that the population structure and environmental variables are important to any explanation of evolution. The difficulties and successes of some of these studies are discussed in your text. There can be no doubt that it is a challenging arena of genetic investigation.

IMPORTANT TERMS

Allopatric population
Cyclical selection
Directional selection
Disruptive selection
Genetic homeostasis
Genetic load
Hybrid zone
Incipient species
Introgressive hybridization
Isolating mechanisms
Segregational load
Selection pressure (intensity)
Stabilizing selection
Sympatric population

PROBLEM SET 22

1. A plant-breeding institute establishes four new strains of winter grain (G_1, G_2, G_3, and G_4). The genetic variation in each of these for resistance to an infectious fungus (rust) is measured by comparing the F_1 and F_2 phenotypic variances after crosses of each to a standard grain strain (G). The following data are collected. Which one of these four strains would be expected to respond most readily to artificial selection for rust resistance?

Strain	F_1 variance	F_2 variance
G_1	0.41	0.57
G_2	0.40	0.62
G_3	0.43	0.55
G_4	0.46	0.66

2. A rancher has just bought a flock of sheep in which a recessive allele for black wool is segregating. At the time he bought the flock, the allele was at a frequency of

.5 ($p = q = .5$). Theoretically, how many generations of complete selection against the black sheep would be required to reduce the allelic frequency to .1?

3. According to Ernst Mayr, sympatric isolating mechanisms can be placed into two broad categories: premating and postmating. Give some examples of each of these. Which of the two is more adaptive, in terms of the metabolic energy investment of the population?

4. What evidence is there to support the concept of sympatric speciation?

5. Would the heritability for a trait that is important for organism survival and reproduction (a "fitness trait") likely be larger than, smaller than, or the same as the heritability for a trait not directly affecting survival or fitness? Please explain your answer and discuss the expected relative effectiveness for directional selection on a fitness trait.

6. Please test whether the following data are in Hardy–Weinberg equilibrium by using a χ^2 test to compare the observed and the expected genotype frequencies. If the data are not in equilibrium, what explanations can you propose to account for any deviation? Phenotypes have been omitted in order to enable you to have full freedom in forming your explanations.

Genotypes	*RR*	*Rr*	*rr*	Total
Numbers of individuals	602	290	108	1,000

7. The initial frequency of a recessive eye-color mutant in a *Drosophila* population is .4. What is the maximum change in frequency that can occur in 10 generations of complete selection against the recessive, if no other factors, such as mutation, are acting in the population?

8. Suppose you have a large herd of horses in which three phenotypes are segregating: chestnut, carmello, and palomino. Both carmello and palomino are the result of a modifier gene (D') influencing the expression of the chestnut-coat color gene. The alternative allele (D) has no effect. $D'D'$ plus chestnut gives carmello, DD' plus chestnut gives palomino, and DD plus chestnut gives chestnut-colored horses. Suppose you would like to have a herd made up primarily of palomino horses, but you would also like to have more chestnuts than carmellos. Therefore you put selection pressure on this herd by selling all carmello horses and 50 percent of the chestnut horses before they mature. The initial allele frequencies of D and D' were equal ($p = q = .5$), and the population was in Hardy–Weinberg equilibrium when you started selection.

(a) What is the relative fitness of each genotype?
(b) What would the gene frequencies be after one generation of selection?
(c) What would the zygotic frequencies be in the second generation?
(d) What is the equilibrium frequency of q?

9. Beginning with the same original population of horses as in problem 8, suppose you decided to remove all the chestnuts as well as all carmello horses from the first generation of progeny. What would the gene frequencies and the zygotic frequencies be in the next generation?

10. Instead of selecting for the palominos, you decide that you want more chestnut and carmello horses than palominos. Starting at the same gene frequencies ($p = q = .5$) and selecting against the palominos ($s = 50$ percent),

(a) what is the relative fitness of each genotype?
(b) what will the frequency of the D′ allele determining carmello be after one generation of selection?
(c) What would the frequency of carmellos be after one generation of selection, if the starting frequency of D′ were .3?

ANSWERS TO PROBLEM SET 22

1. The heritabilities in each strain can be measured as shown in Chapter 9.

$$h^2 = \frac{V_{F_2} - V_{F_1}}{V_{F_2}} = \frac{V_G + V_E - V_E}{V_G + V_E} = \frac{V_G}{V_G + V_E}$$

The heritability of each strain is obtained by substituting into this formula, with the following results:

$G_1 = 0.28$
$G_2 = 0.35$
$G_3 = 0.22$
$G_4 = 0.30$

Strain G_2 has the largest heritability for rust resistance and thus, by definition, has the largest proportion of genetic (heritable) variation. It would respond most successfully to artificial selection for this trait, assuming that all other characteristics of the strains are equivalent.

2. There are two ways to solve this problem. One is to solve for q after each generation of selection. If $p^2 + 2pq + q^2 = 1$, then after one generation of selection against q^2,

	AA	***Aa***	***aa***	**Frequency of *a***
Before	p^2	$2pq$	q^2	q
After	$\frac{p^2}{p^2 + 2pq}$	$\frac{2pq}{p^2 + 2pq}$	0	$\frac{q}{1 + q}$

Repeating this calculation, however, is time-consuming. The more general form is

$$q_n = \frac{q_o}{1 + nq_o}$$

where q_o is the original frequency of the recessive, q_n is the frequency at generation n, and n is the number of generations of selection. This can be solved for n:

$$n = \frac{q_o - q_n}{q_o q_n}$$

Substituting from the problem, $n = (0.5 - 0.1) / (0.5)(0.1) = 8$ generations to make the desired reduction in allele frequency.

3. *Premating barriers* include seasonal or habitat isolation, behavioral isolation, and mechanical isolation. *Postmating barriers* include gametic mortality, zygotic mortality, hybrid inviability, and hybrid sterility. Premating isolation mechanisms are more energetically economical, because gametes are not wasted and doomed developmental activities are not initiated.

4. One of the most generally accepted theories of speciation is allopatric speciation. Many workers believe that populations can accumulate significant genetic differences only when they are separated to prevent gene exchange. Others, including K. Mather and J. M. Thoday, have proposed that selection for adaptation to different niches within a single locality could lead to polymorphism and that, if the disadvantages of the hybrids were severe enough, speciation might occur. Thoday and J. B. Gibson, for example (see reference at end of this answer), selected for high and low bristle number in a single culture of *Drosophila melanogaster,* and they obtained significant divergence between these extremes, in spite of the fact that extremes were given the opportunity to interbreed freely. Clearly, geographic separation is not required for divergence to occur. Many mechanisms could encourage sympatric divergence, such as differences in flowering time in two groups of plants in the same field or localized heterogeneity in the environment (e.g., high levels of toxic waste at a mining site). These and other possibilities are discussed more thoroughly in a review by J. M. Thoday (*Proceedings of the Royal Society of London* (B) 182 [1972]: 109–143).

5. Heritability measures the proportion of phenotypic variation that can be traced to genetic segregation. For heritability to be high, a large number of polygenic heterozygotes must be present; that is, a large number of "increasing" and "decreasing" polygenic alleles must be maintained in the population. This is not likely to be true for a critical fitness trait, since there would seldom be a selective advantage to maintaining anything less than the greatest possible expression of such characters. Heritability would therefore tend to be higher for traits not directly affecting survival or fitness.

6. The 1,000 individuals in this population make up a gene pool of 2,000 copies of each gene. To test the fit to Hardy–Weinberg, we must first calculate the allele frequencies. Since the homozygote has two alleles of one type, and the heterozygote has one (*RR* and *Rr*, respectively),

$$p \text{ (the frequency of } R) = \frac{(2 \cdot 602) + 290}{2{,}000} = .747$$

$$q \text{ (the frequency of } r) = \frac{(2 \cdot 108) + 290}{2{,}000} = .253$$

At Hardy–Weinberg equilibrium,

$RR = p^2 = (.747)^2 \cdot 1{,}000 = (.558) \cdot 1{,}000 = 558$ individuals
$Rr = 2pq = 2(.747)(.253) \cdot 1{,}000 = (.378) \cdot 1{,}000 = 378$
$rr = q^2 = (.253)^2 \cdot 1{,}000 = (.064) \cdot 1{,}000 = 64$

The fit to Hardy–Weinberg can be calculated using a χ^2 analysis.

	RR	*Rr*	*rr*	Total
Observed (*O*)	602	290	108	1,000
Expected (*E*)	558	378	64	1,000
$O - E$	+44	−88	+44	0
$(O - E)^2/E$	1,936/558	7,744/378	1,936/64	

$$\chi^2 = \sum \frac{(O - E)^2}{E} = 3.470 + 20.487 + 30.250 = 54.207$$

The number of degrees of freedom is 1; 3 (the number of classes) minus 1 (for using the sample size to calculate the expectations) minus 1 (for estimating one parameter for the data, the allele frequency p, which automatically establishes q and the genotype frequencies) = 1. The probability value for this magnitude of χ^2 is $< .001$, indicating a significant discrepancy between the observed and expected frequencies. There appears to be a relative deficiency of heterozygotes, suggesting selection favoring the homozygous genotypes. Another way of getting such a pattern is to have this sample of 1,000 individuals really represent a mixture of individuals from two different local populations, each of which had been selected for an alternative allele or in which drift had led to the increase of alternative alleles.

7. From problem 2,

$$q_n = \frac{q_o}{(1 + nq_o)}$$

Substituting $q_o = .4$ and $n = 10$,

$$q_{10} = \frac{.4}{1 + 10(.4)} = \frac{.4}{5} = .08$$

8. From the problem, the frequency of $D = p = .50$, and the frequency of $D = q = .50$. Since $p^2 + 2pq + q^2 = 1$, the phenotypic frequencies in the herd are $p^2 = .25$ for carmellos, $2pq = .50$ for palominos, and $q^2 = .25$ for chestnuts. The selection pressure against carmellos is $s = 1$ and against chestnuts is $t = .5$.

(a) The fitness of

carmellos $= W_1 = 1 - s = 0$
palominos $= W_2 = 1$
chestnuts $= W_3 = 1 - t = .50$

The contribution of each genotype to the next generation is

carmellos $= p^2W_1 = 0$
palominos $= 2pqW_2 = .50$
chestnuts $= q^2W_3 = .125$

(b) After one generation of selection,

$$q_1 = \frac{q - q^2t}{1 - p^2s - q^2t}$$

Substituting $p = q = .5$, $s = 1$, and $t = .5$, we find that $q_1 = .6$, and $p_1 = .4$.

(c) The zygotic frequencies can be calculated by using $p = .4$ and $q = .6$.
For carmellos, $p^2 = .16$; for palominos, $2pq = .48$; and for chestnuts, $q^2 = .36$.

(d) At equilibrium, $ps = qt$. Substituting $1 - q$ for p, $(1 - q)s = qt$

$$(1 - q)s = qt$$
$$s - sq = qt$$
$$q(s + t) = s$$
$$q = \frac{s}{s + t}$$

Correspondingly,

$$p = \frac{t}{s + t}$$

In this problem, at equilibrium $q = .667$ and $p = .333$.

9. The selection intensities would be $s = 1$ and $t = 1$.

$$q_1 = \frac{q - q^2t}{1 - p^2s - q^2t} = .5$$

This makes sense if you recognize that all remaining horses will be heterozygotes, so p must equal q. From this point on, the population will be in genotypic (and therefore phenotypic) equilibrium.

10. **(a)** $W_1 = 1$, $W_2 = 1 - s$, and $W_3 = 1$. Since s = 5, W_2 must be .5.

(b) After one generation of selection,

$$q_1 = \frac{q - spq}{1 - 2pqs} = \frac{.375}{.750} = .5$$

The gene frequencies did not change.

(c) Letting q be the frequency of D' and substituting $q = .3$ into the formula just given in **(b),**

$$q_1 = \frac{0.3 - (0.5)(0.7)(0.3)}{1 - 2(0.7)(0.3)(0.5)} = 0.247$$

Since $q_1 < .5$, the frequency of the D' allele will decrease.

CROSSWORD PUZZLE

SELECTION AND EVOLUTION

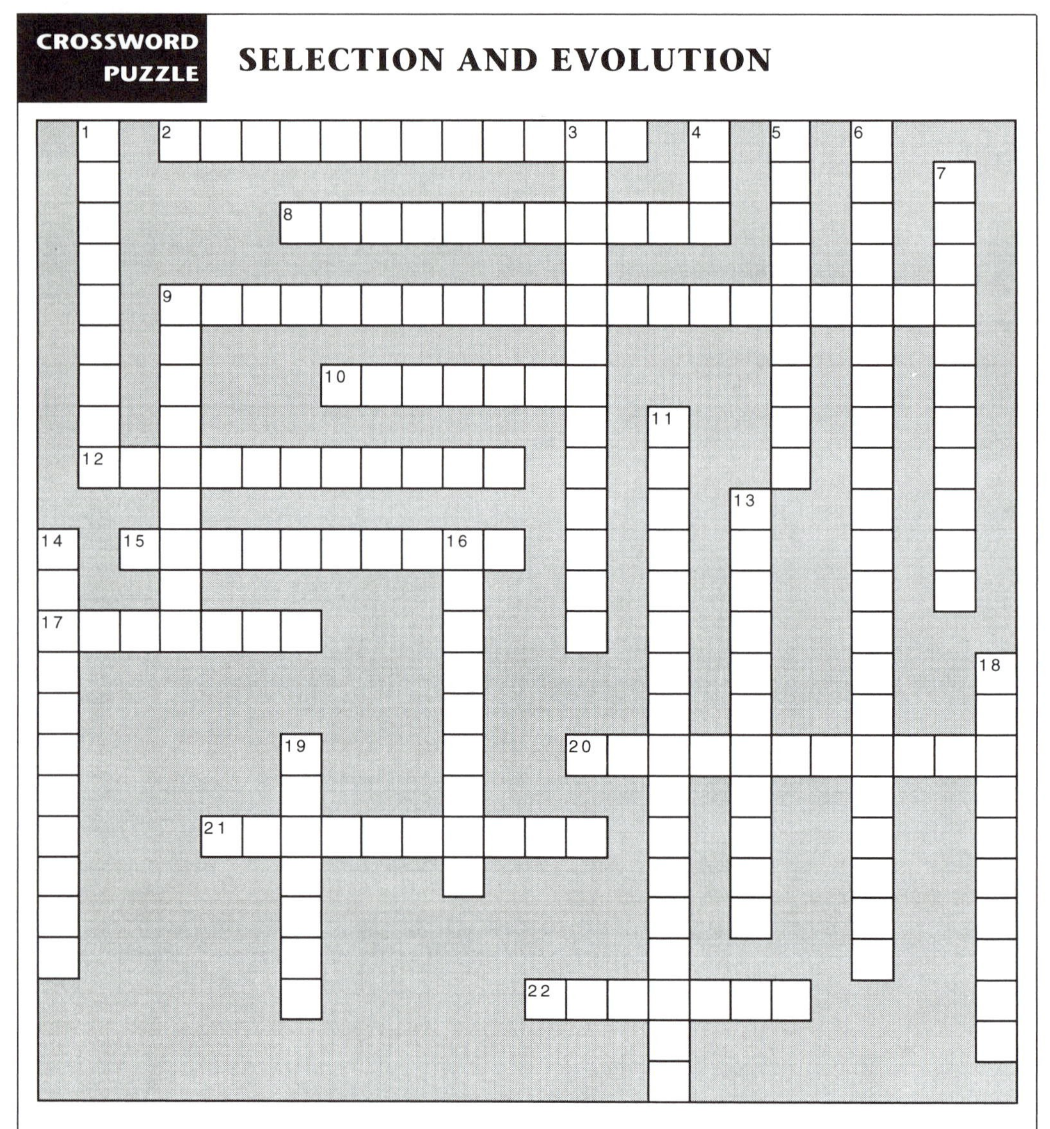

ACROSS

2. ___ evolution: change that splits a single ancestral lineage into two or more branches
8. Evolution in one or more species in response to changes in another associated species
9. Measure of the effects of selection (loss of fitness) of one genotype relative to a best fit genotype that is given a fitness value of 1
10. Small colonizing population that has only a sample of the genetic variation found in the population from which it came
12. Selection against variations from the "normal"
15. Genetic mechanisms that prevent the production of hybrid zygotes
17. Basic taxonomic category, which describes lineages evolving separately; one definition involves reproductive isolation
20. Selection against one extreme and the intermediate, while another extreme phenotype has an advantage
21. Reduction in genetic variation due to the effect of genetic drift on a population that was reduced in size (crashed) and then at a later date increases in numbers
22. Selection for the resemblance of one species to another species (model) that gives the impersonator some form of protection or advantage

DOWN

1. Presence of similar adaptations in widely diverse species having structures with similar function but different embryonic origins
3. ___ hybridization; incorporation of genetic information from one gene pool into another gene pool as a result of hybrid fertility

4. Selection that is often used to explain altruistic behavior
5. Causes differential reproductive survival of genetically variable populations
6. Divergence from the normal gamete ratios produced by heterozygotes (meiotic drive)
7. Reproductive mechanisms that prevent hybrid zygotes from producing fertile offspring
9. Related populations that are not isolated geographically from each other
11. Modification of specific population traits such that the population diverges into different habitats and a production of a large variety of species or groups results
13. Describes the average loss of genetic fitness in individuals of a population because they carry harmful alleles
14. Selection that tends to favor the survival of the extreme phenotypes and eliminate the intermediate phenotypes
16. Mechanisms that act as barriers to genetic exchange between populations
18. Closely related species that are geographically isolated from each other
19. Describes the relative reproductive success of one genotype over others; opposite of selection

23 PRACTICE TESTS

MITOSIS AND MEIOSIS

Multiple Choice

Please select the single best answer for each of the following questions.

1. A human cell at prophase II of meiosis will normally have how many centromeres? **(a)** 46, **(b)** 45, **(c)** 92, **(d)** 23, **(e)** none of the above.

2. An individual carries a pair of homologous chromosomes, *A* and *a*. A cell at mitotic metaphase or at metaphase I of meiosis has **(a)** one *A* and one *a* chromosome, **(b)** two *A* and two *a* chromosomes, **(c)** four *A* and four *a* chromosomes, **(d)** none of the above.

3. A cell from the same individual at metaphase II of meiosis has **(a)** one *A* and one *a* chromosome, **(b)** either one *A* or one *a* chromosome, **(c)** two *A* and two *a* chromosomes, **(d)** either two *A* or two *a* chromosomes.

4. Bivalents are seen **(a)** at prophase I of meiosis, **(b)** at prophase II of meiosis, **(c)** at prophase of mitosis, **(d)** at prophase I of meiosis and at prophase of mitosis, **(e)** at prophase II of meiosis and at prophase of mitosis, **(f)** during the interphase just prior to the beginning of meiosis.

5. Histone proteins are found in **(a)** eukaryotes only, **(b)** bacteria only, **(c)** viruses only, **(d)** both eukaryotes and bacteria, but not viruses.

6. Centromeres associated with duplicated chromosomes (sister chromatids) do *not* divide during **(a)** mitotic anaphase, **(b)** anaphase I of meiosis, **(c)** anaphase II of meiosis.

7. A chromosome with a long arm and a short arm is called **(a)** telocentric, **(b)** egocentric, **(c)** epicentric, **(d)** metacentric, **(e)** acrocentric, **(f)** paracentric, **(g)** chromocentric.

8. A diploid organism has 10 pairs of chromosomes. These 20 chromosomes will all behave independently of each other during **(a)** mitosis, **(b)** meiosis, **(c)** both mitosis and meiosis.

9. The epidermal cells of a certain mouse have 32 chromosomes. This species therefore has how many linkage groups? **(a)** 32, **(b)** 64, **(c)** 16, **(d)** 31, **(e)** none of the above.

10. If parthenogenetic development of an unfertilized egg could produce a viable diploid child in humans, its sex would be **(a)** female, **(b)** male, **(c)** either female or male.

Diagrams and Definitions

1. Please draw and label the main parts of **(a)** a cell at anaphase I and **(b)** a cell at anaphase II, where $2n = 6$ acrocentric chromosomes.

2. Please define each of following terms briefly:

(a) Bivalent
(b) Synapsis
(c) Homologous chromosomes
(d) Chromatids
(e) Linkage group

NUCLEIC ACIDS

Multiple Choice

Please select the single best answer for each of the following questions.

1. In double-stranded RNA, the nucleotides of each strand are connected to those in the other by **(a)** hydrogen bonds, **(b)** phosphodiester bonds, **(c)** peptide bonds, **(d)** none of the above.

2. Each "rung" in the double-stranded "ladder" of nucleic acid contains how many nitrogenous rings: **(a)** one, **(b)** two, **(c)** three, **(d)** two or three, **(e)** two, three, or four?

3. A virus whose genetic material consists of single-stranded RNA has about 22 percent of its RNA nucleotides consisting of uracil. The frequency of adenine in this organism's RNA is **(a)** 22 percent, **(b)** 28 percent, **(c)** 78 percent, **(d)** not ascertainable from this information.

4. In the Meselson and Stahl experiment of DNA replication, *E. coli* whose DNA was fully labeled with ^{15}N were placed in a ^{14}N medium and permitted to undergo a number of replications. After the fourth replication in ^{14}N, the DNA should consist of **(a)** one 14/14 to one 14/15 hybrid, **(b)** three 14/14 to one 14/15 hybrid, **(c)** seven 14/14 to one 14/15, **(d)** none of the above.

5. A virus that contains approximately 29.7 percent adenine and 29.3 percent uracil most likely has as its genetic material **(a)** single-stranded RNA, **(b)** single-stranded DNA, **(c)** double-stranded RNA, **(d)** double-stranded DNA.

6. If a hybrid tobacco mosaic virus (TMV) is made by combining the RNA from strain I with the protein coat from strain II, and this hybrid TMV is used to infect a plant, then the progeny will be **(a)** strain I in both RNA and protein, **(b)** strain II in both RNA and protein, **(c)** hybrid like the hybrid parent.

7. RNA differs from DNA in the nucleotide bases it contains. To be specific, it differs in one of the **(a)** purines, **(b)** pyrimidines.

8. Griffith, in 1928, and Avery, MacLeod, and McCarty in 1944 studied bacterial transformation in *Diplococcus pneumoniae,* where rough (R, nonvirulent) forms were transformed by DNA taken from smooth (S, virulent) forms. Is it possible to transform S forms to R forms by treating them with DNA extracted from R bacteria? **(a)** yes, **(b)** no.

9. Experiments with labeled protein and labeled DNA in viruses helped confirm that DNA was the genetic material. These experiments were carried out by **(a)** Watson and Crick, **(b)** Hershey and Chase, **(c)** Griffith, **(d)** Avery, MacLeod, and McCarty.

10. From Chargaff's rule, we predict that **(a)** $(A + T)/(C + G) = 1$, **(b)** $(A + C)/(T + G) = 1$, **(c)** $(G + A)/(C + T) = 1$, **(d)** answers a and b only, **(e)** answers b and c only, **(f)** answers a, b, and c, **(g)** none of the above.

Diagrams and Definitions

1. Using the following symbols, diagram **(a)** a purine nucleotide, **(b)** a nucleoside of DNA.

2. On the following diagrams designate the 5′ and the 3′ carbon positions, and identify the sugars.

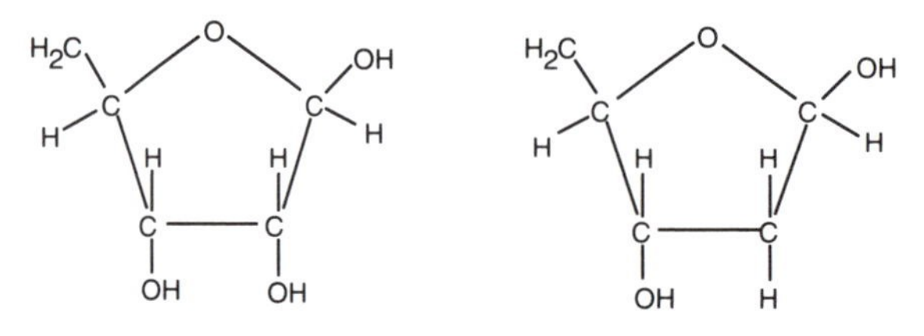

3. Please define each of the following terms briefly:

(a) Semiconservative replication
(b) Polymerase
(c) Transcription
(d) Replication
(e) Ligase

BASIC MENDELIAN GENETICS 1

Multiple Choice

Select the single best answer for each of the following questions.

1. A married couple has children whose blood types include A, B, AB, and O. The blood types of the parents are **(a)** AB and O, **(b)** A and B, **(c)** AB and AB, **(d)** none of the above.

2. Curly (*Cy*), a wing mutant in *Drosophila,* is dominant to the allele for normal (straight) wings and is lethal when homozygous. If two Curly flies are mated, the proportion of wild-type flies among the progeny would be **(a)** 1/4, **(b)** 1/2, **(c)** 3/4, **(d)** 1/3, **(e)** 1/6, **(f)** 2/3.

3. Two genetically different strains of recessive albino mice are crossed. The first strain, *aaWW,* is albino because of *aa,* whereas the second, *AAww,* is albino because of *ww.* The two loci assort independently. Assume that the dihybrid F_1 mice are crossed with each other to produce the F_2. The fraction of the F_2 mice that is expected to be albino is **(a)** 3/4, **(b)** 1/2, **(c)** 9/16, **(d)** 3/16, **(e)** 7/16, **(f)** 1/4, **(g)** 1/16.

4. In a self-fertilizing plant, what proportion of a trihybrid F_2 will breed true (that is, will not segregate for any of the three loci)? **(a)** none, **(b)** 1/64, **(c)** 1/8, **(d)** 1/4, **(e)** 1/2.

5. Camptodactyly (permanently stiff, bent fingers) is an autosomal dominant trait in humans. A normal woman marries an affected man whose father was normal. If the condition is expressed in only 60 percent of the carriers in a hypothetical population, what proportion of the children produced by these parents would be expected to carry the mutant? **(a)** 100 percent, **(b)** 60 percent, **(c)** 50 percent, **(d)** 30 percent, **(e)** 25 percent, **(f)** none of the above.

6. A 9:3:4 phenotypic ratio for F_2 in a dihybrid cross is the result of the action of **(a)** a dominant epistatic gene, **(b)** a recessive epistatic gene, **(c)** an ecstatic gene, **(d)** a dominant pleiotropic gene, **(e)** a recessive pleiotropic gene.

7. Assume that a dihybrid cross in the aardvark is made between the genotypes *IiAa* and *IiAa.* The *i* homozygote is lethal, preventing the embryo from implanting. The *A* allele permits pigmentation, whereas the *a* homozygote is albino. The two loci assort independently. The *I*–genotype is phenotypically normal. The baby aardvarks from this cross will segregate what ratio of pigmented to albino phenotypes? **(a)** 3:1, **(b)** 9: 7, **(c)** 13:3, **(d)** 2: 1, **(e)** none of the above.

8. Rh incompatibility between fetus and mother will be less likely to lead to problems for a future child if the ABO blood types of the fetus and the mother are, respectively **(a)** O and AB, **(b)** O and A, B, or AB, **(c)** A and AB, **(d)** AB and O, **(e)** B and AB.

9. Linkage is an exception to the Mendelian law of **(a)** segregation, **(b)** dominance and recessiveness, **(c)** unit factor inheritance, **(d)** independent assortment.

10. In chickens, two pairs of genes are involved in the inheritance of comb type. One gene, *P,* produces pea-combs, whereas its recessive allele, *p,* produces a single-comb. The second gene, *R,* produces a rose-comb, whereas its recessive allele, *r,* also produces a single-comb. When *R* and *P* occur together (that is, *R* – *P* –), a walnut-comb is formed. Considering both loci together, the only genotype that can produce a single-comb is *rrpp.* Suppose that a walnut-comb chicken is crossed with a rose-comb chicken. They produce two walnut-comb, three rose-comb, one single-comb, and two pea-comb chickens. Which of the following are the genotypes of the parents **(a)** *rrPp* × *RrPp,* **(b)** *RrPp* × *Rrpp,* **(c)** *RrPp* × *RrPp,* **(d)** *RRPP* × *RRpp,* **(e)** *RRPp* × *Rrpp*?

MENDELIAN GENETICS 2

Multiple Choice

Select the single best answer for each of the following questions.

1. A woman with blood type B has a child with blood type O. She claims that a friend of hers is the child's father. He denies it. His blood type is B. The fact that his parents are type A and type AB would be welcome information **(a)** to the woman, **(b)** to the man, **(c)** to neither the woman nor the man.

2. A true-breeding (homozygous) mouse strain with black hair is crossed to a true-breeding strain with white hair. The F_1 are all black. When the F_1 mice are crossed to each other, they give rise to an F_2 consisting of about 9 black: 3 brown: 4 white mice. This is an example of **(a)** recessive epistasis, **(b)** dominant epistasis, **(c)** pleiotropy, **(d)** duplicate genes.

3. A woman with blood type B and a man with blood type A have a child with blood type A. The probability that their next child will be type B is **(a)** either 0 or 1/4, depending upon the genotype of the father; **(b)** either 1/4 or 1/2, depending upon the genotype of the father; **(c)** either 0 or 1/4, depending upon the genotype of the mother; **(d)** either 1/4 or 1/2, depending upon the genotype of the mother; **(e)** 0; **(f)** 1/4; **(g)** 1/2.

4. If you were to flip a coin twelve times, you would expect six heads and six tails (in any order), with a probability of about **(a)** 23 percent, **(b)** 26 percent, **(c)** 30 percent, **(d)** 34 percent.

5. Two normal-sized rats are each heterozygous for a recessive gene *d*, causing dwarfism. The probability that any two of their progeny will include one normal and one dwarf rat is **(a)** 1, **(b)** 3/16, **(c)** 3/8, **(d)** none of the above.

6. During the lifetime of a woman with the genotype *CcHHTtUu*, what proportion of the ova she produces will have the genetic makeup *CHTU*: **(a)** all, **(b)** 1/2, **(c)** 1/4, **(d)** 1/8, **(e)** 1/16, **(f)** none of the above.

7. If both a husband and wife have the genotype given in question 6 (*CcHHTtUu*), what is the likelihood that their child will have the genotype *CcHHttUU*? **(a)** 1/4, **(b)** 3/4, **(c)** 3/16, **(d)** 3/32, **(e)** 1/32, **(f)** none of the above.

8. Mendel interpreted his results with garden peas as demonstrating **(a)** blending inheritance, **(b)** particulate inheritance, **(c)** chromosomal inheritance, **(d)** cytoplasmic inheritance.

9. The pedigree shown in the accompanying figure is for the ability to taste PTC, a dominant trait. The probability that a child of II-3 and II-4 will be a taster is **(a)** 3/4, **(b)** 1/2, **(c)** 1/4, **(d)** 1, **(e)** none of the above.

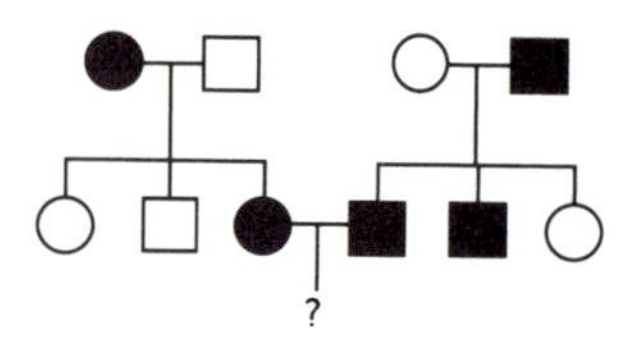

10. A rare dominant trait, when exhibited in men, is transmitted to half their sons and to half their daughters. The gene for this trait is carried on **(a)** the X chromosome, **(b)** an autosome, **(c)** the Y chromosome, **(d)** a cytoplasmic factor.

11. In *Drosophila,* Dichaete is a gene with a dominant effect on the angle at which the wings are held relative to the body. Flies that breed true for Dichaete never occur. A cross between phenotypically Dichaete males and Dichaete females produces approximately twice as many Dichaete progeny as wild-type progeny. The number of progeny from such a cross is also invariably smaller than the number from a cross of Dichaete females or males to wild-type. These observations are consistent with the hypothesis that Dichaete is **(a)** an epistatic gene, **(b)** a hypostatic gene, **(c)** a quantitative trait, **(d)** a recessive lethal trait.

12. What proportion of all six-child families in which one parent is a heterozygote showing polydactyly (a dominant trait) and the other is a normal individual will be expected to produce at least one child with polydactyly? **(a)** 5/6, **(b)** 1/64, **(c)** 1/6, **(d)** 1/2, **(e)** none of the above.

13. In a family, two phenotypically normal parents have a son who suffers from the X-linked form of hemophilia. What is the probability that both of their next two children will also have hemophilia? **(a)** 1/4, **(b)** 1/8, **(c)** 1/16, **(d)** extremely low likelihood of having two such children in a row.

14. A man whose mother had blood type A and whose father had blood type O marries a woman with blood type AB. If the man has blood type A, how many different blood types can be expected among their children? **(a)** 8, **(b)** 3, **(c)** 1, **(d)** 2, **(e)** 4.

15. Assume that you are told that in the F_2 of a cross between tall and short plants there were 77.6 percent tall and 22.4 percent short plants. With this information you **(a)** can, **(b)** cannot test the results by chi-square.

Problems

1. In humans, achondroplastic dwarfism and neurofibromatosis are both dominant conditions. An achondroplastic woman whose father also suffered from the condition marries a man with neurofibromatosis. Please define symbols for these conditions and then do the following:

(a) Determine what phenotypes their children could have, and in what ratios they would be expected.

(b) A son with both traits marries a normal woman, and their first child is normal. What is the probability that if they have a family of five children, three will be normal and two will show both traits?

2. Answer the following questions after examining this pedigree which follows:

(a) What is the probable mode of inheritance of this trait?

(b) What characteristics of the pedigree led you to your answer in part **(a)**?

(c) What is the probability that the proband would have an affected child if mated with another individual from the population at large?

(d) What is the probability that individual III-4 is a carrier?

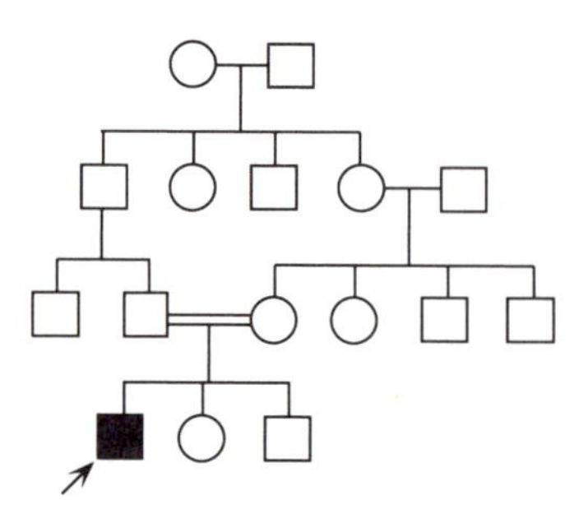

MENDELIAN GENETICS 3

Multiple Choice

Select the single best answer for each of the following questions.

1. In a cross of *RrBbeeFf* × *rrBbEEff*, with complete dominance at all loci, independent assortment, and no gene interaction, the probability that the progeny will show all four dominant traits is **(a)** 1, **(b)** 1/2, **(c)** 3/4, **(d)** 1/4, **(e)** 1/8, **(f)** 3/8, **(g)** none of the above.

2. If the heritability of high milk yield in a dairy herd is 0.12, the dairy farmer should be advised to **(a)** try soft music and better food to increase yield, **(b)** try an intensive program of selected breeding with the herd.

3. A dihybrid cross with codominance at each locus and with the two genes assorting independently and not interacting will give an F_2 phenotypic ratio of **(a)** 9:3:3:1, **(b)** 12:3:1, **(c)** 9:7, **(d)** 9:3:4, **(e)** 1:2:2:4:1:2:1:2:1.

4. A dihybrid cross with incomplete dominance at each locus and with the two genes assorting independently and not interacting will give an F_2 phenotypic ratio of **(a)** 9:3:3:1, **(b)** 12:3:1, **(c)** 9:7, **(d)** 9:3:4, **(e)** 1:2:2:4:1:2:1:2:1.

5. In a pedigree for polydactyly, some individuals have six phalanges on one appendage and five on the other three appendages; some have six phalanges on two appendages and five on the other two appendages; other individuals may have as many as seven phalanges on one or more appendages. This phenomenon is called **(a)** variable penetrance, **(b)** variable expressivity, **(c)** discordance, **(d)** phenocopying, **(e)** cultural modification.

6. A fruit fly that is a phenocopy for winglessness **(a)** will, **(b)** will not transmit this trait to its offspring.

7.

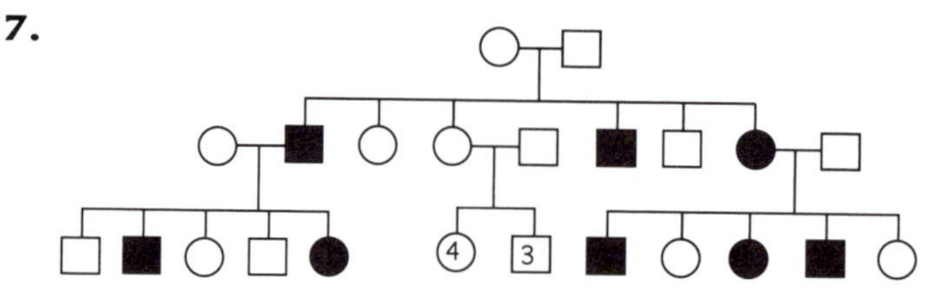

The trait indicated in the accompanying pedigree is most likely inherited as **(a)** an autosomal dominant, **(b)** an autosomal recessive, **(c)** a sex-linked recessive, **(d)** a sex-linked dominant trait.

8. The kappa particles are bacteria inhabiting the cytoplasm in paramecia. Kappa particles **(a)** are, **(b)** are not capable of mutating.

9. The transmission of information that is not coded for by DNA or RNA is seen in **(a)** infectious heredity of CO_2 sensitivity in *Drosophila,* **(b)** inheritance of episomes in bacteria, **(c)** inheritance of the suppressive petite condition in yeast, **(d)** inheritance of cortical structure in paramecia.

10. The accompanying pedigree is for a rare autosomal recessive trait; it also shows a consanguineous marriage. The probability that a child of IV-1 and IV-2 will be affected is **(a)** 1/4, **(b)** 1/8, **(c)** 1/16, **(d)** 1/64, **(e)** 1/128.

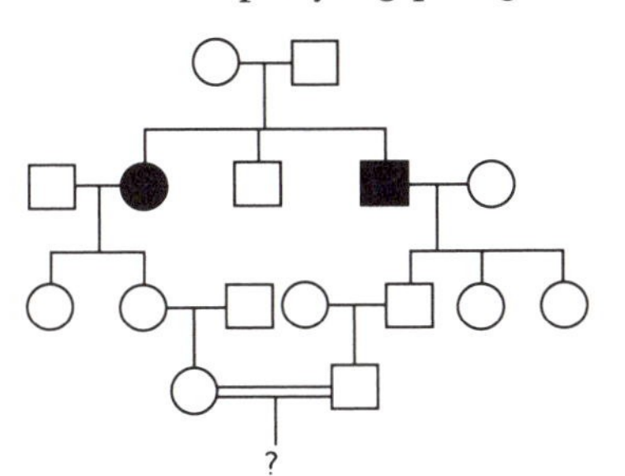

11. In the dihybrid cross I^AI^BMN to I^AI^BMN, the expected number of different phenotypes among the offspring is **(a)** 4, **(b)** 6, **(c)** 8, **(d)** 9, **(e)** 16, **(f)** none of the above.

12. Assume a deleterious recessive trait (*ee*) with a penetrance of 0.6. In a cross of *Ee* × *Ee,* the expected proportion of normal to affected progeny is **(a)** 3:1, **(b)** 6:4, **(c)** 45:55, **(d)** 85:15, **(e)** 90:10, **(f)** none of the above.

13. In the chi-square test, the larger the χ^2 value, the **(a)** better, **(b)** worse the fit of the observed results to the expected results.

14. A sample of ears of corn has a mean weight of 850 grams with a standard deviation of 40 grams. With this information we can predict that 34 percent of the ears will have weights ranging from **(a)** 830 to 870 g, **(b)** 810 to 890 g, **(c)** 830 to 850 g, **(d)** 850 to 890 g.

15. What proportion of five-child families from marriages of one albino parent and one heterozygous parent would you expect to be composed of three albino and two normal children? **(a)** 1/32, **(b)** 3/32, **(c)** 5/32, **(d)** 1/2, **(e)** 10/32, **(f)** none of the above.

Problems

1. The following questions refer to the accompanying pedigree.

(a) Assuming, when no evidence points otherwise, that all individuals marrying into the pedigree are homozygous normal, what is the probable mode of inheritance?

(b) What is the probability that the child shown as "?" will express the trait?

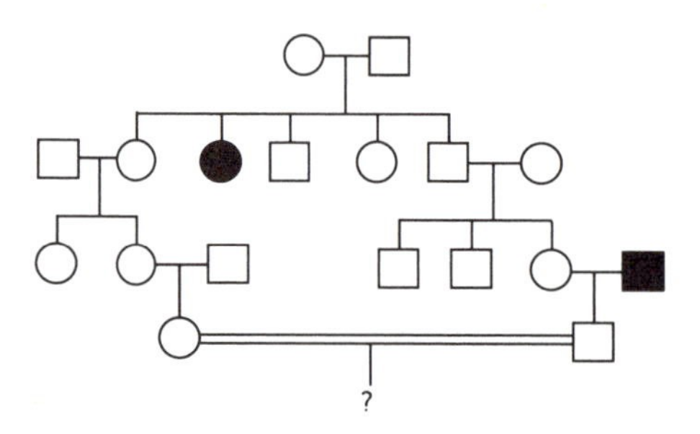

2. A cross is made between a true-breeding shaggy llama strain and a true-breeding short-haired llama strain. The shaggy llamas are also homozygous brown; the short-haired ones are homozygous cream-colored. An F_2 from their progeny gave the following distributions of phenotypes:

Shaggy brown	35
Short cream	31
Shaggy cream	79
Short brown	15
	160

Propose a hypothesis to explain this pattern of phenotypes and determine

(a) The expected ratio
(b) Chi-square for goodness of fit to this hypothesis
(c) The interpretation of these results: **(1)** reject, **(2)** do not reject

LINKAGE AND MAPPING IN DIPLOIDS

Multiple Choice

Select the single best answer for each of the following questions.

1. In a series of two-point testcrosses, the following crossover data were obtained (given in map distances). The distance from locus *S* to *D* is 2; *R* to *S* is 9; *P* to *S* is 8; *D* to *R* is 11; *D* to *P* is 6. The relative positions of these loci are **(a)** *D R S P,* **(b)** *P D S R,* **(c)** *R D S P,* **(d)** *S D P R,* **(e)** none of the above.

2. A three-strand double crossover will give rise to **(a)** 3 double and 1 noncrossover strands, **(b)** 2 double, 1 single, and 1 noncrossover strands, **(c)** 3 single and 1 noncrossover strands, **(d)** 1 double, 2 single, and 1 noncrossover strands, **(e)** none of the above.

3. In a certain plant, 28 percent of meioses have an exchange between the *p* and *rpl* loci. It therefore follows that in a *p rpl* /+ + dihybrid, a + *rpl* chromosome is recovered with a frequency of **(a)** 28 percent, **(b)** 56 percent, **(c)** 14 percent, **(d)** 7 percent, **(e)** none of the above.

4. The pair of chromosomes shown in the following figure contains both genetic markers and cytologically detectable markers. In which region (or regions) will an

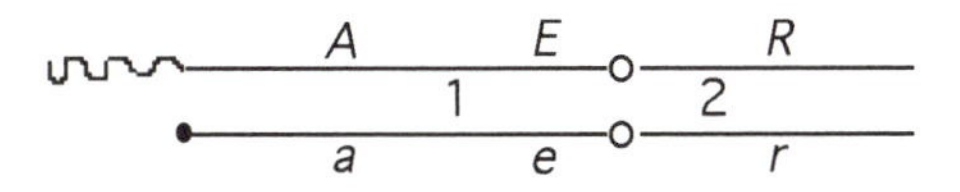

exchange produce chromosomes that can be shown cytologically to be crossovers? **(a)** region 1 only, **(b)** region 2 only, **(c)** either region 1 or region 2, **(d)** neither region 1 nor region 2.

5. Half-tetrad analysis of attached-X females of *Drosophila* provides evidence that crossing over occurs at the four-strand stage of meiosis. Females are heterozygous for white eyes (w) and forked bristles (f). In the genotype illustrated in the following figure, exchange between the forked locus and the centromere would be consistent with crossing over at the four-strand stage if which of the following types of progeny are produced? **(a)** females with white eyes, **(b)** females with forked bristles, **(c)** females with white eyes and forked bristles, **(d)** females with white eyes and females with forked bristles.

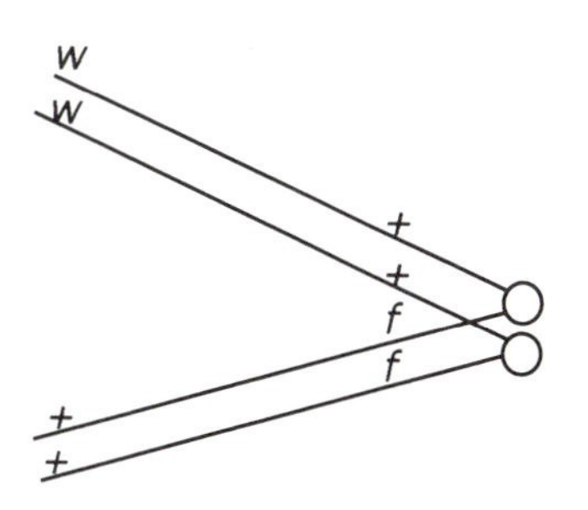

6. In *Drosophila,* a large number of experiments have shown that the vermilion eye-color gene (v) and the miniature-wing gene (m) are separated by approximately 3 crossover map units. In a female that is $v\,m\,/\,v\,m$, the percentage of meioses that *can be expected* to have an exchange in the v—m region is **(a)** approximately 3.0, **(b)** approximately 1.5, **(c)** approximately 6.0, **(d)** not ascertainable in the absence of heterozygosity at the v and m loci.

7. The attached-X chromosome of a female *Drosophila,* shown in the following figure, is drawn at the four-strand stage, and it is heterozygous for the recessive miniature-wing allele (m). A female with miniature wings can be produced by an exchange **(a)** between the centromere and the m locus, **(b)** between the m locus and the end of the chromosome, **(c)** at any point along the chromosome as long as it involves the correct chromatids, **(d)** none of the above.

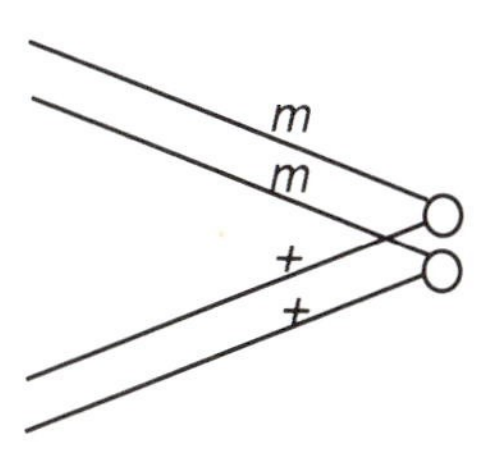

8. Genetic recombination can take place **(a)** only between genes, **(b)** only within genes, **(c)** either between or within genes.

9. Which of the following spore arrangements in *Neurospora* would result from first-division segregation in an ascus heterozygous for spore color at one locus?

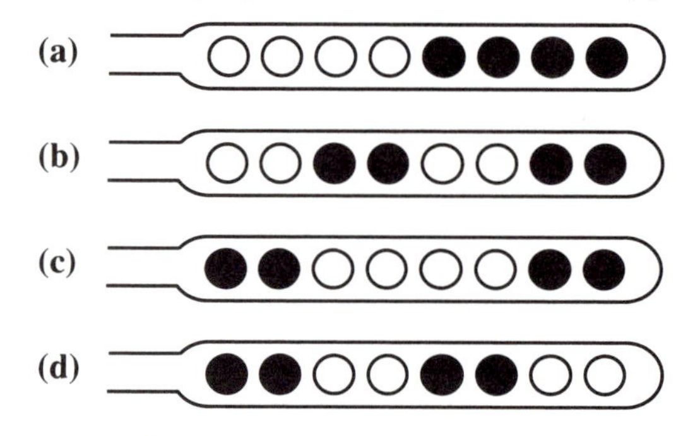

(e) None of the above

10. Three closely linked loci (*X, Y,* and *Z*) are separated by 2.5 and 1.2 map units, respectively. In mapping them, a total of 25 double recombinants were recovered in 100,000 scored progeny. The coefficient of coincidence is **(a)** 1/6, **(b)** 1/3, **(c)** 1/2, **(d)** 5/6, **(e)** none of the above.

Problems

1. Following is a sample of 1,000 gametes produced by a trihybrid individual. Please answer the following questions.

Gametes	No.	Gametes	No.
BRT	74	*BRt*	147
brt	65	*brT*	125
Brt	255	*BrT*	26
bRT	288	*bRt*	20

(a) Label the complementary types as parentals, single crossovers, or double crossovers.

(b) Diagram the chromosomes of the trihybrid, placing the genes in their proper order.

(c) Determine the map distances between the three loci.

(d) Calculate the coefficient of coincidence for the double crossovers. What does this value mean?

2. The following data were obtained from a testcross of *BbEeGg* mice.

Gametes	No.	Gametes	No.
BEG	310	*BEg*	113
beg	328	*beG*	99
Beg	52	*BeG*	18
bEG	60	*bEg*	20

(a) What is the distance from B to E ?
(b) What is the distance from E to G?
(c) What is the *expected* number of double crossovers?

3. The following data were obtained in a mapping experiment that involved three loci. A plus (+) denotes the normal (dominant) allele for each locus.

Gametes	No.	Gametes	No.
$+r+$	359	$w++$	92
$wr+$	47	$w+s$	351
wrs	4	$+++$	6
$+rs$	98	$++s$	43

(a) Construct a map of the three loci, showing their relative distances.
(b) Is there any interference?

CHROMOSOME NUMBER AND STRUCTURE

Multiple Choice

Select the single best answer for each of the following questions.

1. A hexaploid ($6n$) autopolyploid is derived from a diploid species that contains eight pairs of chromosomes. This hexaploid will have how many linkage groups? **(a)** 2, **(b)** 6, **(c)** 8, **(d)** 24, **(e)** 48, **(f)** none of the above.

2. The chromosome aberration that requires breaks in at least two nonhomologous chromosomes is **(a)** inversion, **(b)** deletion, **(c)** duplication, **(d)** translocation, **(e)** centric fusion.

3. A person who has Klinefelter syndrome, Down syndrome, and cri du chat syndrome will have how many Barr bodies in each somatic-cell nucleus? **(a)** 0, **(b)** 1, **(c)** 2, **(d)** 3, **(e)** 4, **(f)** 5.

4. A family history of Down syndrome is reason to suspect **(a)** nondisjunction, **(b)** highly mutable genes, **(c)** chemical or drug sensitivity, **(d)** inherited translocation.

5. An amphidiploid allopolyploid is produced from diploid species A (n = 6) and diploid species B (n = 8). This allopolyploid will have how many bivalents formed during prophase I of meiosis? **(a)** 7, **(b)** 14, **(c)** 28, **(d)** none of the above.

6. A woman who has no Barr bodies in her cell nuclei is **(a)** aneuploid, **(b)** euploid.

7. A chromosome could be shortened by **(a)** a deficiency and a pericentric inversion, **(b)** a translocation, **(c)** a pericentric inversion or a paracentric inversion, **(d)** a duplication and a deficiency.

8. Assume that a specimen of *Drosophila melanogaster* is found to have a chromosome constitution of XXYAAAA (where A refers to a complete haploid set of autosomes).

This fly would be classified as a **(a)** male, **(b)** female, **(c)** intersex, **(d)** metamale, **(e)** metafemale.

9. Nondisjunction of sex chromosomes during spermatogenesis can result in the birth of children with **(a)** either Klinefelter or Turner syndrome, **(b)** Klinefelter but not Turner syndrome, **(c)** Turner but not Klinefelter syndrome.

10. If a woman with blood type AB has a child by parthenogenesis (only a hypothetical possibility in humans), the child will be expected to be **(a)** a daughter with blood type AB, **(b)** a son with blood type AB, **(c)** a daughter with blood type A, B, or AB, **(d)** a son with blood type A, B, or AB, **(e)** either a son or a daughter with blood type AB, **(f)** either a son or a daughter with blood type A, B, or AB.

Problems

1. Assume that a nonsister-chromatid crossover occurs in the inverted region of the inversion heterozygotes in the two figures that follow.

(a) Diagram and describe the products of these crossover tetrads.

(b) Which inversion heterozygote, A or B, is more likely to produce a viable crossover strand that can be recovered in viable progeny? Please explain your answer.

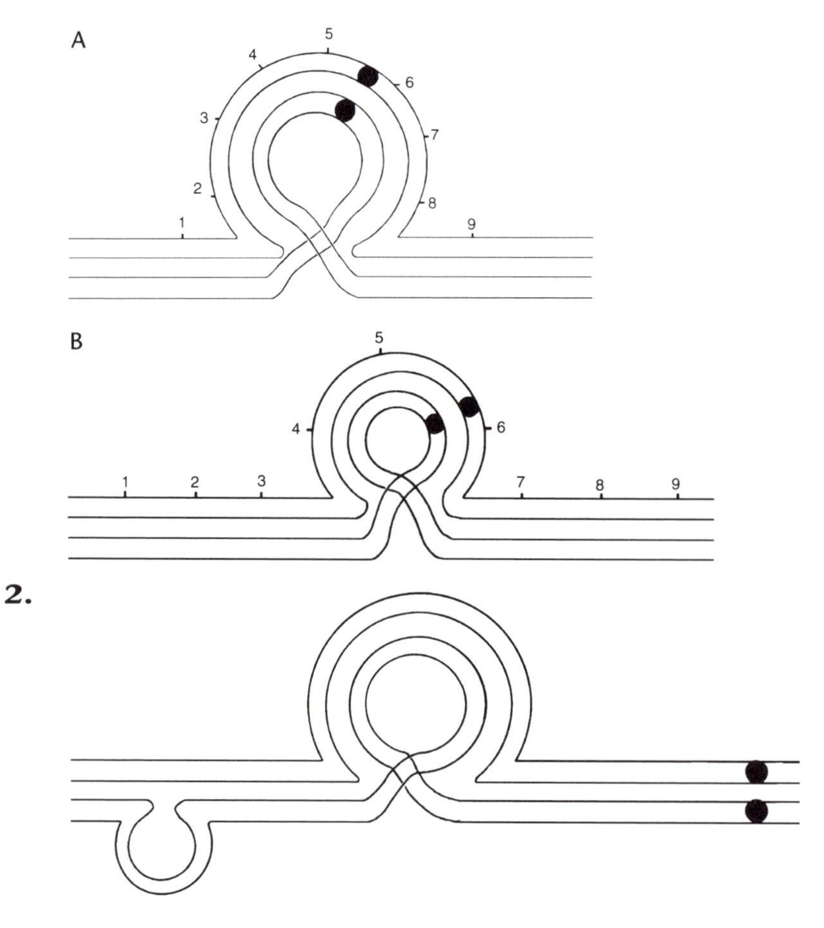

2.

Please describe the products from recombination between nonsister chromatids within the inversion loop in the preceding figure. First draw the chiasma for such a crossover event; then determine how many different chromosome aberrations are involved in this tetrad and how many genetically normal chromosomes will result from meiosis.

MUTATION AND FINE STRUCTURE

Multiple Choice

Select the single best answer for each of the following questions.

1. Ultraviolet radiation is a mutagen for bacteria because it is absorbed and changes the energy state of **(a)** proteins, **(b)** DNA, **(c)** RNA.

2. Genes that are functionally allelic **(a)** do, **(b)** do not show complementation.

3. It is **(a)** surprising, **(b)** not surprising, to find out that a substance is a carcinogen for one strain and is harmless for a second, genetically different strain.

4. At the present time, the total exposure of the population from atomic sources is **(a)** much less than 1 percent, **(b)** about 10 percent, **(c)** about 50 percent, **(d)** much more than 50 percent of the radiation from natural sources or from medical diagnoses and treatment.

5. Ultraviolet radiation in humans can cause **(a)** somatic mutations only, **(b)** germ-line mutations only, **(c)** both somatic and germ-line mutations.

6. Current studies in mutagenesis indicate that **(a)** both radiation and chemicals have a threshold, **(b)** neither radiation nor chemicals have a threshold, **(c)** radiation, but not chemicals, has a threshold, **(d)** chemicals, but not radiation, have a threshold.

7. In support of Benzer's recon concept, crossovers in *Escherichia coli* have been demonstrated **(a)** between cistrons, **(b)** within a triplet codon, **(c)** between pseudoallelic subregions.

8. Mutants induced by the base analogue 5-bromouracil (5-BU) can be caused to revert by treatment with **(a)** 5-BU, but not nitrous acid or acridine dyes; **(b)** both 5-BU and nitrous acid, but not acridine dyes; **(c)** nitrous acid, but not 5-BU or acridine dyes; **(d)** acridine dyes, but not 5-BU or nitrous acid; **(e)** 5-BU, nitrous acid, and acridine dyes.

9. Other factors being equal, which of the following types of mutations is most likely to be severely deleterious? **(a)** frameshift, **(b)** transition, **(c)** transversion.

10. In his fine-structure analysis of the rII region of phage T_4, Benzer wanted to cross rII mutants that were closely linked to each other. To facilitate finding such closely linked genes, Benzer used the techniques of **(a)** deletion mapping, **(b)** directed mutagenesis, **(c)** recombinant DNA.

11. Benzer's work indicated that genetic recombination can occur between any two adjacent **(a)** pseudoalleles, **(b)** nucleotide pairs, **(c)** cistrons, **(d)** codons.

12. Which typically has a higher rate: **(a)** reverse mutation, **(b)** forward mutation, **(c)** they both have the same rate.

13. Normal hemoglobin A has glutamic acid as the amino acid at position 6 in the beta chain. This is replaced by valine in sickle cell hemoglobin and by lysine in hemoglobin C. Considering the codons for these amino acids, one could conclude that the two mutants differ by a base change in **(a)** the same codon, **(b)** the same muton, **(c)** neither of these.

14. A mutation caused when a purine is replaced by a pyrimidine is called a **(a)** frameshift mutation, **(b)** transversion, **(c)** transition, **(d)** tautomeric shift, **(e)** deletion.

15. Repair of a potential mutation when a thymine dimer is produced **(a)** always **(b)** does not always require an endonuclease that nicks the defective strand.

Problems

1. Please define each of the following terms:

(a) Position effect
(b) Complementation
(c) Suppressor mutation
(d) Tautomeric shift
(e) Pseudoalleles

PROTEIN SYNTHESIS AND THE GENETIC CODE

Multiple Choice

Select the single best answer for each of the following questions.

1. The mRNA triplets UAG, UGA, and UAA **(a)** bind to protein releaser factors, **(b)** bind to "nonsense" tRNA molecules, **(c)** do not bind to anything, and thus allow the completed protein to fall off the ribosome.

2. Consider the ribosome shown in the accompanying figure, with the A and P binding sites indicated. The mRNA sequence in this diagram is moving to the **(a)** left, **(b)** right.

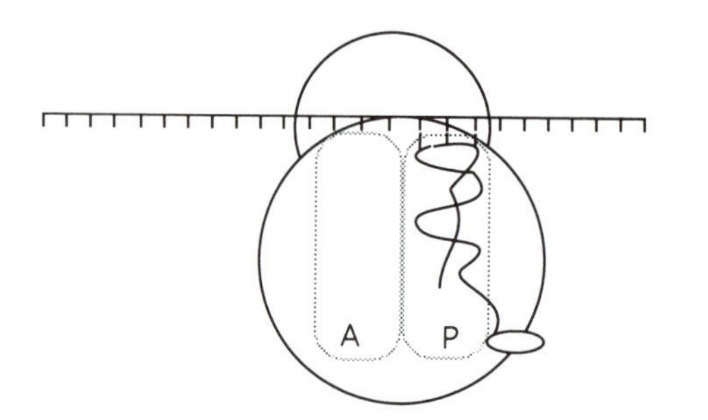

3. Which corresponds to the 5′ end of mRNA, the amino terminal or the carboxyl terminal of a polypeptide? **(a)** the amino terminal, **(b)** the carboxyl terminal.

4. If a particular polypeptide chain is 210 amino acids long, what is the minimum number of nucleotides in the mRNA molecule required for the synthesis of the chain? **(a)** 70, **(b)** 105, **(c)** 210, **(d)** 630, **(e)** 840.

5. A mutation in the gene coding for the sigma factor involved in RNA synthesis would yield a failure in RNA polymerase recognition of **(a)** the initiation point, **(b)** the termination point, **(c)** the ribosome.

6. Assume that the anticodon for an unknown amino acid is 3′ AUG 5′. The DNA sequence that would be consistent with coding for this amino acid is **(a)** 3′ TAC 5′, **(b)** 5′ TAC 3′, **(c)** 3′ UAC 5′, **(d)** 5′ UAC 3′, **(e)** 3′ AUG 3′, **(f)** 5′ AUG 3′, **(g)** 3′ ATG 5′, **(h)** 5′ ATG 3′.

7. Assume a regular copolymer AAGAAGAAGAAGAAG . . . Also assume that the genetic code is a four-base code, rather than a triplet code. The polypeptide that would be consistent with this hypothetical process would be composed of repeating units of how many different amino acids? **(a)** 1, **(b)** 2, **(c)** 3, **(d)** 4, **(e)** 5, **(f)** 6, **(g)** 7, **(h)** 8.

8. A cistron coding for a polypeptide has a base change in a glycine (Gly) codon that results in a nonsense (terminator) mutation. The glycine triplet was **(a)** GGG, **(b)** GGA, **(c)** GGC, **(d)** GGU. (Feel free to refer to the table of genetic codes at the end of the book, Table R.3, if necessary.)

9. Continuing with the example from question 8, this terminator triplet can back-mutate to code once more for glycine. It can also mutate to a missense codon that results in a different amino acid at that site. The number of different single amino acid substitutions that can occur here (with different single-base substitutions) is **(a)** 1, **(b)** 2, **(c)** 3, **(d)** 4, **(e)** 5, **(f)** 6, **(g)** 7, **(h)** 8.

10. The genetic code is said to be unambiguous, which means that **(a)** each amino acid is coded for by one codon, **(b)** each triplet codes for just one amino acid.

Problems

1. Please identify the molecule in the accompanying figure and explain its importance in protein synthesis in prokaryotes.

2. Consider the following nucleotide sequence and then answer the questions listed (please feel free to refer to Table R.3):
GCAACUGGCAUUCGACAC

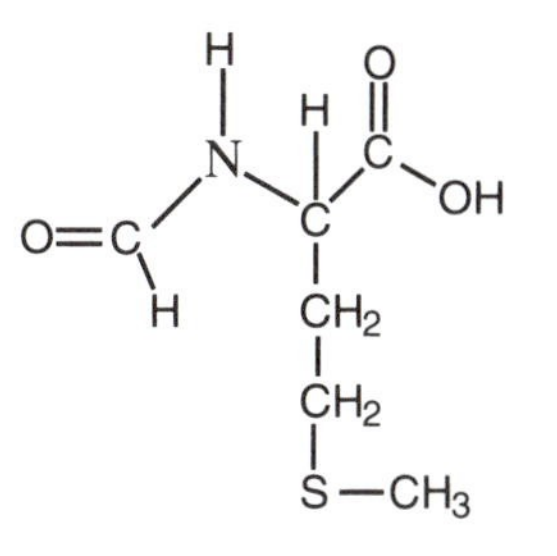

(a) What is the DNA template sequence that coded for this segment of mRNA?
(b) For which amino acid sequence does it code?
(c) A loss of the first DNA pyrimidine **(1)** would, **(2)** would not produce a chain-termination sequence somewhere along this segment.

GENE REGULATION AND DEVELOPMENT

Multiple Choice

Select the single best answer for each of the following questions.

1. The biochemical defect in a mosaic patch of hair cells in a gopher cannot be compensated for by diffusion of material from surrounding cells. On this basis, the mutant in the mosaic patch would be classified as **(a)** nonautonomous, **(b)** autonomous.

2. Posttranslational control of gene expression involves **(a)** masking of mRNA in the cytoplasm, **(b)** modification of a newly synthesized polypeptide, **(c)** synthesizing of mRNA under the control of an inducer substance, **(d)** inactivation of ribosomes in the cytoplasm.

3. The following diagram shows an operon and its regulator gene (i), where P is the promoter, O is the operator, and SG_1 to SG_4 are four structural genes. The operon is repressible. A mutation that causes the operon to function as a constitutive operon could be localized in **(a)** i or P or O, **(b)** P or O but not i, **(c)** i but not P or O, **(d)** i or O but not P, **(e)** i or P but not O, **(f)** none of the above.

i — — P O SG_1 SG_2 SG_3 SG_4

4. In an inducible system, the repressor protein that is coded for by the i gene would be expected to have **(a)** one, **(b)** two, **(c)** three recognition sites.

5. Assume the indicated somatic exchange in *Drosophila,* where y (yellow body color) and f (forked bristles) are recessive to their normal (plus, +) alleles. Given the appropriate chromatid separation at anaphase, this exchange can generate **(a)** a single yellow spot; **(b)** a single yellow, forked spot; **(c)** yellow and forked twin spots.

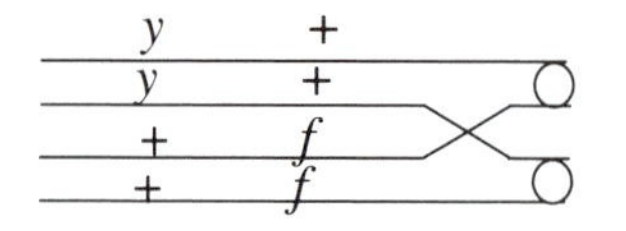

6. Histones probably function in genetic regulation by **(a)** binding to tRNA, **(b)** acting as inducers, **(c)** suppressing amino acid activation, **(d)** binding to ribosomes, **(e)** acting as a general suppressor of RNA synthesis, **(f)** blocking mRNA as it is transcribed from the DNA.

7. Polycistronic messages can be defined as messages **(a)** with several ribosomes attached to them, **(b)** that are from several operons combined together, **(c)** that have information for several different polypeptides.
8. The puffing of bands in polytene chromosomes of Diptera appears to be directly associated with **(a)** gluttony, **(b)** DNA synthesis, **(c)** RNA synthesis, **(d)** polypeptide synthesis.
9. Mutation of the regulator gene of an inducible operon will most likely result in **(a)** the absence of the enzymes, even in the presence of the inducer, **(b)** a change to a repressible system, **(c)** a change to a constitutive system, **(d)** none of the above.
10. Some hormones, such as ecdysone in insects, have been shown to exert their effects upon development by stimulating **(a)** replication, **(b)** translation, **(c)** transcription.

GENETIC MATERIAL IN POPULATIONS

Multiple Choice

Select the single best answer for each of the following questions.

1. Which of the following changes allele frequencies in a population? **(a)** migration, **(b)** inbreeding, **(c)** mutation, **(d)** segregation, **(e)** both inbreeding and mutation, **(f)** both migration and mutation, **(g)** migration, mutation, and segregation, **(h)** all of the above.
2. Assuming a Hardy–Weinberg equilibrium in a population with $p = .2$, what is the expected frequency of the heterozygote two generations later? **(a)** .04, **(b)** .8, **(c)** .32, **(d)** none of the above.
3. The mean and variance in amount of spottedness are monitored in a strain of bean plants over many generations. It is found that the amount of spotting decreases steadily over time but the variance changes only slightly. It is likely that spotting in this plant is under what type of selection: **(a)** directional, **(b)** stabilizing, **(c)** disruptive, **(d)** cyclical.
4. Assume that a certain gene affects gonadal development so that it results in a 20 percent loss of fertility. Selection against this gene will be least effective if it is **(a)** an autosomal dominant with 100 percent penetrance, **(b)** an autosomal dominant with 80 percent penetrance, **(c)** a sex-linked recessive, **(d)** an autosomal recessive, **(e)** a sex-linked dominant with 80 percent penetrance.
5. A sample of 100 people is tested for the trait "hitchhiker's thumb," and 64 of them are found to have this condition. If hitchhiker's thumb is assumed to be dominant to normal straight thumbs (its actual mode of determination is probably more complex than this), what proportion of the sampled population is expected to be

homozygous for hitchhiker's thumb? **(a)** 6 percent, **(b)** 8 percent, **(c)** 10 percent, **(d)** 16 percent, **(e)** none of the above.

6. A population of one million *Drosophila melanogaster* contains 100 flies homozygous for an autosomal recessive mutation causing curled bristles. The chance of picking two homozygous normal flies out of the population is **(a)** .0099, **(b)** .9998, **(c)** .0198, **(d)** $(.0198)^2$, **(e)** none of the above.

7. Assume that the *A* allele occurs with a frequency of .8 and the *a* allele with a frequency of .2 in three different plant populations. One population is a diploid, one is a triploid, and the third is a tetraploid. In samples of 1,000 from each of the three populations, it is more likely that the homozygous recessive phenotype will be found in **(a)** the $2n$ population, **(b)** the $3n$ population, **(c)** the $4n$ population.

8. In humans, a certain sex-linked recessive trait is found in females with a frequency of .09. The frequency with which this trait is expected to occur in males is **(a)** .09, **(b)** .03, **(c)** .0081, **(d)** .18, **(e)** none of the above.

9. If a population begins with 92 percent heterozygotes, what proportion of the population will be homozygous after two generations of self-fertilization? **(a)** 23 percent, **(b)** 46 percent, **(c)** 77 percent, **(d)** 54 percent, **(e)** none of the above.

10. Assuming that mating is random within a city and assuming that there is no selection or migration acting in the populations, the phenomenon of genetic drift would be more likely to occur in **(a)** Muskogee, Oklahoma, **(b)** Chicago, Illinois, **(c)** London, England.

ANSWERS TO PRACTICE TESTS AND CROSSWORD PUZZLES

MITOSIS AND MEIOSIS

Multiple Choice

1. (d) 23. Chromosome number was reduced during anaphase I. The centromeres will not replicate until late metaphase II or early anaphase II.

2. (a) One *A* and one *a* chromosome. You count chromosomes by counting centromeres. There are two *A* chromatids and two *a* chromatids but only one *A* chromosome and one *a* chromosome.

3. (b) Either one *A* or one *a* chromosome. The chromosome number has been reduced in anaphase I, so each cell has only one of the homologous chromosomes.

4. (a) At prophase I of meiosis.

5. (a) Eukaryotes only.

6. (b) Anaphase I of meiosis.

7. (e) Acrocentric.

8. (a) Mitosis.

9. (c) 16. A linkage group is composed of the genes on each kind of chromosome and is equal to the number of homologous pairs.

10. (a) Female. Since the egg is not fertilized, only the female's original X chromosomes are present.

Diagrams and Definitions

1. **(a)** A cell at anaphase I: $2n = 6$.

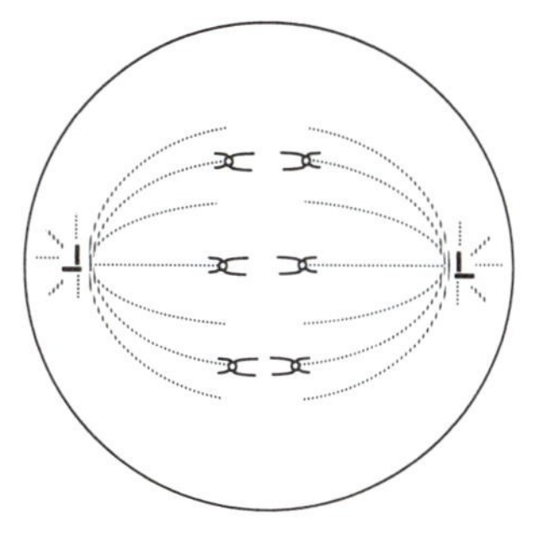

(b) A cell at anaphase II: $2n = 6$.

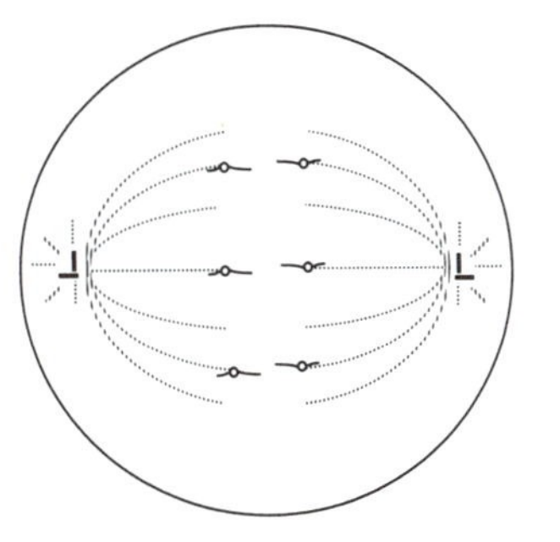

2. The definitions for the terms can be found in the Glossary.

NUCLEIC ACIDS

Multiple Choice

1. **(a)** Hydrogen bonds. In double-stranded RNA and DNA, the bases of one strand are connected to the complementary bases of the other strand by hydrogen bonds.

2. **(c)** Three nitrogenous rings. Each ladder rung contains a purine (a double-ringed nitrogenous base) hydrogen bonded to a pyrimidine (a single-ringed nitrogenous base).

3. **(d)** Not ascertainable from this information. The ratios A = T (or U) and C = G apply only to complementary bases in double-stranded DNA or RNA.

4. **(c)** The experiment demonstrated that replication was semiconservative. The number of hybrid $^{15}N^{14}N$ double-stranded DNA molecules would not change after replication 1, so when 2 molecules of DNA replicate to 4 molecules, 2 will be hybrids and 2 pure ^{14}N. When 4 double-stranded DNA molecules replicate to form 8, 2 will be hybrids, and 6 will be pure ^{14}N. When the 8 molecules replicate to form 16, again 2 will be hybrids, and 14 will be pure ^{14}N (a ratio of 1 hybrid to 7 pure ^{14}N).

5. **(c)** Double-stranded RNA. Uracil is an RNA base. The ratio of adenine to uracil is almost 1:1, which would suggest pairing.

6. **(a)** Strain I in both RNA and proteins. Genetic information is carried by the nucleic acids.

7. **(b)** Pyrimidines. Uracil has replaced thymine.

8. **(a)** Yes. Transformation could occur either way, but it is easier to identify transformation from nonvirulence to virulence or to detect restored biosynthetic activity than loss of activity.

9. **(b)** Hershey and Chase. They radioactively labeled the phosphates in DNA and the sulfur in the proteins of the T_2 virus. They then determined which of these was the infecting agent.

10. **(e)** Answers **(b)** and **(c)** only. Since A = T and C = G, then A + either C or G = T + either C or G.

Diagrams and Definitions

1. **(a)** Adenine or guanine:

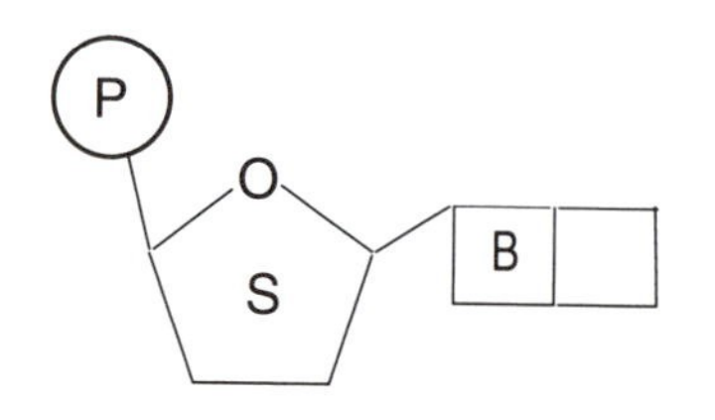

(b) Either of these:

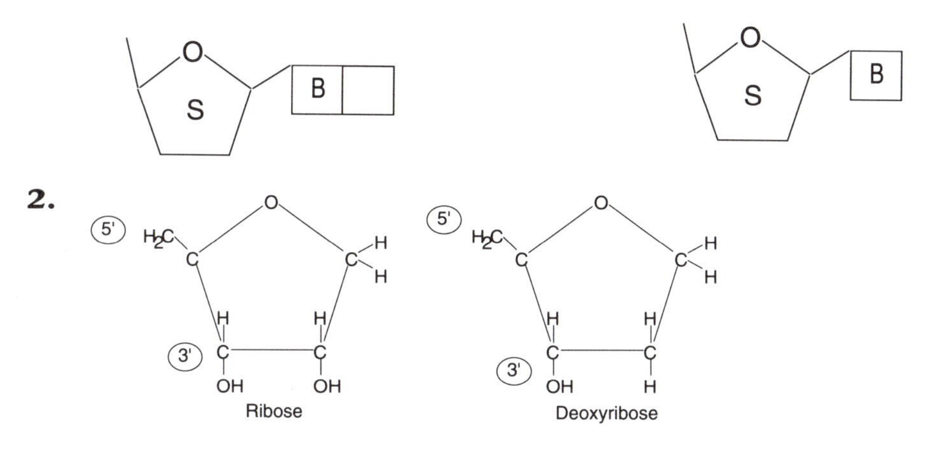

3. The definitions for the terms can be found in the Glossary.

BASIC MENDELIAN GENETICS 1

Multiple Choice

1. **(b)** A and B. Both parents must be carrying a recessive O allele in order for them to have a child with type O blood.

2. **(d)** 1/3. Since the *CyCy* die, the *Cy* parents in this cross must be heterozygous. A cross between them will produce 1/4 *CyCy,* 1/2 *Cy* +, and 1/4 + +, but the *CyCy* die, so the only offspring available to count would be the heterozygous *Cy* and the wild-type. Of these, 2/3 should be *Cy,* and 1/3 should be wild-type.

3. **(e)** 7/16. The expected ratios in a dihybrid cross are 9:3:3:1, but these genes are not phenotypically independent of each other. Thus all genotypes that are homozygous recessive *aa* or *ww* are albino.

4. **(c)** 1/8. All genotypes that are homozygous will breed true. These genotypes are *AABBCC* (1/64), *AABBcc* (1/64), *AAbbCC* (1/64), *AAbbcc* (1/64), *aaBBCC* (1/64), *aaBBcc* (1/64), *aabbCC* (1/64), and *aabbcc* (1/64) = 8/64 or 1/8.

5. **(c)** 50 percent. Fifty percent would be expected to *carry* the mutation, whereas 30 percent would be expected to express the mutant.

6. **(b)** A recessive epistatic gene.

7. **(a)** 3:1. Since the genes assort independently, the only interaction would be the death of those *A* or *a* alleles that were associated with the homozygous *i*. This would be random assortment, so we are really considering a single gene in a heterozygous monohybrid cross.

8. **(d)** AB and O. The mother's O blood has anti-A and anti-B antibodies that will lead to the destruction of fetal blood cells before an Rh immune reaction can be initiated in the mother.

9. (d) Independent assortment.

10. (b) *RrPp* x *Rrpp*. The only way the single comb could have been produced was for each parent to carry at least one recessive allele for each gene.

MENDELIAN GENETICS 2

Multiple Choice

1. (a) To the woman. Since the man's AB parent would have to give an I^A or I^B allele, he got his I^B allele from this parent. The other parent has type A blood and in this case also has the recessive allele *i*. The *i* allele is the one the man received, so his genotype is I^Bi. Although this is consistent with his being the father, it does not, of course, prove it.

2. (a) Recessive epistasis. Two loci (*A* and *B*) are needed to produce the color phenotypes in this cross. The 4/16 that are white are the result of one of the loci (*aa*) being homozygous recessive and masking segregation at the second locus (*aaB–* or *aabb*). In the presence of the dominant allele at the *A* locus, coat color can be either black (*A–B–*) or brown (*A–bb*).

3. (a) Either zero or 1/4, depending upon the genotype of the father. In the first child, the I^A allele came from the father. The mother did not give a I^B allele to the child, so her genotype must be I^Bi. However, from these data we cannot tell the father's genotype. If he is homozygous I^AI^A, there is no chance; if he is a carrier of the *i* allele, there is a 1/4 chance.

4. (a) About 23 percent. Since the six heads and six tails can be obtained in numerous ways, one would thus use the formula

$$\frac{n!}{s!t!}(p)^s(q)^t$$

Here, n (the number of flips) = 12, s (the number of heads) = 6, t (the number of tails) = 6, p (the probability of heads) = 1/2, and q (the probability of tails) = 1/2.

$$\frac{12!}{6!6!}\left(\frac{1}{2}\right)^6\left(\frac{1}{2}\right)^6 = \frac{924}{4{,}096} = .226$$

5. (c) 3/8. The probability of a normal rat is 3/4. The probability of a dwarf rat is 1/4. There are two ways to obtain a normal and a dwarf rat in any two offspring, so 3/4 • 1/4 = 3/16, 3/16 + 3/16 = 3/8. This could also be done with the probability formula, as in problem 4.

6. (d) 1/8 (= 1/2[*C*] • 1[*H*] • 1/2[*T*] • 1/2[*U*]).

7. (e) 1/32 (= 1/2[*Cc*] • 1[*HH*] • 1/4[*tt*] • 1/4[*UU*]).

8. (b) Particulate inheritance. Mendel did not know about chromosomes.

9. (a) 3/4. The parents are heterozygous tasters.

10. (b) An autosome. If it were on the X chromosome, only their daughters would receive it; on the Y, only their sons; and if cytoplasmic, neither the sons nor the daughters would receive it from their father.

11. (d) A recessive lethal trait. All the Dichaete parents and offspring are heterozygous, since Dichaete homozygotes die.

12. (e) None of the above. The correct answer is 63/64. There is only one situation in which *at least* one child with polydactyly is not found: $(1/2)^6$ or 1/64 is the probability for a family with all normal children. All the other families (1 – 1/64 = 63/64) have at least one affected child.

13. (c) 1/16. The mother is a carrier; hemophilia is sex-linked. Only her sons have a chance of expressing the gene. So there is 1/2 chance of the progeny's being a boy and 1/2 chance of its being hemophiliac if a boy. The chance of its being a hemophiliac boy is 1/2 • 1/2 = 1/4. The chance of two such boys is 1/4 • 1/4 = 1/16.

14. (b) 3. The man's genotype is $I^{A}i$, so their children could be phenotypically A, AB, or B.

15. (b) Cannot. Chi-square cannot be applied to percentages.

Problems

1. (a) Suppose that *A* – = achondroplastic dwarf, *aa* = normal size, *N* – = neurofibromatosis, *nn* = normal skin. The cross is female *A* – *nn* × male *aaN* – . Assuming that the genes are relatively rare, these parents are probably heterozygous for the two dominant alleles, so the cross is *Aann* × *aaNn*. Half of the offspring will be dwarfs, and 1/2 will be normal size; 1/2 will have neurofibromatosis, and 1/2 will not. So the ratios are 1/2 • 1/2 = 1/4, dwarfs with normal skin; 1/2 • 1/2 = 1/4, normal size with neurofibromatosis; 1/2 • 1/2 = 1/4, normal; and 1/2 • 1/2 = 1/4, dwarfs with neurofibromatosis.

(b) The son is a heterozygote for both traits (*AaNn*). The cross is *AaNn* × *aann*. The probability of a normal child is 1/4. The probability of a child showing both traits is also 1/4. The probability of three normal and two with both achondroplastic dwarfism and neurofibromatosis is

$$\frac{4!}{2!2!}\left(\frac{1}{4}\right)^2\left(\frac{1}{4}\right)^2 = 6\left(\frac{1}{4}\right)^4 = \frac{6}{256}$$

(Note: One child is normal, so the probability only has to be calculated for the remaining four.)

2. (a) Recessive.There is no evidence to help one decide whether the trait is autosomal or sex-linked.

(b) No one in the pedigree expresses the gene until a consanguineous marriage occurs.

(e) Although we do not know the frequency in the population at large, if we assume that it is a rare trait the probability is essentially zero.

(d) 1/2. We know that III-4's mother is a carrier, since III-3 is a carrier.

MENDELIAN GENETICS 3

Multiple Choice

1. **(g)** None of the above. Half will be *Rr,* 3/4 *B–,* all *Ee,* and 1/2 Ff, so 1/2 • 3/4 • 1 • 1/2 = 3/16 are dominant for all four traits.

2. **(a)** Try soft music and better food to increase yield. A heritability of 0.12 is low, so most of the variation is environmental.

3. **(e)** 1:2:2:4:1:2:1:2:1. Specifically, (1) *AABB:* (2) *AABb:* (2) *AaBB:* (4) *AaBb:* (1) *AAbb:* (2) *Aabb:* (1) *aaBB:* (2) *aaBb:* (1) *aabb.*

4. **(e)** 1:2:2:4:1:2:1:2:1. In both this question and the preceding one, all heterozygotes will be distinguishable from the homozygotes.

5. **(b)** Variable expressivity.

6. **(b)** Will not. Phenocopies are environmentally induced changes in somatic tissue.

7. **(a)** Autosomal dominant. Note that there are several examples of father-to-son transmission, so the trait cannot be X-linked.

8. **(a)** Are capable of mutating. Since kappa contains DNA (it is a bacterium), it is capable of mutating.

9. **(d)** Inheritance of cortical structure in paramecia. All the others involve DNA or RNA. The cortex of paramecia apparently does not contain nucleic acids, and preexisting structure appears to act as a guide or template for the structure of the offspring.

10. **(c)** 1/16. The probability that IV-1 is a carrier is 1/2, since III-2 must be *Aa.* The same is true for IV-2. If they are heterozygous, then the probability of an affected child is 1/4, so the total probability is 1/2 • 1/2 • 1/4 = 1/16.

11. **(d)** 9. A M, A N, A MN, B M, B N, B MN, AB M, AB N, AB MN.

12. **(d)** 85:15. Penetrance of 0.6 means the *ee* genotype would give an expectation of 25.0 • 0.6 (that is, 60 percent of 1/4 of 100 progeny) = 15 per 100 progeny. The rest have the dominant phenotype.

13. **(b)** Worse. Chi-square measures the *deviation* between the observed and the expected results.

14. **(d)** 850 to 890 g. This is one of two possible answers. One standard deviation below the mean would be the range from 810 to 850 g, but this was not listed.

15. **(e)** 10/32. The probability of an albino child is 1/2, as is the probability of a normal child. In a family of five, then

$$\frac{5!}{3!2!}\left(\frac{1}{2}\right)^3\left(\frac{1}{2}\right)^2 = \frac{10}{32}$$

Problems

1. **(a)** Autosomal recessive.

(b) 1/24. Parent IV-1 has a probability of being a carrier, which is (2/3 • 1/2 • 1/2)

= 2/12. Parent IV-2 is a heterozygote, since one parent was a homozygous recessive, and he himself is not a homozygote. The probability that the child will express the trait is thus 2/12 • 1 • 1/4 = 2/48, or 1/24.

2. From the phenotypes produced this would appear to be a dihybrid cross in which shaggy is dominant over short and cream is dominant over brown.

(a) The expected ratio: 9:3:3:1.

(b) Chi-square for goodness of fit: $\chi^2 = 4.71$, with 3 degrees of freedom.

	Observed	**Expected**	**$(O - E)^2/E$**
Shaggy brown	35	30	0.833
Short cream	31	30	0.033
Shaggy cream	79	90	1.344
Short brown	15	10	2.500
			$\chi^2 = 4.710$

(c) Since $.2 > p > .1$, do not reject the hypothesis. The answer is **(2).**

LINKAGE AND MAPPING IN DIPLOIDS

Multiple Choice

1. **(b)** PDSR.

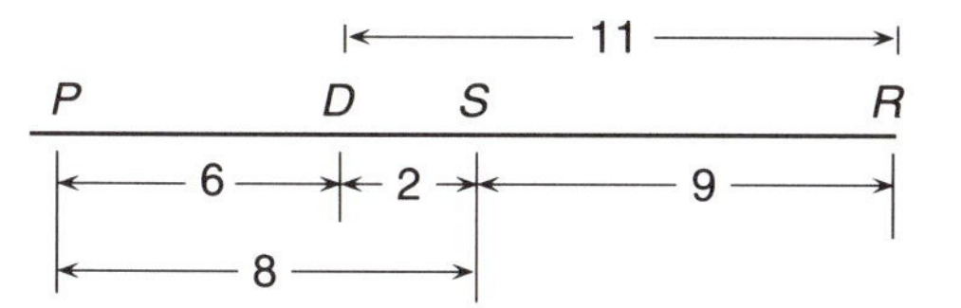

2. **(d)** 1 double, 2 single, and 1 noncrossover strand.

3. **(d)** 7 percent. Fourteen percent of the chromosomes from these meioses will be + *rpl* and *p* +; therefore, one-half (or 7 percent) will be of each type. See the explanation given in the answer to problem set 12, question 2.

4. **(d)** Neither region 1 nor region 2. It would have been possible to identify the crossovers cytologically only if the cytological markers had been at different ends of the regions in question.

5. **(d)** Females with white eyes and females with forked bristles. The two resulting attached-X chromosomes would be

6. **(e)** Approximately 6 percent. Although these crossovers could not be detected, they could be predicted from the information given, so (**d**) is *not* correct.

7. **(a)** Between the centromere and the *m* locus. The resulting attached-X chromosomes would be

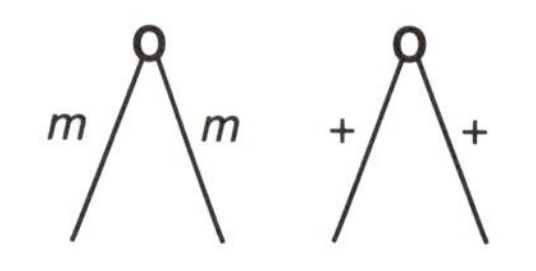

8. **(c)** Either between or within genes.

9. The answer is **(a).**

10. **(d)** 5/6. The coefficient of coincidence is the ratio of observed doubles (25 out of 100,000) to expected doubles (30 out of 100,000).

Problems

1. **(a)**

Brt	= 255	Parental	*BRT*	= 75	Single crossovers
bRT	= 288		*brt*	= 65	
BRt	= 147	Single crossovers	*BrT*	= 26	Double crossovers
brT	= 125		*bRt*	= 20	

(b) $\frac{B\ t\ r}{b\ T\ R}$

(c)

Region 1		Doubles		Region 2	
Gametes	**No.**	**Gametes**	**No.**	**Gametes**	**No.**
BTR	74	*Btr*	26	*BtR*	147
btr	65	*btR*	20	*bTr*	125
	139		46		272

$$\text{Region 1}\ \frac{(139 + 46)}{1{,}000} = .185$$

$$\text{Region 2}\ \frac{(272 + 46)}{1{,}000} = .318$$

B *T* *R*
18.5 31.8

(d) Observed = 46/1,000 = .046
Expected = .185 • .318 = .059
Coincidence = .046/.059 = .78

The crossover events are not independent; the occurrence of one event interferes with the occurrence of the second.

2. **(a)** B to E = (112 + 38)/1,000, or 15 percent.
   (b) E to G = (212 + 38)/1,000, or 25 percent.
   (c) Expected doubles = .0375(= .25 • .15) • 1,000 = 38.

3. Total count = 1,000
   (a)

Region 1		Doubles		Region 2	
Gametes	**No.**	**Gametes**	**No.**	**Gametes**	**No.**
wr+	47	*wrs*	4	+*rs*	98
++*s*	43	+++	6	*w*++	92
	90		10		190

$$w \text{ to } r = \frac{(90 + 10)}{1{,}000} = .10$$

$$r \text{ to } s = \frac{(190 + 10)}{1{,}000} = .20$$

w *r* *s*
10 20

(b) The expected frequency of doubles is .20 • .10 = .02. The observed frequency of doubles is 10/1,000 = .01. Therefore, there is interference.

CHROMOSOME NUMBER AND STRUCTURE

Multiple Choice

1. **(c)** 8. Since the multiple-chromosome sets are derived from within the same organism, there will be 8 linkage groups, 6 chromosomes in each.
2. **(d)** Translocation. There has to be a break in both chromosomes because the telomeres are "nonsticky." The other choices either can or must involve a single chromosome.
3. **(b)** 1. The presence of Barr bodies is associated with the sex chromosome and with Klinefelter syndrome. The other syndromes are associated with autosomal changes.
4. **(d)** Inherited translocation, involving chromosome 21.
5. **(b)** 14.
6. **(a)** Aneuploid. She would have lost an X chromosome and would have Turner syndrome.
7. **(b)** Translocation. All other choices include at least one aberration that would not reduce chromosome size.
8. **(a)** Male. The ratio of autosome sets to sex chromosomes determines the sex of the organism in *Drosophila*.

9. **(a)** Either Klinefelter syndrome or Turner syndrome.

10. **(c)** A daughter with blood type A, B, or AB. The offspring would be either homozygous at all loci or heterozygous like the mother, depending upon the type of parthenogenesis that produced the child.

Problems

1. **(a)** The products would be duplications and deficiencies of genes that are outside the inversion, and rearrangements of genes within the inversion.

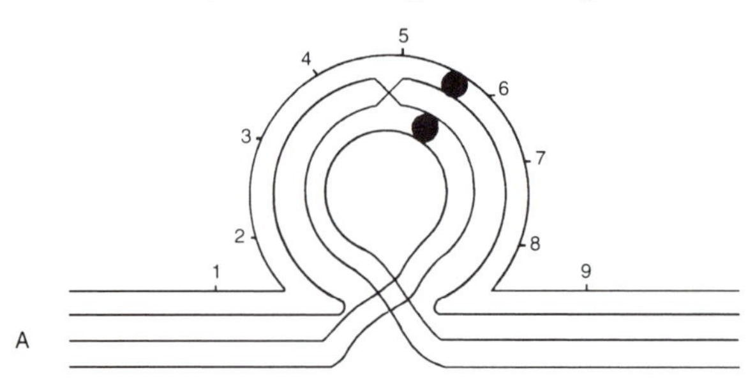

1 8 7 6 $_0$ 5 4 3 2 1; 9 2 3 4 5 $_0$ 6 7 8 9; plus one normal 1 2 3 4 5 $_0$ 6 7 8 9 and one inverted chromosome, 1 8 7 6 $_0$ 5 4 3 2 9.

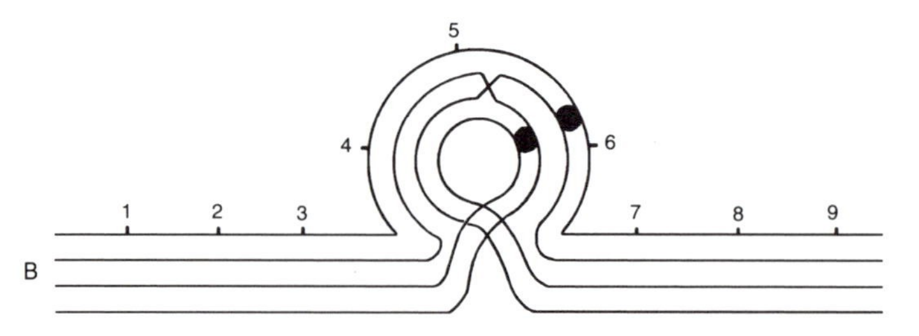

1 2 3 6 $_0$ 5 4 3 2 1; 9 8 7 6 $_0$ 5 4 7 8 9; plus one normal 1 2 3 4 5 $_0$ 6 7 8 9 and one inverted chromosome, 1 2 3 6 $_0$ 5 4 7 8 9.

(b) The larger inversion (A). The larger the inversion, the smaller the deficiency and duplications will be. If there is no position effect, one would be more likely to obtain a viable offspring from the large inversion. If the problem had not stipulated that the crossover occurred within the inverted region, however, the smaller inversion would be less severe. The smaller the inversion, the less often a random crossover occurs within the inversion. Crossovers outside the inversion produce no duplications or deficiencies.

2. Chromatids 1 and 3 are noncrossover products. The two chromosomes are heterozygous for a paracentric inversion. Either chromatids 1 and 2 are in normal sequence, or 3 and 4 are in normal sequence. In addition, the small distal loop indicates that either (a) chromatids 3 and 4 carry a small duplication, whereas 1 and 2 are normal, or (b) chromatids 1 and 2 carry a small deficiency, whereas 3 and 4 are normal. Chromatids 2 and 4 will give crossover products that include a dicentric bridge and an acentric fragment with associated duplication and deficiency. Depending on the reason for the small distal loop, either noncrossover chromatid 1 or 3 will be the only normal chromosome.

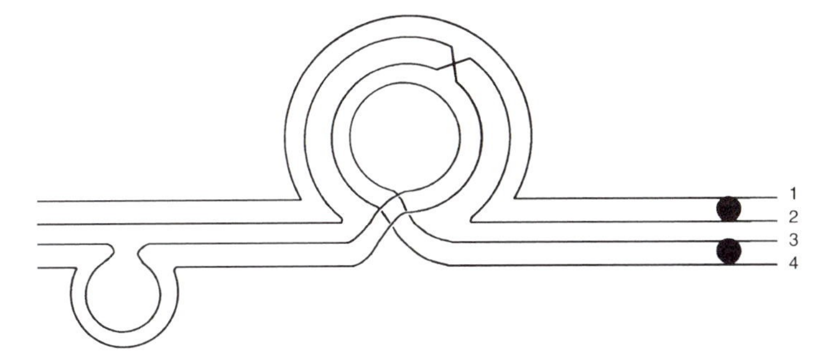

MUTATION AND FINE STRUCTURE

Multiple Choice

1. **(b)** DNA. Absorption of ultraviolet light in cells is confined primarily to compounds with organic rings, such as purines and pyrimidines. The genetic material in bacteria is DNA, not RNA.

2. **(b)** Do not show complementation. The cis–trans complementation test is one of the methods used to determine functional alleles.

3. **(b)** Not surprising. Genetically determined physical differences in the way in which chemicals are absorbed and turned over in the body can lead to major variations in response. Indeed, this is one of the basic problems studied in the medical field of pharmacogenetics.

4. **(a)** Much less than 1 percent.

5. **(a)** Somatic mutations only. Ultraviolet light affects only those compounds that can absorb it. The rays are too long to penetrate more than a cell layer or so.

6. **(d)** Chemicals, but not radiation, have a threshold.

7. **(b)** Within a triplet codon.

8. **(b)** Both 5-BU and nitrous acid. Mutations that are induced by base analogues can be induced to revert by base analogues. Both nitrous acid and 5-BU produce transitions and are bidirectional in their action.

9. **(a)** Frameshift.

10. **(a)** Deletion mapping.

11. **(b)** Nucleotide pairs. The term *recon* is used for the smallest unit of recombination.

12. **(b)** Forward-mutation. Back mutation requires a specific sequence of events to correct a previous mutation, whereas forward mutation just requires a change.

13. **(a)** Codon. The codons for the three amino acids are glutamic acid–GAA, GAG; valine–GUA, GUG; and lysine–AAA, AAG. Different bases (mutons) are changed in these two mutants.

14. **(b)** Transversion. Transversions are believed to be caused by error-prone pathways of repair or by gaps resulting from replication before dimer repair can be accomplished.

15. **(b)** Not always. Light repair does not require endonuclease nicks.

Problems

1. Answers to the definitions can be found in the Glossary.

PROTEIN SYNTHESIS AND THE GENETIC CODE

Multiple Choice

1. **(a)** Protein releaser factors.

2. **(b)** Right.

3. **(a)** The amino terminal. The polypeptide grows carboxyl to amine, so the first amino acid has a free amino end.

4. **(d)** 630 (= 210 • 3 nucleotides per codon).

5. **(a)** The initiation point.

6. **(g)** 3′ ATG 5′. The sequence is

tRNA 3′ AUG 5′
mRNA 5′ UAC 3′
DNA 3′ ATG 5′

7. **(c)** 3.

8. **(b)** GGA. The first G is changed to U, forming nonsense codon UGA.

CGA, AGA } = Arg
UCA = Ser
UUA = Leu
UGG = Trp
UGU, UGC } = Cys

9. **(e)** 5. UGA → GGA = back mutation to glycine

10. **(b)** Each triplet codes for just one amino acid.

Problems

1. The amino acid is *N*-formyl-methionine. Its codon represents the starting site for protein synthesis. New amino acids cannot attach to the amine end of this molecule.

2. **(a)** CGT TGA CCG TAA GCT GTG.
(b) Ala Thr Gly Ile Arg His.
(c) The answer is **(2).** It would not produce a termination sequence. The first pyrimidine is cytosine.

GENE REGULATION AND DEVELOPMENT

Multiple Choice

1. **(b)** Autonomous. The cells were not affected by the environment around them.
2. **(b)** Modification of a newly synthesized polypeptide. Note that the question asked for *posttranslational,* not *posttranscriptional* control.
3. **(d)** *i* or *O* but not *P.*
4. **(b)** Two: one for the operator and one for the inducer.
5. **(c)** Yellow and forked twin spots:

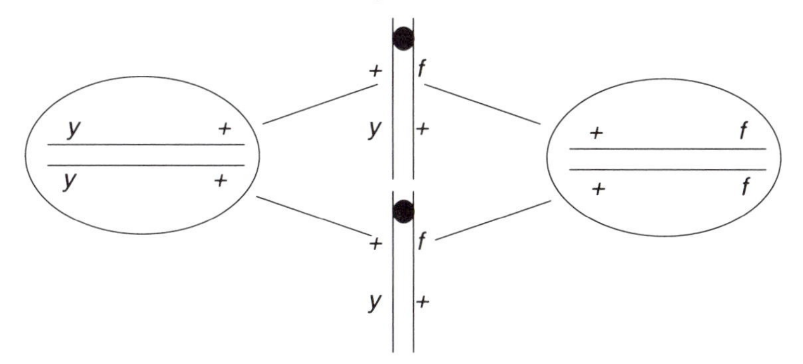

6. **(e)** Acting as a general suppressor of RNA synthesis.
7. **(c)** Messages that have information for several different polypeptides.
8. **(c)** RNA synthesis. Note: The phrase "directly associated" is important.
9. **(c)** An inducible changed to a constitutive system.
10. **(c)** Transcription.

GENETIC MATERIAL IN POPULATIONS

Multiple Choice

1. **(f)** Both migration and mutation. Inbreeding will only change genotype frequencies.
2. **(c)** .32. The stipulation of two generations can be ignored, since in equilibrium, frequencies are stable from one generation to the next.

3. (a) Directional selection.

4. (d) An autosomal recessive. In an autosomal recessive, the proportion of alleles expressed in the phenotype (for selection to act upon) is the smallest, and selection would thus be least effective.

5. (d) 16 percent. If 64/100 are phenotypically dominant, 36/100 are homozygous recessive, $q^2 = .36$, $q = .60$, $p = .40$, and $p^2 = .16$.

6. (e) None of the above. The answer is $(.99)^4$, since the probability of one being a homozygote is $p^2 = (.99)^2$. The probability of two such independent events is the product of their likelihoods $(.99)^2(.99)^2 = .9606$.

7. (a) The $2n$ population. Since each chromosome represents an independent event, the probability of being homozygous decreases as the number of homologues increases.

8. (e) None of the above. The answer is .3. The frequency in males is the square root of the frequency in females, since females have two X chromosomes, but males have only one.

9. (c) 77 percent. The proportion of homozygotes is 1 minus the proportion of heterozygotes. The heterozygotes will decrease by one-half each generation, being reduced to 23 percent after two generations.

10. (a) Muskogee, Oklahoma. Drift is greatest in the smallest population.

ANSWERS TO CROSSWORD PUZZLES

Chapter 2 (p. 14)

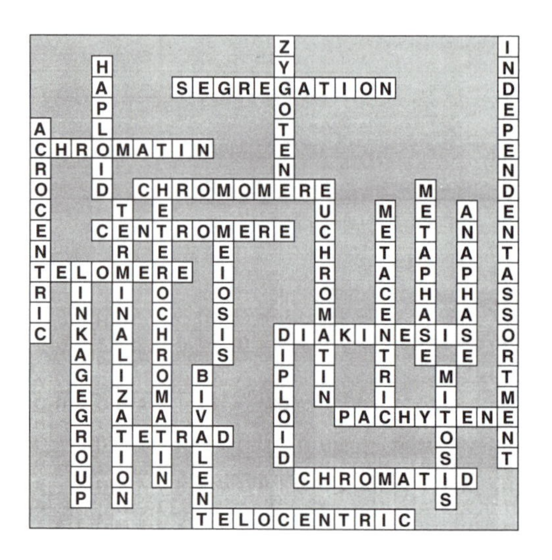

Chapter 3 (p. 24)

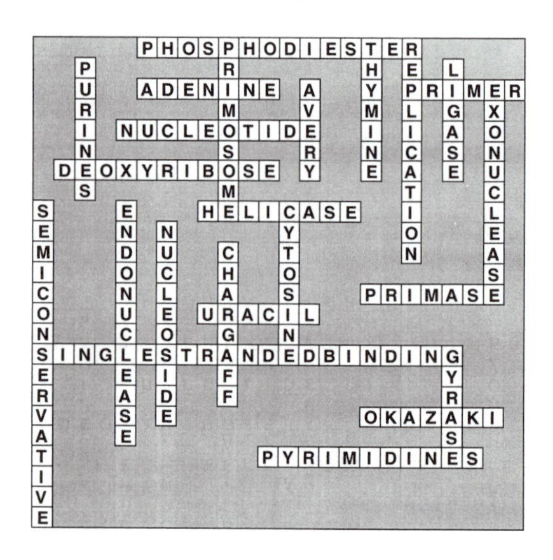

Chapter 4 (p. 35)

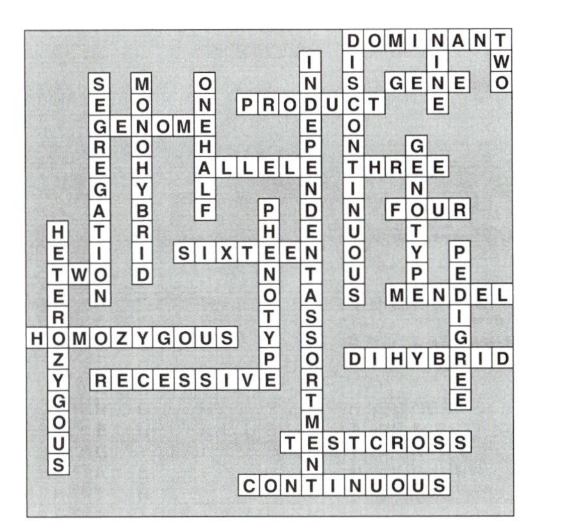

Chapter 6 (p. 61)

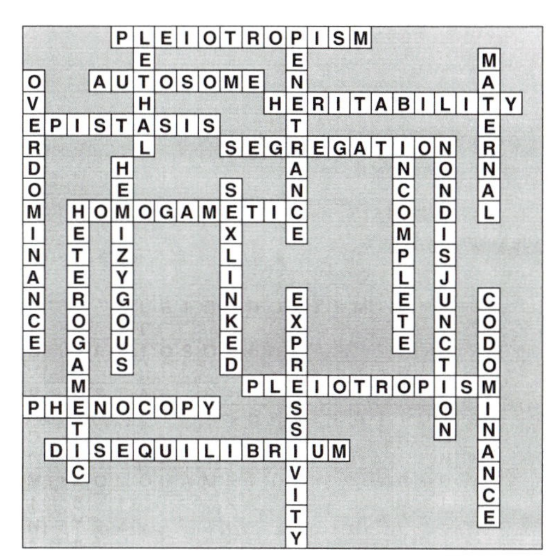

Chapter 9 (p. 85)

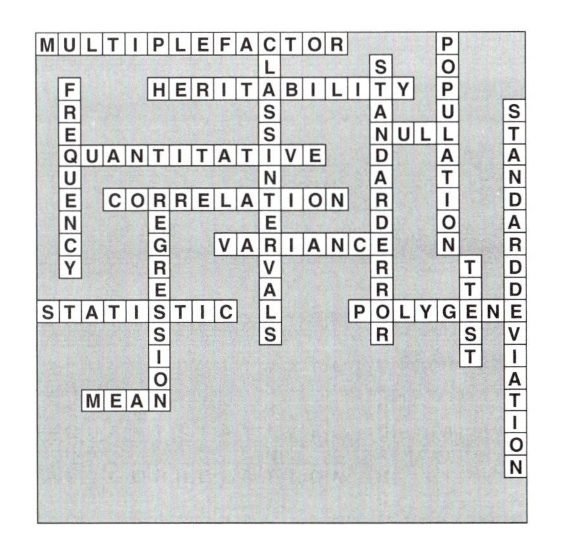

Chapter 13 (p. 117)

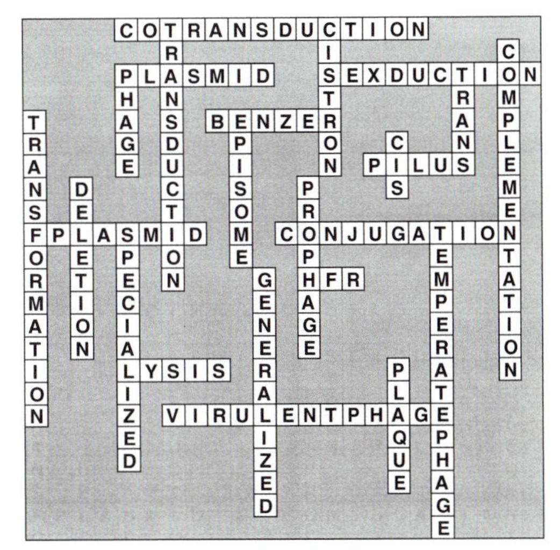

Chapter 16 (p. 137)

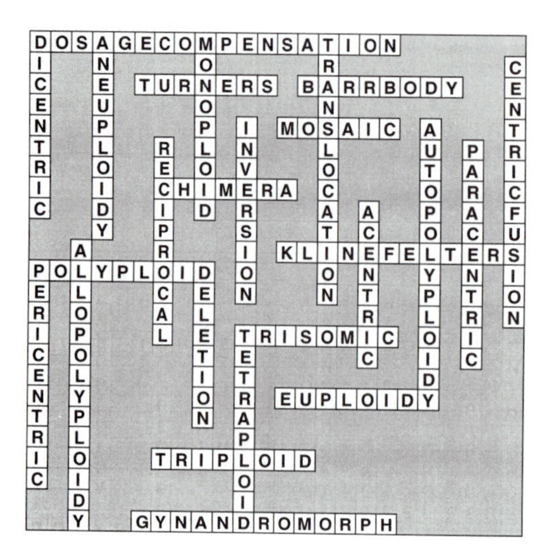

Chapter 17 (p. 157)

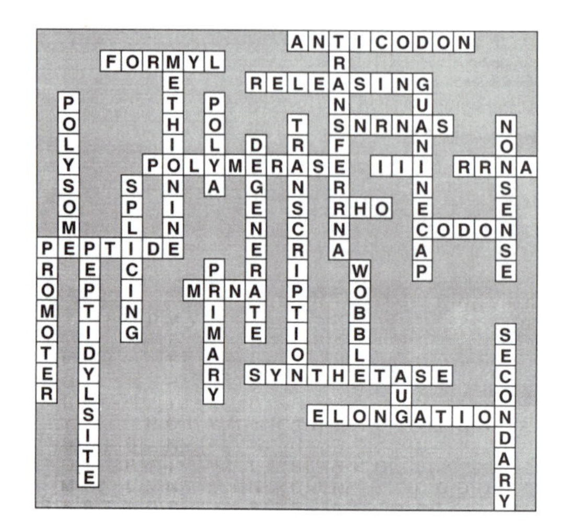

Chapter 21 (p. 194)

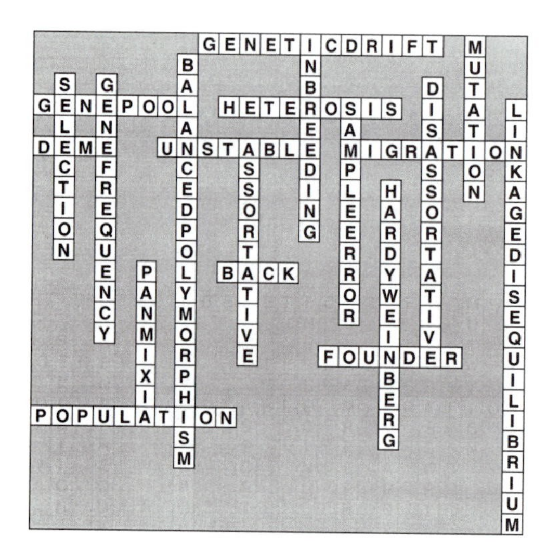

Chapter 22 (p. 204)

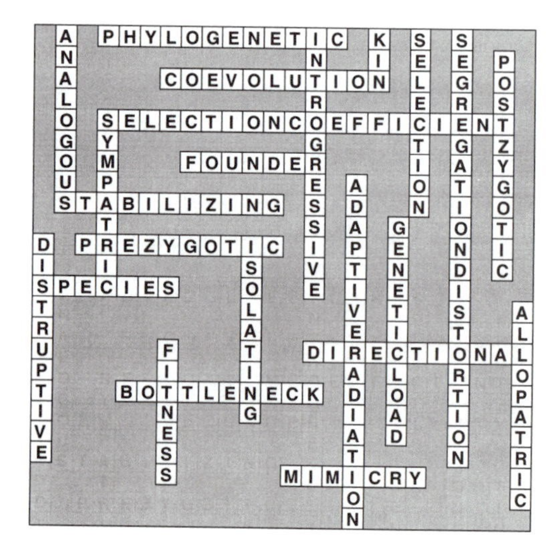

25 LANDMARKS IN THE HISTORY OF GENETICS

The science of genetics has been the recipient of information from many unrelated fields of science as well as those closely related, such as cytology and evolutionary biology. Genetics, cytology, and evolutionary biology are endeavors that tend to link biochemistry, geology, all of biology, and many other sciences in an all-encompassing theory of life on this earth. Correspondingly, change in our knowledge in any of these areas advances our knowledge in the others. Thus, it is quite hard to create a chronological list of the historic events that have had an impact on genetics. Even an attempt to describe the most important or most directly associated events is difficult. This list is not meant to be inclusive. The chronology in R. C. King's *Dictionary of Genetics* (New York: Oxford University Press, 1990) is an excellent source of information. Our list is not extensive for the last few years. This is not because of a lack of important research, but rather an inability to step back and observe from a distance the numerous events as they unfold. The history of genetics continues to be written at a dazzling pace.

The formal rules of genetic transmission and knowledge of *DNA* and gene action are fairly modern advances. Yet several examples indicate that an appreciation of inheritance has a long history. Clay tablets suggest that Babylonians bred horses according to pedigrees 6,000 years ago. A law in the Jewish Talmud written before 600 A.D. recognized the familial inheritance of hemophilia by excusing certain male relatives from ritual circumcision (from M. W. Strickberger, *Genetics,* 3rd ed. New York: Macmillan, 1985). But it was not until the mid- to late 1800s that growing knowledge of biological organization, geological history, and challenges to ideas like Lamarck's inheritance of acquired characters set the stage for establishing a new branch of biology.

Although the science of genetics perhaps could be said to begin with C. Darwin and G. Mendel, there were numerous events that predate these men and were important to the sciences of evolution and genetics. For example, in 1668, F. Redi disproved the theory of spontaneous generation of maggots. L. Spallanzani demonstrated in 1769 that "spontaneous generation" of microorganisms was preventable if containers were heated and sealed. In 1780, just four years after the United States declared its independence from England, L. Spallanzani performed artificial insemination experiments on amphibians, demonstrating the need for spermatic fluid for fertilization and development.

Experiments like these that showed the continuity among generations was also aided by technical advances. The use of simple microscopes by Hooke (1635–1703), Leeuwenhoek (1632–1723), and others to study biological materials added a critical level of precision to knowledge about cells and early development.

We have decided to start this history of genetics in the 1800s. This is about the time that science began to understand the importance of microorganisms in disease. Travel was resulting in a new understanding of geology and fossils were being examined from a different perspective. A social conscience was forming and biology was on the doorstep of a revolution.

1818	W. C. Wells	Suggests selection was responsible for African populations that were relatively resistant to local diseases (thus the first to suggest natural selection)
1820	C. F. Nasse	Suggests a *sex-linked* mode of inheritance for hemophilia
1831	R. Brown	Notes *nuclei* within cells
1838–39	M. J. Schleiden and T. Schwann	Develop the theory that plants and animals are composed of cells (*cell theory*)
1858	C. Darwin and A. Wallace	Present abstracts to the Linnean Society of London on the theory of evolution based on natural selection, Darwin publishes *On the Origin of Species* one year later
1866	G. Mendel	Publishes his genetic studies on garden peas, *Versuche über Pflanzenhybriden* (Experiments on Plant Hybridization)
1871	F. Miescher	Publishes a method for the isolation of a cell nucleus; isolates "nuclein," which is now known to be a nucleic acid and protein mixture
1875	E. Strasburger	Describes *cell division* in plants
1876	F. Galton	Uses twin studies to describe the relative influence of heredity and the environment (*nature vs. nurture*) on behavioral traits
1879	W. Flemming	Demonstrates that nuclear division involves splitting of the chromosome and migration of sister *chromatids;* later in 1882, he will coin the term *mitosis*
1899	M. W. Beijerinck	Demonstrates tobacco mosaic disease is the result of a self-reproducing subcellular form of life, the *virus*
1900	H. de Vries, C. Correns, and E. von Tschermak	Independently perform experiments that parallel Mendel's studies and arrive at similar results, discover Mendel's paper, recognize its significance, and stress its importance
1900	K. Landsteiner	Discovers human blood groups
1901	H. de Vries	Uses the term *mutation* to describe the sudden, spontaneous changes in hereditary material
1902	T. Boveri and W. Sutton	Propose the *chromosome theory of inheritance*
1905	W. Bateson and R. C. Punnett	Using the sweet pea as an experimental model, report the first example of genes linked to a chromosome (*chromosome linkage*)
1908	G. H. Hardy and W. Weinberg	Independently formulate the Hardy–Weinberg law of population genetics

1909	A. E. Garrod	With the publication of *Inborn Errors of Metabolism,* the earliest to discuss biochemical genetics
1909	W. Johannsen	While studying the inheritance of seed size, realizes the distinction between appearance of an organism and its actual genetic composition; coins the terms *phenotype, genotype,* and *gene*
1909	H. Nilsson Ehle	Proposes the multiple-factor hypothesis to explain *quantitative inheritance*
1910	T. H. Morgan	In discovering the white eye mutant in *Drosophila,* describes sex-linkage in this fly; *Drosophila* genetics begins
1911	T. H. Morgan	Demonstrates several genes are linked on the X chromosome in *Drosophila*
1912	T. H. Morgan	Discovers a *sex-linked lethal* in *Drosophila;* demonstrates that male *Drosophila* do not have recombination
1913	A. H. Sturtevant	Experimentally demonstrates the linkage concept in *Drosophila* and produces the *first genetic map*
1914	C. B. Bridges	Demonstrates meiotic *nondisjunction* in *Drosophila*
1916	H. J. Muller	Discovers *interference* with recombination in *Drosophila*
1917	O. Winge	Discusses the importance of *polyploidy* in the evolution of angiosperms
1917	C. B. Bridges	Finds the first chromosome *deficiency* in *Drosophila*
1918	H. Spemann and H. Mangold	Demonstrate embryonic induction
1919	T. H. Morgan	Calls attention to the relationship between the *haploid number of chromosomes* and the number of *linkage groups* in *Drosophila*
1923	J. K. Santos, H. Kihara, T. Ono, and O. Winge	Demonstrate the *XX–XY sex determination* in certain dioecious plants: Santos for *Elodea,* Kihara and Ono for *Rumex,* and Winge for *Humulus*
1926	S. S. Chetverikov	Begins the genetic analysis of wild populations of *Drosophila*
1927	B. O. Dodge	Initiates genetic studies on *Neurospora*
1927	H. J. Muller	Demonstrates that mutations can be induced by x-rays
1928	L. J. Stadler	Demonstrates the *dose–frequency* curve is linear in artificially induced mutations

1928	F. Griffith	Discovers transformation in pneumococci
1929	C. D. Darlington	Suggests that chiasmata function to hold homologues together during metaphase I of meiosis
1930–32	R. A. Fisher, J. B. S. Haldane, and S. Wright	Develop the mathematical foundations for population genetics
1931	C. Stern	Provides the cytological proof of crossing over in *Drosophila*
1931	H. B. Creighton and B. McClintock	Independently of C. Stern, provide the cytological proof of crossing over in maize
1932	M. Knoll and E. Ruska	Make the prototype of the electron microscope
1933	T. S. Painter	Begins cytogenetic studies of *Drosophila* salivary gland chromosomes
1934	A. Følling	Discovers phenylketonuria, the *first hereditary metabolic disorder* associated with mental retardation
1934	H. Bauer	Suggests that the giant chromosomes found in the salivary gland cells of fly larvae are *polytene*
1935	J. B. S. Haldane	Calculates the spontaneous mutation frequency of a human gene
1935	C. B. Bridges	Publishes the first salivary gland chromosome maps for *Drosophila*
1937	Th. Dobzhansky	Publishes *Genetics and the Origin of Species,* a landmark in the study of evolutionary genetics
1939	E. L. Ellis and M. Delbrück	Invent the "*one-step growth*" method of experimenting with bacterial phages
1939	E. Knapp and H. Schreiber	Demonstrate the correspondence between the effectiveness of ultraviolet light in inducing mutation and the absorption spectrum of nucleic acid
1940	E. B. Ford	Defines genetic polymorphism
1941	G. W. Beadle and E. L. Tatum	Introduce the one gene–one enzyme hypothesis
1944	O. T. Avery, C. M. MacLeod, and M. McCarty	In describing the pneumococcus transforming principle, suggest that DNA and not protein is the hereditary material
1948	H. K. Mitchell and J. Lein	Demonstrate that in certain mutant strains of *Neurospora* tryptophan synthetase is missing: the first evidence for the one gene–one enzyme theory
1948	P. A. Gorer, S. Lyman, and G. D. Snell	Discover H2, the first major histocompatibility locus found in mice

1948	G. D. Snell	Formulates the laws of transplantation acceptance and rejection; introduces the term *histocompatibility gene*
1949	A. D. Hershey and R. Rotman	Demonstrate genetic recombination in bacteriophage
1950	B. McClintock	Proposes transposable elements in maize
1950	E. Chargaff	Demonstrates that the numbers of adenine and thymine groups are always equal and the numbers of cytosine and guanine groups are likewise equal in DNA
1951	Y. Chiba	Demonstrates the presence of DNA in chloroplasts
1952	W. Beermann	Suggests that the puffing patterns of polytene chromosomes reflect differential gene activities
1952	A. D. Hershey and M. Chase	Demonstrate that DNA is the genetic material in phages
1953	J. D. Watson and F. H. C. Crick	Propose the *double-helix model* for DNA
1953	A. Howard and S. R. Pelc	Demonstrate the cell cycle (G_1, S, and G_2 periods preceding mitosis)
1955	S. Benzer	Coins the terms *cistron, recon,* and *muton* while working out the *fine structure map* of the rII region of phage T_4
1956	H. B. D. Kettlewell	Studies industrial melanism in the pepper moth: the first well-documented change in gene frequency by natural selection
1956	F. Jacob and E. L. Wollman	Experimentally interrupt mating in *E. coli* and demonstrate DNA is inserted from the donor bacterium into the recipient
1956	J. H. Tjio and A. Levan	Demonstrate the diploid number of humans is 46
1956	M. J. Moses and D. Fawcett	Independently observe synaptonemal complexes
1957	V. M. Ingram	Demonstrates that sickle-cell and normal hemoglobin differ by one amino acid
1958	F. Jacob and E. L. Wollman	Demonstrate that the DNA of *E. coli* is circular and suggest that different linkage groups in *Hfr* strains result from different insertion points of a factor that ruptures the circular DNA
1958	F. H. C. Crick	Predicts the discovery of tRNA in suggesting that during protein formation the amino acid is carried to the template by an adapter molecule of nucleotides
1958	M. Meselson and F. W. Stahl	Demonstrate the *semiconservative replication* of DNA in *E. coli*

1959	S. Ochoa	Discovers the first RNA polymerase; with A. Kornberg receives Nobel Prize for work with the in vitro synthesis of nucleic acids
1959	E. Freese	Suggests that mutation can occur by changes in single base-pairs in DNA; uses the terms *transition* and *transversion*
1960	P. Doty, J. Marmur, J. Eigner, and C. Schildkraut	Demonstrate that separation and later recombining of complementary strands of DNA are possible
1961	F. Jacob and J. Monod	Propose the *operon theory* of gene regulation; also suggest the existence of mRNA
1961	S. Brenner, F. Jacob, and M. Meselson	Demonstrate the presence of *mRNA* with F. Gros, W. Gilbert, H. Hiatt, C. G. Kurland, and J. D. Watson
1961	M. F. Lyon and L. B. Russell	Independently find evidence suggesting *deactivation of one of the X chromosomes* in female mammals
1961	B. D. Hall and S. Spiegelman	Demonstrate a technique for producing hybrid molecules containing one strand of DNA and one of RNA that leads to the isolation and characterization of mRNAs
1961	F. H. C. Crick, L. Barnett, S. Brenner, and R. J. Watts-Tobin	Demonstrate that the genetic language is a *three-letter code*
1962	J. B. Gurdon	Demonstrates that *the somatic and germinal nuclei are qualitatively alike;* his experiment on frogs involved enucleating an egg and replacing the nucleus with an intestinal cell nucleus; normal fertile frogs develop from the modified egg
1963	E. Margoliash	Sequences cytochrome c polypeptides from a variety of organisms and produces the *first phylogenetic tree utilizing a specific gene product*
1963	L. B. Russell	Demonstrates that a piece of an autosomal chromosome translocated to an X chromosome would be deactivated with the X chromosome
1964	A. S. Sarabhai, A. O. W. Stretton, S. Brenner, and A. Bolle	Demonstrate *colinearity of gene and protein product* in the virus T_4
1964	C. Yanofsky, B. C. Carlton, J. R. Guest,	Demonstrate the *colinearity of gene and protein product* (tryptophan synthetase) in the bacterium *E. coli*

	D. R. Helinski, and U. Henning	
1964	D. D. Brown and J. B. Gurdon	Illustrate the nucleolus is involved in the production of 18S and 28S rRNAs
1965	L. Hayflick	Discovers human diploid cells in tissue culture have about *50 doubling cycles*
1965	R. W. Holley	Completely *sequences alanine tRNA* from yeast
1965	A. J. Clark and A. D. Margulies	Find bacteria mutants that are abnormally sensitive to UV light; this suggests enzyme systems for *repairing damaged DNA*
1966	H. G. Khorana and M. W. Nirenberg	Working independently complete the genetic code
1966	F. H. C. Crick	Proposes the *wobble theory* to explain the degeneracy pattern found in the genetic code
1966	M. Waring and R. J. Britten	Demonstrate the presence of repetitious nucleotide sequences (*repetitive DNA*) in vertebrates
1966	V. A. McKusick	Publishes a catalogue listing about 1,500 genetic disorders of *Homo sapiens* (*Mendelian Inheritance in Man*)
1966	R. C. Lewontin and J. L. Hubby	Use electrophoretic techniques to demonstrate heterozygosity of proteins in natural populations
1966	H. Harris	Uses electrophoretic techniques to demonstrate human enzyme polymorphisms
1967	K. Taylor, S. Hredecna, and W. Szybalski	Demonstrate genes on the same chromosome may have different orientations of transcription: one gene may be read 3′–5′ on one strand, another read 3′–5′ on the other strand of the double helix
1968	R. T. Okazaki	Reports that short lengths of DNA are synthesized during replication discontinuously; pieces are later spliced together (*Okazaki fragments*)
1968	H. O. Smith, K. W. Wilcox, and T. J. Kelley	Isolate the first restriction endonuclease (*Hin*d II)
1968	S. Wright	Publishes the first of four volumes of *Evolution and the Genetics of Populations*
1968	E. H. Davidson, M. Crippa, and A. E. Mirsky	Demonstrate a long-lived form of *mRNA* is stored in the egg for use in early embryogenesis
1968	J. E. Cleaver	Demonstrates that xeroderma pigmentosum in humans is the result of a *defective DNA repair mechanism*

1969	C. Boon and R. Ruddle	Use somatic hybrid cell line containing human and mouse chromosomes to correlate the loss of human chromosomes with loss of phenotypic characters; this leads to the use of hybrid lines to assign specific loci to particular human chromosomes
1969	H. A. Lubs	Demonstrates a fragile site on the human X chromosome in some mentally retarded males
1970	T. Caspersson, L. Zech, and C. Johansson	Use quinacrine dyes to demonstrate specific fluorescent banding patterns in human chromosomes
1970	R. Sager and Z. Ramanis	Publish a genetic map of eight genes residing on the chloroplast chromosome of *Chlamydomonas:* the first non-Mendelian genetic map
1971	M. L. O'Riordan, J. A. Robinson, K. E. Buckton, and H. J. Evans	Discover that all 22 pairs of human autosomes are visually identifiable by staining with quinacrine hydrochloride
1972	G. H. Pigott and N. G. Carr	Hybridize DNA from cyanobacteria to the chloroplasts of *Euglena gracilis;* this genetic homology supports the theory that chloroplasts are descendants of endosymbiotic cyanobacteria
1972	J. Mendlewicz, J. L. Fleiss, and R. R. Fieve	Demonstrate a psychosis (manic-depression) is genetic and a dominant gene located on the short arm of the X chromosome is involved
1972	D. E. Kohne, J. A. Chisson, and B. H. Hoyer	Use DNA–DNA hybridization techniques to study the evolution of primates; conclude that the chimpanzee is closely related to humans
1972	P. Berg	Produces the *first recombinant DNA in vitro*
1973	H. Boyer and S. Cohen	Use a *plasmid to clone DNA;* this led to recombinant DNA techniques
1974	A. Tissieres, H. K. Mitchell, and U. M. Tracy	Discover six new proteins are synthesized in *Drosophila* when given heat shocks
1974	B. Ames	Develops a rapid method for detecting mutagenic compounds
1975	Asilomar meetings	*Historic meeting* where molecular biologists from all over the world meet to write *rules to guide research in recombinant* DNA; in the USA, the NIH Recombinant DNA Committee issues guidelines to minimize any potential risks of this research

1975	D. Pribnow	Determines the nucleotide sequences of two bacteriophage promoters; forms a model of promoter function
1975	E. M. Southern	Demonstrates a method for transfer of DNA fragments from an agarose gel to nitrocellulose filters
1975	G. Morata and P. A. Lawrence	Demonstrate a mutant (*engrailed*) whose normal function defines the boundary between wing compartments as the wing develops in *Drosophila;* normal cells recognize anterior versus posterior compartments
1976	B. G. Burrell, G. M. Air, and C. A. Hutchison	Report the presence of overlapping genes in the phage ϕX174
1976	Genentech	First genetic engineering company
1977	A. M. Maxam, W. Gilbert, and F. Sanger	Work out methods for nucleotide sequencing of DNA
1977	F. Sanger, et al.	Sequence the DNA genome of bacteriophage ϕX174
1977	W. Gilbert	Synthesizes insulin and interferon in bacteria
1977	P. Sharp, R. Roberts, et al.	Demonstrate *introns* in eukaryotic genes
1979	J. C. Avise, R. A. Lansman, and R. O. Shade	Using restriction endonucleases and mitochondrial DNA, measure the relationships of organisms in natural populations
1979	B. G. Barrell, A. T. Bankier, and J. Drouin	Discover that the genetic code of human mitochondria has some atypical characteristics
1979	D. V. Goeddel, et al.	Synthesize the human growth hormone gene
1980	U.S. Supreme Court	Rules that patents can be awarded for genetically modified microorganisms
1980	L. Olsson and H. S. Kaplan	Manufacture a pure antibody in a laboratory culture
1981	L. Margulis	Summarizes the evidence for the symbiosis theory for the origin of such organelles as mitochondria and chloroplasts, in *Symbiosis in Cell Evolution*
1981	J. D. Kemp and T. H. Hall	Transfer a gene from beans to sunflowers via a plasmid of the crown gall bacterium
1981	T. R. Cech, A. J. Zaug, and P. J. Grabowski	Demonstrate self-splicing in rRNA: first evidence of molecules other than proteins acting as biological catalysts
1981	S. Anderson, B. G. Barrell, F. Sanger, et al.	Completely sequence the human mitochondrial genome

1982	Eli Lilly International Co.	Market the first drug (human insulin) made by recombinant DNA techniques
1982	E. P. Reddy, R. K. Reynolds, E. Santos, and M. Barbacid	Report the genetic changes in a line of human bladder carcinoma cells that activate an oncogene
1984	W. McGinnis, C. P. Hart, W. J. Gehring, and F. H. Ruddle	Demonstrate that the homeobox of *Drosophila* is also found in mice, suggesting a developmental function for the homeobox
1985	K. Mullis	Develops the polymerase chain reaction (PCR) for amplifying small amounts of DNA; receives the Nobel Prize in 1993
1985	A. J. Jeffries, V. Wilson, and S. L. Thien	Devise DNA fingerprint techniques
1987	C. Nüsslein-Volhard, H. G. Frohnhüfer, and R. Lehmann	Demonstrate a small group of maternal effect genes determine the polarized pattern of development in *Drosophila*
1987	E. P. Hoffman, R. H. Brown, and L. M. Kunkel	Isolate the protein (dystrophin) produced by the muscular dystrophy gene
1987	D. C. Page et al.	By cloning a section of the human Y chromosome, discover a factor influencing testis differentiation and thus illuminate the mechanism of sex determination in humans
1987	R. L. Cann, M. Stoneking, and A. C. Wilson	Using mtDNA erect a genealogical tree tracing all human mtDNAs to a common African maternal ancestor
1988	N. Wexler, M. Conneally, and J. Gusella	Associate Huntington disease with human chromosome 4
1988	P. Leder and T. Stewart	Develop a genetically altered animal (oncomice) patented by Harvard University: the first U.S. patent for genetically altered animals
1989	L-C. Tsui, et al.	Identify the cystic fibrosis gene and predict its product's amino acid sequence
1990	W. F. Anderson et al.	First to treat patients with gene therapy
1993	J. Hall and R. Stillman	Report producing genetically identical embryos from cells fertilized in vitro; although the initial cell was an abnormally fertilized egg, this experiment drew attention to the possibility of cloning humans

GLOSSARY

7-Methyl guanosine cap a modification of eukaryotic mRNA, where a methylated guanine nucleotide is attached to the end of the mRNA molecule in a 5′–5′ linkage. The cap appears important in ribosome assembly at the appropriate initiation codon in the mRNA

Acentric chromosome a chromosome without a centromere

Acridine dye a chemical mutagen known to produce shift mutations

Acrocentric chromosome a chromosome with the centromere near one end

Alkylating agents mutagens such as nitrogen mustard that produce mutations by adding methyl or ethyl groups to the bases (particularly guanine), causing loss of bases, transitions, or chromosome breaks

Allele any of the various forms in which a gene can exist (e.g., a^1, a^2, A)

Allopatric populations geographically separated populations

Allopolyploidy a euploid in which the multiple chromosome sets are of different origin, from either different species, races, or populations

Ames test analysis for detecting mutagenic chemicals using bacterial strains

Aminoacyl attachment site (A) the site on the ribosome where an entering tRNA molecule is bound by its anticodon to the mRNA codon

Aminoacyl-tRNA synthetases enzymes that activate tRNA by attaching the correct amino acid to its specific tRNA molecule

Amniocentesis a diagnostic procedure for determining genetic abnormalities in utero; involves removal of amniotic fluid containing fetal cells, which are cultured and used to screen for chromosomal aberrations, DNA markers, and certain protein defects

Amphidiploid or *allotetraploid;* tetraploid species in which the genome contains the diploid chromosome complement of both parent species

Analysis of variance (ANOVA) the statistical breakdown of observed variation into its components

Aneuploidy change in chromosome number involving loss or gain of less than a complete set of chromosomes (for example, $2n + 1$, $2n - 1$)

Antibody a protein (immunoglobulin) that reacts with specific substances (antigens) that are foreign to the organism

Anticodon a three-base sequence of nucleotides on tRNA that complements a codon on mRNA

Antigen a substance that elicits the production of antibodies when introduced into an animal

Ascospores sexual spores in the pink bread mold *Neurospora crassa,* occurring as eight haploid spores confined in a sac called an *ascus;* result of one meiotic cycle that produces four haploid spores, which then divide by mitosis, producing the eight spores

Assortative mating a nonrandom mating; see *negative assortative mating, disassortative mating,* and *positive assortative mating*

Attached-X chromosome in *Drosophila,* refers to two X chromosomes attached by one centromere

Attenuation control of transcription termination at specific regulatory sequences in bacterial operons

Autopolyploidy a euploid condition in which the extra chromosome sets are derived from within the same species

Autoradiography a technique for detecting the presence of radioactivity within a sample by overlaying the sample with a photographic emulsion (e.g. X-ray film) and allowing the emitted radiation to expose the film. The location of the radioactive substance (in a cell, on a chromosome, in a gel, etc.) is revealed by developing the film

Autosome any chromosome other than the sex chromosomes

Auxotroph mutant a mutant microorganism that is dependent on specially supplemented media to supply molecules it is unable to synthesize

Back mutation (see *reverse mutation*)

Bacteriophage also termed *phage:* viruses that infect bacterial cells

Balanced lethal a method of retaining recessive lethals at high frequencies in a population; a situation in which linkage assures the maintenance of the recessive lethals in a system of permanent heterozygotes, in which only the heterozygote is viable

Balanced polymorphism the maintenance of several different genotypes in a population, due to different selection pressures on each type; common example: heterozygous individuals have superior fitness

Barr body a dark-staining body in the nucleus of mammals that is the result of inactivation or heterochromatinization of all but one of the X chromosomes

Base analogue a chemical that is functionally similar to a base but structurally different; may undergo tautomeric shifts more often than the normal bases

Binomial distribution a probability distribution in which the expected frequencies of classes are obtained by binomial expansion

Bivalent a pair of homologous chromosomes during synapsis at prophase I of meiosis

Canalized characters characters in which the phenotype is normally unaffected by environmental stresses or by underlying genetic variability

Catabolite repression the lowered expression of many bacterial operons when glucose is present in the growth medium; mediated by decreased levels of cyclic AMP

cDNA DNA that has been enzymatically synthesized from an RNA template (i.e., "complementary" DNA) (see *reverse transcriptase*)

cDNA library a collection of numerous cDNAs (spliced into a vector) that is representative of an mRNA population at a given time or in a given tissue during development

CentiMorgan the genetic unit of crossing over, equal to 1 percent crossing over; abbreviated cM

Centric fusion the translocation of two nonhomologous acrocentric or telocentric chromosomes with the loss of one of the centromeres; leads to reduction in chromosome number

Centriole a self-replicating cytoplasmic body found in animal cells, located at each of the poles during mitosis and meiosis, and appearing to have spindle fibers radiating from it

Centromere a chromosomal region to which the spindle fibers attach during mitosis and meiosis

Chi-square test a statistical test for measuring the significance of the discrepancy between observed and expected results; also written χ^2

Chiasma (pl. chiasmata) the visible evidence of exchange (crossing over), seen in prophase I of meiosis

Chimera another term for *mosaic*

Chromatid a strand of a replicated chromosome that is attached by a centromere to an identical sister strand

Chromatin the interphase DNA–protein complex, identifiable by its staining properties

Chromomere local coiled regions of the linear DNA thread that give the eukaryotic chromosome a beaded appearance. In *Drosophila's* polytene chromosomes, these coils are in register and produce the banded appearance of the chromosomes

Cis-acting factor a site in DNA, such as a promoter or regulatory region (e.g., Lac operator), that must be adjacent or physically linked to a gene to exert an effect

Cis-configuration linkage arrangement in which the alleles of two linked genes are on the same homologue, and the recessive alleles are on the other (for example, *ab* versus *AB*) or are *coupled;* opposite is *trans-configuration,* or *noncoupling* (*Ab* versus *aB*)

Cistron a sequence of DNA coding for a single polypeptide (or tRNA or rRNA product) as defined by the complementation test

Class intervals subdivisions arbitrarily made in frequency distributions for a quantitative trait to facilitate presentation and analysis; for example, one-inch class intervals used to subdivide human height

Codominance type of dominance in which alleles express their product independently in heterozygotes, so that a heterozygote for two such alleles expresses products from both; seen, for example, in blood-group antigens and protein electrophoretic alleles

Codon a section of mRNA, three bases in length, that codes for an amino acid

Coefficient of coincidence the ratio of the observed frequency of double crossovers to the expected frequency of double crossovers

Coefficient of regression (*b*) the slope of the line of y on x

Coefficient of variation (*CV*) a statistic used to compare the variability of different samples; defined as a percentage, $CV = (s/\bar{x}) \cdot 100$, where s = standard deviation and $\bar{x}$ = the mean; produces an estimate of variability that is independent of the sample measurement

Complementary gene action refers to the interaction of nonallelic genes in the production of a specific phenotype. Specific alleles from each gene are necessary for the expression of the phenotype

Complementation the ability of two recessive mutations to restore the normal (wild) phenotype in trans-linkage

Complementation test test used to identify functional alleles; based on the fact that in trans-linkage a mutant phenotype is produced if two mutants are alleles, but a wild-type phenotype (*complementation*) is produced if they are functionally different genes

Concatemer a tandemly repeating nucleotide sequence found in viruses, in which the entire virus genome is multiply repeated

Conditional mutation an allele that produces a mutant phenotype in certain environments, but a wild phenotype under other conditions. The environment is called *permissive* if wild-type is produced, and *restrictive* if it produces a mutant phenotype. The mutants are often lethal (conditional lethals)

Confidence limits intervals, established around the observed frequency, within which we can be relatively sure that the true frequency is found; set by using the normal distribution curve and standard deviation intervals (usually the 95 percent and 99 percent confidence intervals)

Conjugation (in bacteria) a unidirectional transfer of genetic information, following the pairing of a donor and a recipient cell

Consanguinity the sharing of at least one common ancestor; mating between close relatives

Constitutive the production of gene products continuously, without regard to the cell's requirements; often referring to an operon in which there is a mutation in one of the control genes, so that the cell no longer responds to repression or to induction

Constitutive heterochromatin heterochromatin that is identical on a given chromosome pair and forms a permanent structural component of homologous chromosomes

Contingency chi-square test a statistical test used to compare one set of observations to another set to see whether the results are dependent or independent of the conditions under which the observations were made

Continuous variation variation among phenotypes, measured on some scale, that cannot be divided into a few distinct categories (see *discontinuous variation*)

Copy choice a hypothesis to explain intrachromosomal recombination in which pairing of two parental chromosome sequences allows the copying process to begin on one strand and then switch to the other parental strand

Correlation when two or more characteristics are dependent on, and vary with, each other; *positive correlation:* one characteristic increases as the other increases; *negative correlation:* one characteristic increases as the other decreases

Correlation coefficient a statistical measure of the relationship between two characteristics; measures the extent to which variations in one variable are related to variations in another

Cosmid a recombinant DNA vector that can "infect" a bacterial cell packaged within a bacteriophage protein coat, but is capable of replicating as a plasmid after entrance. Vector DNA sequences (*cos* sites) required for packaging are minimized, permitting large amounts of foreign DNA to be included in phage heads and subsequently introduced into cells

Coupling linkage see *cis-configuration*

Covariance a variance component used in calculating the correlation coefficient

Cri du chat syndrome a human condition associated with a deletion at the tip of chromosome 5 and characterized by congenital abnormalities and a thin "mewing" cry in newborn infants

Crossing over the events leading to genetic recombination between linked genes

Cyclical selection a shifting selection that results in selection in one direction for one or a few generations or seasons, then selection in the opposite direction for one or a few generations or seasons

Cytokinesis the division of the cytoplasm during cell division

Cytoplasmic inheritance inheritance of genetic traits from determinants that are not located on the chromosomes: extranuclear inheritance

Degenerate code a code in which there is more than one codon for the same amino acid

Degrees of freedom (d.f.) the number of independently variable classes in an experiment (for example, in a typical χ^2 analysis, where d.f. equals the number of classes minus one)

Deletion a loss of a part of a chromosome

Deme a local inbreeding group in which individuals have an equal chance of mating with all members of the opposite sex

Deoxyribonucleic acid (DNA) Organic molecule that is the basis of heredity. A double-stranded structure composed of nucleotides (phosphate group, 5-carbon deoxyribose sugar, and purine or pyrimidine base)

Deoxyribose the 5-carbon sugar found in a DNA nucleotide

Dicentric a chromosome with two centromeres; often the result of recombination within a heterozygous paracentric inversion

Dideoxyribonucleotides analogues of the normal DNA precursor molecules that lack 3′ hydroxyl residues. Incorporation of dideoxyribonucleotides into DNA terminates chain elongation and is the basis for the DNA sequencing technology developed by Frederick Sanger

Diester bond the bond in the DNA backbone between the 3′ carbon of the sugar of one nucleotide and the phosphate attached to the 5′ carbon of the adjacent nucleotide

Diesterase an enzyme that breaks diester bonds

Differentiation the developmental process by which apparently unspecialized cells form tissues and structures of adult forms

Dihybrid cross a cross between two individuals having different alleles at two loci (for example, *AaBb*)

Diploid having two sets of chromosomes; found in the somatic cells of most higher organisms; represented as $2n$

Directional selection selection favoring an extreme phenotype

Disassortative mating the mating of unlike phenotypes or genotypes (see *assortative mating*)

Discontinuous characters phenotypic variation that can be divided into a few distinctive classes (for example, brown fur and white fur)

Disruptive selection selection of two or more different genotypes or phenotypes in a single population or species; results in an increase in phenotypic variance

DNA see *deoxyribonucleic acid*

DNA sequencing a technique that permits the base-by-base determination of the linear array of nucleotides within a DNA chain

Dominant an allele that manifests itself phenotypically in heterozygotes

Dosage compensation a term used to describe the fact that phenotypic effects of sex-linked genes are similar, whether an individual carries one (for example, X chromosome in a man) or two (X chromosome in a woman) gene doses

Down syndrome or trisomy 21; human condition associated with aneuploidy for chromosome 21 and characterized by mental retardation; a flat, round face; and simian crease in the palm

Duplication the presence of a section of a chromosome in excess of the normal amount

Dyad a chromosome having two sister chromatids attached at a single centromere

Effective population size the number of mating pairs contributing to the next generation

Electrophoresis a technique used to detect variation in proteins, involving the use of an electric field to cause the proteins to migrate along a gel and then observing their relative positions on the gel by protein-specific staining reactions

Elongation factors proteins that are necessary for translation to proceed

Endonuclease an enzyme that produces an internal cut in the sugar–phosphate backbone of DNA (for example, in repair of thymine dimers, in which a specific endonuclease cuts the backbone at a nucleotide adjacent to a thymine dimer)

Enhancer a specific DNA sequence, distinct from the promoter, that can bind a transcription factor, leading to an increase in gene transcription. Enhancers may be located either upstream or downstream of the promoter, sometimes separated by distances of thousands of nucleotides

Epigenesis the theory that during development new tissues and organs appear that were not present in the original zygote; replaced *preformation*

Episome a circular piece of self-replicating double-stranded DNA, additional to the normal genome of the bacterial cell; can either exist independently in the cytoplasm or become an integral part of the host's chromosome; there is not full agreement as to the definition of *plasmid* and *episome*

Epistasis the interaction between two nonallelic genes, such that one interferes with the phenotypic expression of the other

Euchromatin light-staining chromatin that carries most of the typical genes of an organism

Eugenics controlled breeding of humans to eliminate unfavored genotypes or select for favored ones

Eukaryote an organism that has chromosomes composed of DNA and protein, nuclear membranes, and other membrane-bound organelles

Euploidy a normal chromosome complement or a complete multiple of a haploid set

Exons the coding regions of a gene that are separated from each other by intervening sequences (*introns*)

Exonuclease an enzyme that removes the terminal nucleotide of nucleic acids

Expressivity the degree of phenotypic expression in those individuals showing a certain trait

F factor the sex or fertility episome in bacterial donor cells; gives bacteria that have it the ability to form conjugation tubes

Facultative heterochromatin heterochromatin that varies in different cell types at different stages of development, or even from one homologous chromosome to another (for example, the Barr body)

Fitness in the genetic sense, a term referring to the ability to produce offspring in a particular environment; the relative reproductive success of a genotype

Founder principle a term referring to the small sample of genetic information carried by a group of individuals establishing a new population; one form of genetic drift

Frameshift a mutation caused by deletion or addition of nucleotides in DNA, which has the effect of changing the reading frame of codons, causing the wrong sequence of amino acids to be produced

Frequency distribution an arrangement of statistical data that demonstrates the number of occurrences of each value in a data set

G-banding banding patterns of chromosomes that are produced by treating chromosomes with trypsin and then staining with Giemsa. Most heterochromatin stains darkly, and most euchromatin stains lightly with this technique

Gamete a haploid cell; in multicellular organisms is usually called the egg or the sperm

Gene a nucleotide sequence coding for mRNA, tRNA, or rRNA

Gene conversion the process by which, in a heterozygote, one allele of a gene is lost and is replaced by the other allele; the consequence of partial heteroduplexes and the use of either strand to repair the helix

Gene frequency a measure of the commonness of an allele in a population; the proportion of each allele of a gene in a population

Gene pool the total genetic information in a population, as represented, for example, by the genes in the gametes of a population

Genetic drift changes in gene frequency due to random sampling of the gene pool each generation; caused by populations being limited in size rather than being infinite

Genetic homeostasis the maintenance of genetic variability in a population in the face of all forces acting to reduce it; probably arises from the necessity for the maintenance of certain levels of heterozygosity to ensure normal development

Genetic load a complex concept that one might describe as the extent to which a population departs from a genetic constitution producing an optimum phenotype in all members

Genome the total complement of genes characteristic of a species; has different usages, which should be clear from the context: (1) for viruses and bacteria, the entire set of genes; (2) for eukaryotes, all the genes present in the monoploid set of chromosomes (in diploids, sometimes called the *haploid genome*): for example, the allopolyploid *Raphanobrassica,* derived from the radish (*Raphanus*), and the cabbage (*Brassica*), which contains both the *Raphanus* and *Brassica* (diploid genomes); (3) the genetic makeup of an individual diploid organism

Genotype the total genetic information contained in an organism or the genetic constitution of an individual with respect to the genes under consideration

Gynandromorph sexual mosaic, usually in insects, in which male and female tissues are present in the adult; in *Drosophila,* usually the result of the loss of an X chromosome in a developing female; in the parasitic wasp *Bracon hebetor,* the result of the fertilization of one nucleus of an infrequent binucleate egg, producing diploid (female) and monoploid (male) tissues

Gyrase jargon for a type II topoisomerase that relaxes closed-circle DNA supercoils in *E. coli.* Important in DNA replication

Haploid having one set of unpaired chromosomes; found in cells such as gametes; the haploid chromosome number ($1n$) is the same as the number of linkage groups; also termed *monoploid*

Haplotype the genetic makeup of a single chromosome; commonly used to describe the genotype of a set of closely linked genes such as the histocompatibility complex. A genetic pattern that arises from polymorphisms at any given genetic locus or closely linked group of loci, such as differences in restriction fragment sizes due to individual chromosomal variations in the location of restriction enzyme sites

Hardy–Weinberg equilibrium rule stating that gene frequencies do not change from one generation to the next if no outside forces are acting upon them

Helicase an enzyme that unwinds the DNA helix ahead of the replicating DNA polymerase

Helix-destabilizing protein (HDP) a protein that binds to single-stranded DNA at the replication fork and prevents the parental strands from reannealing (single-strand binding protein)

Helix-relaxing protein (gyrase) relaxes the tension caused by the unwinding of the double helix during replication; acts by breaking a phosphodiester bond in one of the parental strands ahead of the replication fork

Helix-unwinding protein a protein that causes the unwinding of the DNA strands during replication, creating the replication fork

Hemizygous genes present only once in the genotype, such as the genes on the X chromosome in male flies or mammals

Heritability the proportion of phenotypic variation in a given trait in a specified population that can be accounted for by genetic segregation

Heterochromatin dark-staining chromatin that is believed to be primarily structural and to include condensed, inactivated regions of the chromosome and only a few functioning genetic loci

Heteroduplex a double strand of DNA in which the bases within the helix are not totally complementary; may be the result of recombination between single strands of the homologues

Heterogametic sex the sex that produces two kinds of gametes relative to the sex chromosomes: in humans, the male (XY), since his gametes contain either an X or a Y chromosome; in birds, the female

Heterogeneous nuclear RNA (HnRNA) precursor mRNA, which must be processed before it becomes mRNA and leaves the nucleus; processing involves splicing, capping on the 5′ end, and adding a repeat adenine chain (poly A) on the 3′ end

Heterosis superiority of a hybrid to either parental type; superiority of heterozygous genotypes

Heterozygote an organism having different alleles at a given locus on homologous chromosomes

Hfr (high-frequency recombination) a strain of *E. coli* in which the F factor has incorporated into the bacterial chromosome inducing conjugative transfer of the bacterial chromosome

Histocompatibility gene a gene that belongs to either the major histocompatibility (MHC) system or other minor complexes. The systems are responsible for the histocompatibility molecules (antigens)

Histocompatibility molecules cell-surface antigens that cause rejection of cells or tissue that does not have them. They are genetically encoded glycoproteins

Histones basic proteins (of five major types) that complex with the phosphate groups of DNA and are involved in the packing of DNA

Holandric a gene that is found only on the Y chromosome, and thus will never be passed from a father to his daughters

Homeotic mutation a mutation that causes one body structure to be replaced by another body structure during development

Homogametic sex the sex that produces one kind of gamete relative to the sex chromosomes; contrast with *heterogametic* sex

Homologous chromosomes chromosomes that pair during meiosis and contain the same gene loci; diploid organism normally receives one chromosome of the pair from each parent

Homozygote an organism having the same allele at a given locus on homologous chromosomes

Hybrid zones zones where species barriers have broken down and viable fertile hybrids are being produced

Hydrophilic molecules molecules that interact with water

Hydrophobic molecules molecules that are insoluble in water

Hypothesis a generalization made to explain a set of observations; in science, should be *heuristic* – that is a type that leads to predictions beyond the observations that generated it

Imaginal disc a small group of cells in insect larvae that are destined to develop into adult parts

In situ hybridization the ability to detect a specific nucleic acid sequence within a biological structure (e.g., genes on a chromosome, mRNA within a cell) through complementary base pairing with a nucleic acid probe (see *nucleic acid hybridization*)

Inbreeding mating between relatives or genotypically similar individuals

Inbreeding coefficient a value that measures the probability that two genes in a zygote are identical

Inbreeding depression deterioration of fitness, due to inbreeding

Incipient species superspecies; a monophyletic group with differences distinct enough to warrant separation above the species level, but not yet reproductively isolated

Incomplete dominance type of dominance observed when the phenotype of the heterozygote is intermediate between, and distinguishable from, those of the homozygotes

Independent assortment the random combination of alleles of two or more genes, due to the behavior of nonhomologous chromosomes or effectively unlinked chromosome segments during meiosis, since the orientation of one pair of homologous chromosomes at the equatorial plate in cell division is independent of the orientation of other pairs

Independent event event whose occurrence or nonoccurrence does not affect the occurrence or nonoccurrence of another event; probability of two or more independent events occurring together calculated by multiplying their individual probabilities

Informosomes cytoplasmic particles of protein and mRNA found in the egg, which are then activated in some way during differentiation

Initiation factors enzymes that produce the complex of mRNA, ribosomal subunits, and the start tRNA, thus initiating translation

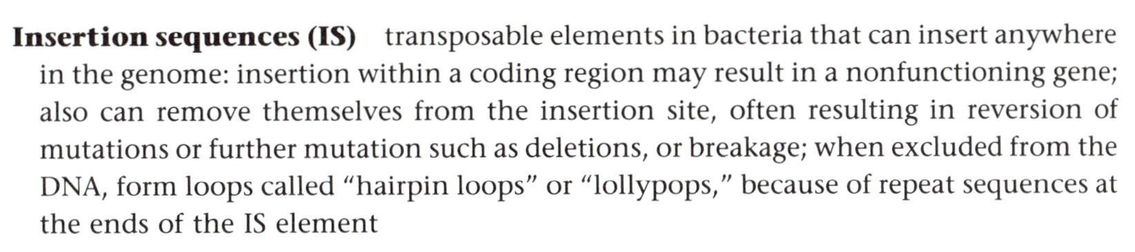

Insertion sequences (IS) transposable elements in bacteria that can insert anywhere in the genome: insertion within a coding region may result in a nonfunctioning gene; also can remove themselves from the insertion site, often resulting in reversion of mutations or further mutation such as deletions, or breakage; when excluded from the DNA, form loops called "hairpin loops" or "lollypops," because of repeat sequences at the ends of the IS element

Interference (chromosomal interference) the effect that one crossover event has on the occurrence of a second crossover event

Intergenic suppressors mutations in one gene that suppress other mutations in different genes

Intervening sequence noncoding nucleotide sequence in eukaryotic DNA, which separates coding portions of the gene; also called *intron*

Intragenic suppressors mutations that suppress other mutations in the same gene

Introgressive hybridization hybridization that results from backcrossing to one of the parental populations

Intron an intervening sequence

Inversion a change in gene order within a chromosome, involving two breaks and a turning around of the segment between the breaks

Isoallele allelic difference that can only be detected under special conditions; for example, elevated temperature, which can uncover thermal lability differences among gene (or allele) products that appear identical under normal conditions

Isochromosome a chromosome in which two genetically identical arms are attached to the same centromere; arms are mirror images of each other

Isolating mechanisms mechanisms that prevent gene exchange between populations

Karyotype the descriptive morphologic characteristics and number of chromosomes of a particular organism; usually, description of chromosomes as they appear in mitotic metaphase

Klinefelter syndrome a human condition associated with sexual aneuploidy; occurs in individuals bearing one Y chromosome and two or more X chromosomes, who are externally male, although usually sterile and phenotypically feminized

Lethal a mutation that causes death before reproductive age (may be recessive or dominant, a point mutation, or a chromosomal mutation)

Ligase an enzyme that joins nucleotides together

Linkage disequilibrium occurrence of alleles of two genes together more often than predicted by their individual frequencies

Linkage group genes located on the same chromosome; same in number as the haploid chromosome number

Locus the physical position of a gene on a chromosome

Lod score a statistical treatment of pedigree data used to establish the probability of linkage

Lysis the bursting of a cell by the destruction of the cell wall and/or membrane

Maternal effect type of development seen in offspring that, instead of expressing its own genotype, expresses the genotype of the mother, for example, because of products that are laid down in the egg and that act early in development

Mean ($\bar{x}$) the average value of a set of numerical data; the sum of all values, divided by the total number of data points

Meiosis the single replication of chromosomes, followed by a sequence of two cell divisions, the first a reduction division and the second similar to a mitotic division; result: four nuclei that are haploid in genetic makeup

Messenger RNA (mRNA) the RNA that carries the code sequence for polypeptide chains

Metacentric chromosome a chromosome in which the centromere is located near the center

Minisatellite a simple DNA sequence (15–100 nucleotides) repeated about 10–50 times and located at more than one chromosomal site. Restriction enzyme polymorphisms surrounding these sites provide the basis for performing DNA typing using hybridization probes complementary to minisatellite sequences

Mitosis the replication of chromosomes and their orderly division into a pair of nuclei having the same number and kind of chromosomes as the original parent nucleus

Monohybrid cross a cross between two individuals heterozygous at a single locus

Monoploid see *haploid*

Monosomic a term applied to individuals lacking one of a pair of homologous chromosomes

Morgan unit a unit of genetic map length, equal to 100 centiMorgans

Mosaic also termed *chimera;* gene or chromosomal difference in tissues that occurs after zygote formation

Multiple-factor inheritance inheritance in which several genes influence the phenotype of one quantitative character; also called *polygenic inheritance*

Mutation any detectable and heritable change in the genetic material that is not caused by segregation or recombination

Mutation equilibrium ($\hat{q}$) situation in which gene frequencies have become stable; forward and back mutations no longer cause a change in them (see *unstable equilibrium*)

Mutation frequency the proportion of alleles in a population that are mutant

Mutation rate the probability of a particular mutational event taking place per biological entity per generation

Muton the basic unit of gene mutation; the smallest alterable element of a gene (a nucleotide)

Mutually exclusive event an event whose occurrence precludes the occurrence of another event (for example, a flipped coin will be either a head or a tail)

Negative assortative mating the mating of unlike phenotypes or genotypes; experimentally, mating of parents chosen and sorted on the basis of their dissimilarity

Nondisjunction the irregular distribution of chromosomes or chromatids during anaphase of either mitosis or meiosis; results in either addition or loss of chromosomes

Nonhomologous chromosomes chromosomes that contain different genes and that do not pair during meiosis

Nonsense codon codons that do not code for an amino acid; specifically, UAA, UAG, and UGA; result in termination of translation, and hence are also called *terminator* or *stop codons*

Normal distribution also called the *normal curve* or *Gaussian distribution;* defined by the equation: $y = (1/\sigma\sqrt{2\pi})e^{-2(x - \bar{x})2/2\sigma 2}$ where $\bar{x}$ is the mean and σ is the standard deviation

Northern blot the identification of a specific RNA within a biological sample by separating the RNA molecules by gel electrophoresis, transfer of the gel contents to a filter membrane, and detection and quantification of the filter-bound RNA by hybridization to a nucleic acid probe

Nucleic acid hybridization the detection or characterization of a particular nucleic acid sequence in a biological sample via its ability to base-pair or hybridize to a complementary nucleic acid probe molecule through Watson–Crick hydrogen-bonding interactions

Nucleolus a dark-staining nuclear organelle of eukaryotes; the location of the genes coding for ribosomal RNA; rich in rRNA

Nucleoside the combination of a nitrogenous base and a 5-carbon sugar; can be either a ribonucleoside or a deoxyribonucleoside

Nucleotide the monomeric unit that makes up DNA or RNA, formed by the addition of a phosphate group to a nucleoside

Null hypothesis a statistical hypothesis that assumes that two samples are drawn from the same population or that a sample fits a specified model (for example, a 3:1 ratio); accumulation of observations or measurements and application of appropriate statistical tests permit the null hypothesis to be rejected

Okazaki fragments small fragments of DNA produced during replication along one strand of a replication fork because of the requirement that DNA nucleotides be added in the 5′ to 3′ direction; may be found in both strands when there are numerous replicating points

Oocyte a female sex cell from which the haploid egg is produced

Operator a regulatory DNA region at one end of an operon; acts as the binding site for repressor proteins

Operon a set of contiguous genes that transcribe a single mRNA unit; function is controlled by regulatory genes

Overdominance type of dominance observed when the phenotype of the heterozygote falls outside the range of either homozygote

***p* arm** the term for the short arm of a human chromosome

Panmixia (or panmixis) random mating

Paracentric inversion an inversion in a chromosome that does not involve the centromere

Parameter values of a population from which a sample has been taken

Parthenogenesis the production of offspring from a female gamete without fertilization by a male gamete; occurs in many different forms, some of which involve nongenetic participation by a male

Pedigree a diagram that shows ancestral relationships among members of a family over several generations

Penetrance the frequency with which a gene manifests itself in the phenotype of the heterozygote (if dominant) or of the homozygote (if recessive)

Peptide bond the formation of a covalent bond between the carboxyl end of one amino acid and the amino end of another amino acid

Peptidyl site (P) one of two tRNA binding sites on the ribosome; the site where peptide bonds are formed between the amino acids carried by the two tRNA molecules attached to the ribosome and mRNA complex

Peptidyl transferase the enzyme that forms a peptide bond between amino acids during translation

Pericentric inversion an inversion in a chromosome that includes the centromere

Petite mutations mutations in strains of yeast that result in lack of mitochondrial function

Phenocopy an alteration of a phenotype by environmental factors such that the phenotype mimics that produced by a specific gene

Phenotype the observed characteristic of an individual; the result of the interaction of the genotype with the environment

Phosphodiester bond a covalent bond between two of a phosphate group's oxygens and two carbon-containing groups. 5′C—O—P—O—C3′

Photoreactivation repair of thymine dimers, which requires the cells to be exposed to visible light

Plaque the formation of an area of lysed cells on a confluent lawn of bacteria, resulting from the clonal propagation of a single infection through adjacent host cells in the population

Plasmid a circular piece of self-replicating DNA that is additional to the normal genome of the bacterial cell and does not become integrated in the host genome (cf. *episome*)

Pleiotropism one gene with more than one phenotypic effect

Point mutation a mutation within a gene, in contrast to chromosomal mutation

Polar body any of three nuclei produced during meiosis in the female that get little cytoplasm and will not become an egg

Polar mutation a mutation in one gene that affects the expression of downstream genes in a prokaryotic polycistronic transcription unit

Poly(A) tail a posttranscriptional modification of eukaryotic mRNA, in which a series of adenine bases are enzymatically added to the 3′ end of the mRNA molecule

Polygene gene of relatively small effect; contributes, along with environmental influences, to the determination of a quantitative trait phenotype

Polymerase chain reaction (PCR) the amplification of a unique segment of DNA, often from a heterogeneous DNA population, through progressive cycles of primer annealing, DNA polymerization, and strand denaturation. The specificity of the amplification is controlled by using two oligonucleotide primers complementary to the DNA flanking the target DNA sequence

Polymerases enzymes that catalyze the assembly of nucleotides into RNA or DNA

Polyploid any organism having a number of complete chromosome sets that exceeds that of the diploid (for example, $3n$, $4n$)

Polysome several ribosomes aligned along a single mRNA strand

Population (1) *Mendelian population:* a group of sexually interbreeding or potentially interbreeding individuals in a defined area; (2) *statistical population:* a large group of potential observations from which a sample is taken

Position effect a change in the phenotypic expression of a gene, caused by a change in its position in relation to other genes

Positive assortative mating the mating of like phenotypes or genotypes; experimentally, mating of parents chosen and sorted for similarity

Preformation theory that there was an exact miniature replica of an adult in either the egg or sperm; supplanted by the epigenetic view in the eighteenth and nineteenth centuries

Pribnow box in *E. coli,* a region of the promoter found 10 nucleotides upstream (−10 position) of the transcription start site

Primary oocyte the original diploid egg cell that produces the polar bodies and the ovum by meiotic divisions

Primary spermatocytes a group of diploid cells, produced mitotically from spermatogonial cells of the testes, that undergo meiosis to produce the haploid sperm cells

Primase an RNA polymerase that is part of the primosome. It produces the RNA primers for DNA replication on the "lagging" strand

Primer a substrate that is necessary for polymerization (for example, during DNA synthesis) and that is similar to, and later replaced by, the subunits of the reaction product (for example, RNA nucleotides replaced by DNA nucleotides)

Primosome protein complex, including primase, required for the initiation and synthesis of the Okazaki fragments used in DNA replication. Without the primase it is often called a *preprimosome*

Proband the affected person who first calls a genetic condition to the attention of a physician; important in pedigree analysis; see *propositus, proposita*

Prokaryotes organisms that lack nuclear membranes and do not have histones associated with their DNA (for example, bacteria and blue-green algae)

Promoter a regulatory region at one end of an operon that acts as the binding site for RNA polymerase

Prophage the DNA of a temperate bacteriophage that is integrated into the genome of the host bacterium

Proposita the term for *proband,* if the person is female

Propositus the term for *proband,* if the person is male

Protein primary structure the sequence of the amino acids in the polypeptide chain

Protein quaternary structure two or more polypeptide chains bound together by one or more bonds

Protein secondary structure the configuration imposed on the polypeptide chain by its peptide linkages; the alpha helix

Protein tertiary structure the bending and folding of a protein secondary structure due to the interactions of specific amino acids (creating disulfide bonds, for example)

Prototroph a nutritionally independent cell that can live on a chemically defined minimal medium

Pseudoalleles genetic loci that are functionally allelic but structurally nonallelic; produce similar phenotypic effects but are capable of recombining with each other

Pseudodominance type of dominance that results when a recessive allele expresses itself in a heterozygote, caused by a deletion of the dominant allele on the homologous chromosome

Purine an adenine or guanine nitrogenous base found in RNA and DNA nucleotides

Pyrimidine nitrogenous base found in RNA and DNA nucleotides; in DNA the two pyrimidines are cytosine and thymine; in RNA they are cytosine and uracil

***q* arm** the term for the longer arm of a human chromosome

Quantitative characters traits in which the phenotype is measured on a continuous scale, caused by polygenic segregation and environmental influences (for example, height, IQ, or amount of pigment)

Recessive gene an allele that expresses itself only when it is homozygous (or hemizygous)

Reciprocal translocation exchange of chromosomal material between two nonhomologous chromosomes

Recombination a process occurring during normal meiosis that gives rise to linked combinations of alleles that differ from those found in the parents; see *crossing over*

Recon the unit of genetic subdivision at which recombination occurs (between one nucleotide and the next)

Regression a statistical method for predicting one character by using another that is correlated to it; on a graph, represented by a straight line, drawn with the formula $y = \bar{y} + b(x - \bar{x})$, where y and x are values of the correlated characters and b is the slope of the line or coefficient of regression of y on x

Releasing factors proteins that bring about the cleavage of the translated protein from the tRNA molecule and the ribosome

Repetitive DNA type of DNA occurring in eukaryotes, with nucleotide sequences in multiple copies ranging from a few short sequences to as many as a million copies; may be composed of a tandem sequence or be dispersed among the unique sequences of the genome

Replication the process by which the two strands of DNA separate and each serves as a template for the synthesis of a new complementary strand

Repulsion linkage when nonallelic mutants are on different homologous chromosomes, $C\ f\ /\ c\ F$ (see *cis-configuration*)

Restriction endonuclease an enzyme that internally cleaves a double-stranded DNA molecule after binding to a specific nucleotide recognition sequence.

Restriction fragment length polymorphism (RFLP) variation in the length of a discrete segment of genomic DNA following restriction endonuclease digestion, due to population differences in the distribution of enzyme sites flanking the sequence. Variations are inherited in a codominant manner and are useful in linkage studies (see *haplotype*) and DNA typing

Reverse mutation a heritable change in a mutant gene that restores the original nucleotide sequence or the original amino acid

Reverse transcriptase RNA-dependent DNA polymerase; an enzyme capable of polymerizing a DNA chain in the 5′ to 3′ direction, copying from a single-stranded RNA template

Rho the factor that binds with RNA polymerase during RNA synthesis, causing the release of a newly formed RNA chain, as well as of the RNA polymerase and the DNA template

Ribose the 5-carbon sugar found in RNA nucleotides

Ribosomal RNA (rRNA) any of several forms of RNA that are a part of the structural unit called a *ribosome*

Ribosomes cellular organelles composed of complex proteins and rRNA; the site of protein synthesis

Roentgen unit (*r*) measurement of ionizing radiation; each roentgen produces 1 electrostatic unit of charge in 1 cubic centimeter of air

Rolling circle replication the method of replicating circular DNA in which one strand is used as a template and is copied many times, producing concatemers

Sampling error variation due to the limited size of a sample

Secondary oocyte a product of first-division meiosis in which reduction has caused one nucleus (the secondary oocyte) to receive most of the cytoplasm and the other to become the first polar body

Secondary spermatocytes male haploid sex cells produced by meiotic division of the primary spermatocytes

Segregation the separation of homologous chromosomes (and thus alleles) during the first meiotic division

Segregation distortion unequal representation of alternate alleles in the gametes, due to abnormalities of segregation during meiosis

Segregational load also termed *balanced load:* refers to the situation in which the heterozygote is superior (has higher relative fitness) to both homozygotes; results in the retention of all three genotypes in the population

Selection the preferential reproduction and survival of some genotypes, along with the elimination of other genotypes

Selection coefficient (*s*) a measure of the force acting on each genotype to reduce its adaptive value

Selection pressure (*I*) also termed *selection intensity;* the difference in survival rates between the optimal (s_o) and suboptimal (s_s) phenotypes, multiplied by the frequency of the suboptimal type in the population (f_s); $I = (s_o - s_s) \cdot f_s$

Semiconservative replication refers to DNA replication in which each of the strands in the double helix is used as a template to produce a new complementary strand, so that after replication the two DNA molecules are each composed of one old and one new strand

Semilethal a lethal mutation causing death before reproductive age in more than 50 percent, but less than 100 percent, of the carriers

Sex-influenced autosomal characters whose expression is conditional on the sex of the organism. For example, the index finger in humans is shorter than the ring finger. The gene that has the major effect on this trait is recessive in females and dominant in males

Sex-limited autosomal or sex-linked genes, that are present in both sexes, but phenotypically express themselves in only one sex

Sex-linked genes that are found on the sex chromosomes; usually used interchangeably with the term *X-linked,* referring to genes on the X chromosome

Sexduction the incorporation of bacterial genes by sex factors and their transfer by conjugation

Shine–Dalgarno sequence a specific sequence in *prokaryotic* mRNA, located just 5′-ward of the start codon, that can base-pair with 16S ribosomal RNA and is important in initiating protein synthesis at the appropriate starting point in translation

Sibling species species that are morphologically very similar but that normally do not mate or, if forced to mate, usually do not produce viable offspring

Sigma factor involved in RNA synthesis that helps *E. coli* RNA polymerase to recognize the initiation point for transcription

Single-strand binding protein (SSBP) protein that binds to single-strand DNA during replication. It prevents the reannealing of the DNA molecule

SnRNPs or small nuclear ribonucleoprotein; some is involved in the removal of intron sequences during mRNA nuclear processing

Somatic cell genetics a subdivision of genetics that studies asexually reproducing body cells. The field uses cell fusion, somatic cell assortment, and crossing-over techniques. *Somatic cell genetic engineering* deals with insertion of genes and attempts to correct hereditary defects

Somatic cell genetics an artificial segregation technique resulting from the loss of human chromosomes from rodent–human hybrid cell fusions. The presence or absence of a human characteristic (protein, enzyme activity, human RFLP) is correlated with the human chromosome complement retained or lost in the hybrid cells

Southern blot a technique, developed by E. M. Southern, whereby DNA fragments generated by restriction enzyme digestion are separated by electrophoresis, transferred to a filter membrane, and detected by nucleic acid hybridization to a specific nucleic acid probe

Species groups of phenotypically similar organisms that can interbreed and produce viable offspring

Spermatocyte germinal cell produced in the testes; see *primary spermatocyte* and *secondary spermatocyte*

Spindle a set of fibers or microtubules that attach to the centromeres during nuclear division and move the chromosomes to opposite poles of the cell

Spliceosome enzymatic complex involved in the elimination of introns and joining of exons in eukaryotic mRNA

Stabilizing selection selection against extreme phenotypes, with corresponding selection for the intermediate type

Standard deviation (s.d.) the square root of the variance; symbols are s.d. or *s* for a sample, and s for the population from which the sample was taken; in a normally distributed population, 68 percent of the population falls within $\pm$ 1 *s* of the mean, 95 percent within $\pm$ 1.96 *s*, and 99 percent within $\pm$ 2.58 *s*

Standard error (s.e.) the square root of the variance of the mean: $s.e. = s_{\bar{x}} = s/\sqrt{n}$ where *s* is the standard deviation and *n* is the sample size

Start codon the codon on mRNA that is recognized as the starting point for the formation of a polypeptide chain, AUG, which codes for methionine

Statistics all values computed from sample data

Structural alleles two functional alleles (that is, two forms of a given gene) are structural alleles if they involve changes in exactly the same nucleotide

Suppressor mutation a second mutation, at a different site, that totally or partially restores a function lost as a result of an earlier mutation

Sympatric populations populations occupying the same geographic area

Synapsis the pairing of homologous chromosomes during prophase I of meiosis

Synaptonemal complex a structure present during prophase I of meiosis that produces the pairing of homologous chromosomes

***t*-test** a statistical test used to compare the differences between two distributions

Tandem repeats short DNA sequences that are duplicated numerous times and lie adjacent to each other

TATA box in eukaryotes, a region of the promoter found approximately 25 nucleotides upstream of the start site (−25), with a consensus sequence of TATAAA; also know as *Goldberg–Hogness box*

Tautomeric shift a reversible change in the location of a proton in a molecule; for example, in DNA nucleotide in amino form (NH_2) is changed to an imino (NH), resulting in mispairing of bases

Telocentric chromosome a chromosome with the centromere at one end

Telomere region found at the tips of chromosomes; unique in being "nonsticky" – that is, chromosome pieces will not attach to it

Temperate phage a bacteriophage that can either incorporate its DNA into the genome of the bacteria or cause the lytic response

Template DNA DNA that serves as a sort of macromolecular mold for the correct synthesis of a complementary DNA or RNA strand

Terminalization the movement of the chiasmata toward the ends of the bivalents during prophase I of meiosis

Testcross a cross between an individual having a dominant phenotype and a homozygous recessive individual; used to determine whether the dominant phenotype is homozygous or heterozygous

Tetrad a group of four associated chromatids of a bivalent; a synapsed pair of dyads

Tetrad analysis the study of crossing over in organisms that keep the products of a single meiotic division together (the tetrad). Useful in the analysis of crossing over

Tetraploid organism containing four sets of chromosomes ($4n$)

Three-point testcross a genetic cross using three linked marker alleles to map the linkage distance between these genes; involves crossing a heterozygote for the three markers to a homozygous recessive individual

Thymine dimer two adjacent thymine bases connected by a double covalent bond; often results when DNA is hit by ultraviolet radiation

Time of entry mapping using the times of transfer of genes from one bacterial cell to another during conjugation as a measure of relative genetic map positions

Trans-acting factor a diffusible molecule, such as a protein transcription factor, capable of binding to a promoter or enhancer region in DNA and regulating gene expression

Trans-configuration linkage arrangement in which the mutant alleles of two linked genes are on different homologues (*Ab* and *aB*); also termed *repulsion* (*noncoupling*), as opposed to *cis-configuration,* in which the alleles are on the same homologue and demonstrate coupling

Transcription the process by which RNA is produced from a DNA strand through the action of DNA-directed RNA polymerase

Transcription factor regulatory proteins that will bind to specific DNA sequences, affecting control of RNA transcription. Generic factors are required for basal transcription; and specialized factors are required for gene expression in response to particular environmental or developmental conditions

Transduction the transfer of genes from one cell to another cell by means of a viral vector; *generalized transduction:* a random fragment of the bacterial chromosome is incorporated into a phage coat and can be transferred; *specialized transduction:* the result of improper excision of a prophage, so that the defective phage carries a specific fragment of the bacterial chromosome

Transfer RNA (tRNA) the RNA molecules that carry amino acids to the ribosomes

Transformation the picking up of extracellular "naked" DNA by a bacterium and the incorporation of this DNA into its genome by recombination

Transition a base-pair substitution in which the purine–pyrimidine orientation is preserved; that is, a purine is replaced by a purine or a pyrimidine by a pyrimidine

Translation the process by which an mRNA sequence is used to produce a polypeptide chain

Translocation the transfer of a section of one chromosome to a nonhomologous chromosome

Transposon a transposable DNA sequence bounded by insertion sequences

Transversion a base-pair substitution in which the purine–pyrimidine orientation is not preserved; a purine is replaced by a pyrimidine, or vice versa

Triploid an organism containing three sets of chromosomes ($3n$)

Trisomic an organism that possesses three homologues of one chromosome

Turner syndrome a human condition in which an X chromosome is missing (XO); affected individuals are externally female, although they do not mature sexually

Two-point-testcross mating of a dihybrid (heterozygous for two genes, *AaBb*) to the homozygous recessive (*aabb*)

Unstable equilibrium a situation in which gene frequencies are not stable; a disturbance will cause the genes to adopt a new frequency or cause one of the genes to go toward fixation (see *mutation equilibrium*)

Variable number of tandem repeat (VNTR) locus an individual genetic locus that shows extreme allelic polymorphism in the population due to variations in the number of tandem repeats of a minisatellite core sequence. With appropriate probes and hybridization conditions, these loci can be distinguished from other minisatellite members and used as genetic markers in linkage studies

Variance a measure of the extent to which values within a distribution depart from the mean; symbol is s^2, for a sample, and σ^2 or $\hat{s}^2$ for the population from which the sample was taken: $s^2 = (\Sigma x^2 - \bar{x}^2[n])/n - 1$ where x represents each measurement and n is the sample size

Vector in recombinant DNA technology, a DNA molecule that can replicate independently and is used to propagate a foreign DNA insert in a heterologous prokaryotic or eukaryotic system

Virulent phage a bacteriophage that provokes a lytic response in which the bacterial cell lyses and releases new virus particles

Wobble hypothesis the idea that certain bases at the 5′ end (third base position) of the anticodon in tRNA are capable of forming hydrogen bonds with several different nitrogen bases at the 3′ end of the codon

Yates's correction factor the subtraction of 1/2, or 0.5, from the absolute value of the observed deviation minus the expected deviation in the chi-square test; corrects for continuity and adds to the accuracy of chi-square determination whenever expected classes are small; formula for chi-square thus becomes $\chi^2 = \Sigma([|O - E| - 0.5]^2/E)$

Yeast artificial chromosome (YAC) a recombinant DNA vector in which a very large foreign DNA insert (> 10^6 base pairs) is coupled to an origin of replication and a centromere and is flanked by telomeres, allowing clonal propagation in yeast

Zygote a diploid cell produced by the union of the egg and sperm

REFERENCE TABLES

TABLE R.1. CRITICAL VALUES OF THE χ^2 DISTRIBUTION

	Probabilities					
d.f.	**0.90**	**0.50**	**0.10**	**0.05**[a]	**0.01**[b]	**0.001**[c]
1	0.02	0.46	2.71	3.84	6.64	10.83
2	0.21	1.39	4.61	5.99	9.21	13.82
3	0.58	2.37	6.25	7.82	11.35	16.27
4	1.06	3.36	7.78	9.49	13.28	18.47
5	1.61	4.35	9.24	11.07	15.09	20.52
6	2.20	5.35	10.65	12.59	16.81	22.46
7	2.83	6.35	12.02	14.07	18.48	24.32
8	3.49	7.34	13.36	15.51	20.09	26.13
9	4.17	8.34	14.68	16.92	21.67	27.88
10	4.87	9.34	15.99	18.31	23.21	29.59
11	5.58	10.34	17.28	19.68	24.73	31.26
12	6.30	11.34	18.55	21.03	26.22	32.91
13	7.04	12.34	19.81	22.36	27.69	34.53
14	7.79	13.34	21.06	23.69	29.14	36.12
15	8.55	14.34	22.31	25.00	30.58	37.70
20	12.44	19.34	28.41	31.41	37.57	45.32
30	20.60	29.34	40.26	43.77	50.89	59.70
50	37.69	49.34	63.17	67.51	76.15	86.66

[a] $.05 > p > .01$.
[b] $.01 > p > .001$.
[c] $p < .001$.

Source: R. A. Fisher and F. Yates, *Statistical Tables for Biological, Agricultural and Medical Research* (Edinburgh: Oliver & Boyd, 1958).

TABLE R.2. CRITICAL VALUES OF STUDENT'S t-DISTRIBUTION

	Probabilities					
d.f.	0.90	0.50	0.10	0.05[a]	0.01[b]	0.001[c]
1	0.16	1.00	6.31	12.70	63.66	636.62
2	0.14	0.82	2.92	4.30	9.92	31.60
3	0.14	0.76	2.35	3.18	5.84	12.92
4	0.13	0.74	2.13	2.78	4.60	8.61
5	0.13	0.73	2.02	2.57	4.03	6.87
6	0.13	0.72	1.94	2.45	3.71	5.96
7	0.13	0.71	1.90	2.36	3.50	5.41
8	0.13	0.71	1.86	2.30	3.36	5.04
9	0.13	0.70	1.83	2.26	3.25	4.78
10	0.13	0.70	1.81	2.23	3.17	4.59
11	0.13	0.70	1.80	2.20	3.10	4.44
12	0.13	0.70	1.78	2.18	3.05	4.32
13	0.13	0.69	1.77	2.16	3.01	4.22
14	0.13	0.69	1.76	2.14	2.98	4.14
15	0.13	0.69	1.75	2.13	2.95	4.07
16	0.13	0.69	1.75	2.12	2.92	4.02
17	0.13	0.69	1.74	2.11	2.90	3.96
18	0.13	0.69	1.73	2.10	2.88	3.92
19	0.13	0.69	1.73	2.09	2.86	3.88
20	0.13	0.69	1.72	2.09	2.84	3.85
21	0.13	0.69	1.72	2.08	2.83	3.82
22	0.13	0.69	1.72	2.07	2.82	3.79
23	0.13	0.68	1.71	2.07	2.81	3.77
24	0.13	0.68	1.71	2.06	2.80	3.74
25	0.13	0.68	1.71	2.06	2.79	3.72
26	0.13	0.68	1.71	2.06	2.78	3.71
27	0.13	0.68	1.70	2.05	2.77	3.69
28	0.13	0.68	1.70	2.05	2.76	3.67
29	0.13	0.68	1.70	2.04	2.76	3.66
30	0.13	0.68	1.70	2.04	2.75	3.65
40	0.13	0.68	1.68	2.02	2.70	3.55
60	0.13	0.68	1.67	2.00	2.66	3.46
120	0.13	0.68	1.66	1.98	2.62	3.37
∞	0.13	0.67	1.64	1.96	2.58	3.29

[a] $.05 > p > .01$.
[b] $.01 > p > .001$.
[c] $p < .001$.
Source: F. J. Rohlf and R. R. Sokal, *Statistical Tables* (San Francisco: W. H. Freeman, 1969).

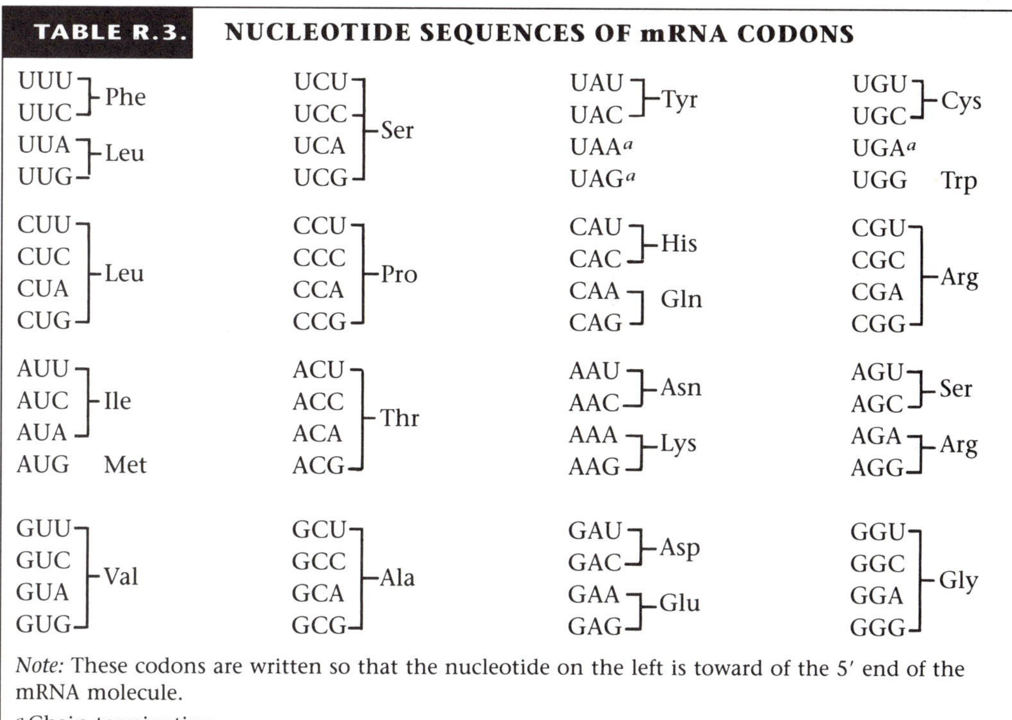

TABLE R.3. NUCLEOTIDE SEQUENCES OF mRNA CODONS

Codon	Amino acid	Codon	Amino acid	Codon	Amino acid	Codon	Amino acid
UUU	Phe	UCU	Ser	UAU	Tyr	UGU	Cys
UUC	Phe	UCC	Ser	UAC	Tyr	UGC	Cys
UUA	Leu	UCA	Ser	UAA[a]		UGA[a]	
UUG	Leu	UCG	Ser	UAG[a]		UGG	Trp
CUU	Leu	CCU	Pro	CAU	His	CGU	Arg
CUC	Leu	CCC	Pro	CAC	His	CGC	Arg
CUA	Leu	CCA	Pro	CAA	Gln	CGA	Arg
CUG	Leu	CCG	Pro	CAG	Gln	CGG	Arg
AUU	Ile	ACU	Thr	AAU	Asn	AGU	Ser
AUC	Ile	ACC	Thr	AAC	Asn	AGC	Ser
AUA	Ile	ACA	Thr	AAA	Lys	AGA	Arg
AUG	Met	ACG	Thr	AAG	Lys	AGG	Arg
GUU	Val	GCU	Ala	GAU	Asp	GGU	Gly
GUC	Val	GCC	Ala	GAC	Asp	GGC	Gly
GUA	Val	GCA	Ala	GAA	Glu	GGA	Gly
GUG	Val	GCG	Ala	GAG	Glu	GGG	Gly

Note: These codons are written so that the nucleotide on the left is toward of the 5′ end of the mRNA molecule.

[a] Chain termination.

TABLE R.4. THE AMINO ACIDS

Symbols		Amino acid	Symbols		Amino acid
Ala	A	Alanine	Leu	L	Leucine
Arg	R	Arginine	Lys	K	Lysine
Asn	N	Asparagine	Met	M	Methionine
Asp	D	Aspartic acid	Phe	F	Phenylalanine
Cys	C	Cysteine	Pro	P	Proline
Gln	Q	Glutamine	Ser	S	Serine
Glu	E	Glutamic acid	Thr	T	Threonine
Gly	G	Glycine	Trp	W	Tryptophan
His	H	Histidine	Tyr	Y	Tyrosine
Ile	I	Isoleucine	Val	V	Valine

TABLE R.5. CHARACTERISTICS OF AMINO ACIDS

Nonpolar (= hydrophobic = internal; the bulky hydrocarbon side chains are sequestered on the interior of the protein)

- Aliphatic (inert carbon side chains): alanine, valine, leucine, isoleucine
- Aromatic (unsaturated carbon ring): phenylalanine
- Sulfur-containing aliphatic: methionine

Ambivalent (may be external or internal, depending on position in polypeptide chain)

- Aliphatic and polar: glycine (R group = H), serine and threonine (contain –OH groups)
- Aromatic and polar: tyrosine and tryptophan
- Side groups that affect backbone conformation: cysteine (can form intrastrand covalent linkages via disulfide bonds), proline (aliphatic side chain bonds back to amino group of main chain, producing bend in peptide backbone)

Polar (= hydrophilic = external; the charged or polar side chains can interact with water on the surface of the protein)

- Acidic (carry negative charge): aspartic acid and glutamic acid
- Basic (carry positive charge): lysine, arginine, and histidine (*his* is charged, depending on local environment)
- Neutral (amides of acidic amino acids): asparagine and glutamine